Hydrology in Practice

Hydrology in Practice

Second Edition

Elizabeth M. Shaw
Formerly of the Department of Civil Engineering
Imperial College of Science and Technology

VNR

International

Van Nostrand Reinhold (International)

First published in 1983 by
Van Nostrand Reinhold (International) Co. Ltd
11 New Fetter Lane, London EC4P 4EE

Reprinted 1984, 1985
Second edition 1988
Reprinted 1989

Typset in 10 on 11 pt Baskerville by
Colset Private Limited, Singapore
Printed and Bound in Great Britain by
T. J. Press (Padstow) Ltd, Padstow, Cornwall

ISBN 0 278 00061 4

British Library Cataloguing in Publication Data

Shaw, Elizabeth M. (Elizabeth Mary)
    Hydrology in practice.—2nd ed.
    1. Hydrology
    I. Title
    551.48

    ISBN 0-278-00061-4

Contents

PART II   HYDROLOGICAL ANALYSIS

Preface

When the writing of a new textbook in hydrology was proposed, the initial reaction was that there was perhaps a sufficient number of suitable books already available. However, few of the many texts form comprehensive introductions to the subject for British engineers. The content of this book attempts to bridge the gap between the text concerned with scientific processes and the applied numerical text. It is based on lecture material given to undergraduate and postgraduate students coming newly to the subject.

Hydrology in Practice is addressed primarily to civil engineering undergraduates. It aims to give an essentially practical approach to the various facets of the subject and emphasizes the application of hydrological knowledge to solving engineering problems. However, the style of the book should also be attractive to undergraduates in other disciplines hoping to make a career in the water industry. Additionally, it could be a useful reference book to the junior engineer in the design office of a consulting engineer or water authority concerned with water engineering problems. Civil engineers in the developing countries could also benefit from this introduction to a specialist subject.

Although presented from a British viewpoint, the subject is set in a global context; hydrological conditions in the different climatic regions are described; incidences of hydrological extremes are necessarily taken from worldwide situations; and examples of major engineering applications from many countries are noted appropriately in the text.

Among the 20 chapters, there are three main groups related to particular themes. The book begins with a brief introduction to the hydrological cycle coupled with greater detail on hydrometeorology; then Part I, Hydrological Measurements, follows with eight chapters on the measurement of water in the different phases of the hydrological cycle and on introductions to the important subjects of water quality and data processing. The second group of six chapters, Part II, Hydrological Analysis, deals with the preparation of basic data into forms required for engineering applications. Included in this group is the core of hydrological study, the derivation of river flow from rainfall, embodied in Chapters 13 and 14. Stochastic Hydrology explains how required statistical information can be gleaned from sequences of hydrological data. Part III, Engineering Applications, contains five chapters demonstrating how hydrological information and analytical techniques are used in solving problems in water engineering. The text concludes with a selection of problems for the student and an appendix of statistical formulae.

This arrangement of the subject matter of engineering hydrology is different from any other text. The treatment of hydrological processes with the emphasis on methods of measurement in Part I is separated from the analytical techniques applied to the major hydrological variables in Part II. The final Part III takes a selection of civil engineering spheres of operation requiring hydrological expertise and demonstrates some of the well tried methods and some of the new techniques applied in current practice.

In the presentation of the material, a good grounding in mathematics is assumed, including calculus, differential equations and statistics. Reference is made where necessary to standard texts on subjects such as fluid mechanics, open channel flow and statistics. Wherever it is relevant the reader is referred to original publications on specialized topics. SI units are given throughout, but where non-standard units are used for convenience in practice, these are retained; certain comparisons with Imperial units have also been included.

Preface to the Second Edition

The preparation of this second edition was assisted greatly by the constructive criticisms of former colleagues, reviewers and users of the text. The initial aims of *Hydrology in Practice* to be an introduction of the subject to students and to be a reference book for junior engineers appear to have been realized.

However, the rapid technological advances in the past five years have necessitated the rewriting of much of the data processing chapter and considerable modifications to the important chapters on precipitation and river flow measurement. In response to several requests, further worked examples have been included but simple applications of catchment models are difficult to find or to invent and these have been left for the research activities of the student or practitioner. Various omissions from the original text have been made good: mention has been made of pressure transducers for water level recording, the Penman-Monteith evaporation formula, and the evaluation of crop water requirements. In the modelling field, the coupled component SHE model has been described and further models added to the Urban Hydrology chapter. An extra example of real time forecasting – flood forecasting – completes the last chapter.

The opportunity has been taken to correct some unfortunate errors in formulae and computation. The added bibliography should assist readers still concerned over remaining shortcomings of some chapters.

Elizabeth M Shaw
(1987)

Acknowledgements

The author wishes to acknowledge the help given over the years by present and former colleagues in the learning and teaching of hydrology in the university. Sources of illustrative and teaching material are cited where pertinent. In the preparation of this book, Professor T. O'Donnell, University of Lancaster, made valuable contributions to Chapters 13, 14, 16 and 19, and without his overall appraisal, the book would never have been completed. The advice of several readers of individual chapters (R.T. Clarke, T.G. Davis, A.E. McIntyre, A. Scott-Moncrieff and H.S. Wheater) is also gratefully acknowledged. However, responsibility for the final version rests solely with the author.

During the writing of the book, the library staff of the Department of Civil Engineering, Imperial College, were tireless in their assistance with elusive references. Thanks are also given to Hazel Guile for all the new drawings, to Anna Hikel for typing the text and to Clare Rogers for assistance with problems. The author is grateful to the publishers for their encouragement and valued suggestions on composition and presentation.

September 1982   *Elizabeth M. Shaw*
                                                                    Imperial College

Note

Since the publication of *Hydrology in Practice* in 1983, Miss Shaw has taken early retirement from the Department of Civil Engineering, Imperial College, but she continues to enjoy the support of her former colleagues, particularly of the ever-helpful librarians. Special thanks are afforded to Kenneth Woodley, formerly of the Meteorological Office, for his detailed appraisal of all things meteorological in the first edition.

1

The Hydrological Cycle and Hydrometeorology

The history of the evolution of hydrology as a multi-disciplinary subject, dealing with the occurrence, circulation and distribution of the waters of the Earth, has been presented by Biswas (1970). Man's need for water to sustain life and grow food crops was well appreciated throughout the world wherever early civilization developed. Detailed knowledge of water management practices of the Sumarians and Egyptians in the Middle East, of the Chinese along the banks of the Hwang-Ho and of the Aztecs in South America continues to grow as archaeologists uncover and interpret the artefacts of such centres of cultural development. It was the Greek philosophers who were the first serious students of hydrology, and thereafter, scholars continued to advance the understanding of the separate phases of water in the natural environment. However, it was not until the 17th century that the work of the Frenchman, Perrault, provided convincing evidence of the form of the hydrological cycle which is currently accepted; measurements of rainfall and stream flow in the catchment of the upper Seine published in 1694 (Dooge, 1959) proved that quantities of rainfall were sufficient to sustain river flow. Since then, scientists have continued to study the processes involved in the movement of water round the hydrological cycle. The main concern of the hydrologist today is with the quantities and time distribution of water passing through the various phases.

1.1   The Hydrological Cycle

The natural circulation of water near the surface of the Earth is portrayed in Fig. 1.1. The driving force of the circulation is derived from the radiant energy received from the sun. The bulk of the Earth's water is stored on the surface in the oceans (Table 1.1) and hence it is logical to consider the hydrological cycle as beginning with the direct effect of the sun's radiation on this largest reservoir. Heating of the sea surface causes *evaporation*, the transfer of water from the liquid to the gaseous state, to form part of the atmosphere. It remains mainly unseen in atmospheric storage for an average of 10 days. Through a combination of circumstances, the water vapour changes back to the liquid state again through the process of *condensation* to form clouds and, with favourable atmospheric conditions, *precipitation* (rain or snow) is produced either to return directly to the ocean storage or to embark on a more devious route to the oceans via the land surface. Snow may accumulate in Polar regions or on high mountains and consolidate into ice,

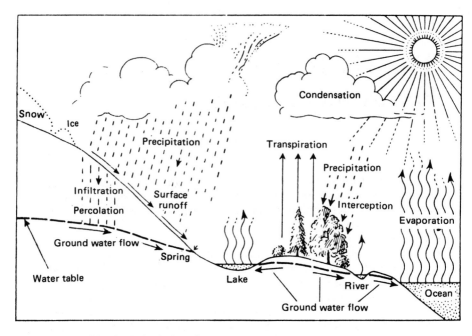

Fig. 1.1   The hydrological cycle. (Reproduced from A.D.M. Phillips &   B.J. Turton (Eds.) (1975) *Environment, Man and Economic Change*, by permission of the Longman Group.)

in which state water may be stored naturally for very long periods (Table 1.1). In more temperate lands, rainfall may be *intercepted* by vegetation from which the intercepted water may return at once to the air by evaporation. Rainfall reaching the ground may collect to form *surface runoff* or it may *infiltrate* into the ground. The liquid water in the soil then *percolates* through the unsaturated layers to reach the *water table* where the ground becomes saturated, or it is taken up by vegetation from which it may be *transpired* back into the atmosphere. The surface runoff and *groundwater* flow join together in surface streams and rivers which may be held up temporarily in lakes but finally flow into the ocean. The hydrological processes are described more fully in the scientific texts (Ward, 1975, Kirkby, 1978).

The land phases of the hydrological cycle have an enhanced importance in nature since evaporation is a purifying process; the salt sea water is transformed into fresh precipitation water and therefore water sources and storages on the continents consist largely of fresh water. The exceptions include groundwater storages with dissolved salts (brackish water) and surface waters polluted by man or natural suspended solids. This general, brief description of the hydrological cycle serves as an introduction to Part I of the text in which each of the various processes from open water evaporation and transpiration through to river flow into the oceans will be described separately in greater detail in conjunction with their quantitative considerations. However, the hydrological processes within the atmosphere, the

descriptive physics of evaporation and precipitation, will form the basis of
the rest of this chapter under the heading of hydrometeorology.

Table 1.1   Estimates of the World's Water

Reproduced from M.I. L'vovich (1979) *World Water Resources*, translated by
R.L. Nace, by permission of The American Geophysical Union.

	Volume (10^6 km^3)	Percentage	Rate of exchange (years)
Oceans	1370	94.2	3000
Groundwater	60	4.13	5000
Ice sheets and glaciers	24	1.65	8000
Surface water on land	0.28	0.019	7
Soil moisture	0.08	0.0055	1
Rivers	0.0012	0.00008	0.031
Atmospheric vapour	0.014	0.00096	0.027

1.2   Hydrometeorology

The science of meteorology has long been recognized as a separate discipline,
though students of the subject usually come to it from a rigorous training in
physics or mathematics. The study of hydrometeorology has evolved as a
specialist branch of hydrology linking the fundamental knowledge of the
meteorologist with the needs of the hydrologist. In this text, hydrometeorology is taken to be the study of precipitation and evaporation, the
two fundamental phases in the hydrological cycle which involve processes in
the atmosphere and at the Earth's surface/atmosphere interface.

The hydrologist will usually be able to call upon the services of a professional meteorologist for weather forecasts and for special studies, e.g. the
magnitude of extreme rainfalls. However, a general understanding of precipitation and evaporation is essential if the hydrologist is to appreciate the
complexities of the atmosphere and the difficulties that the meteorologist
often has in providing answers to questions of quantities and timing. A
description of the properties of the atmosphere and of the main features of
solar radiation will provide the bases for considering the physics of evaporation and the formation of precipitation. A more detailed treatment of these
subjects will be found in the numerous meteorological texts (e.g. McIntosh &
Thom, 1969; Meteorological Office, 1978).

1.2.1   The Atmosphere

The atmosphere forms a distinctive, protective layer about 100 km thick
around the Earth. Its average structure is shown in Fig. 1.2. Note that
although both air pressure and density decrease rapidly and continuously
with increasing altitude, the temperature varies in an irregular but characteristic way. The layers of the atmosphere, 'spheres', are defined by this

temperature profile. After a general decrease in temperature through the *troposphere*, the rise in temperature from heights of 20 to 50 km is caused by a layer of ozone, which absorbs short-wave solar radiation, releasing some of the energy as heat.

To the hydrologist, the troposphere is the most important layer because it contains 75% of the weight of the atmosphere and virtually all its moisture. The meteorologist however is becoming increasingly interested in the stratosphere and mesosphere, since it is in these outer regions that some of the disturbances affecting the troposphere and the Earth's surface have their origins.

The height of the *tropopause*, the boundary zone between the troposphere and the stratosphere, is shown at about 11 km in Fig. 1.2, but this is an average figure, which ranges from 8 km at the Poles to 16 km at the Equator. Seasonal variations also are caused by changes in pressure and air temperature in the atmosphere. In general, when surface temperatures are high and there is a high sea level pressure, then there is a tendency for the tropopause to be at a high level. On average, the temperature from ground level to the tropopause falls steadily with increasing altitude at the rate of 6.5 °C km^{-1}. This is known as the *lapse rate*. Some of the more hydrologically pertinent characteristics of the atmosphere as a whole are now defined in more precise terms.

Atmospheric pressure and density.   The meteorologist's definition of atmospheric pressure is 'the weight of a column of air of unit area of cross-section from the level of measurement to the top of the atmosphere'. More specifically, pressure may be considered to be the downward force on unit horizontal area resulting from the action of gravity (g) on the mass (m) of air vertically above.

At sea level, the average atmospheric pressure (p) is 1 bar (10^5 Pa or 10^5 N m^{-2}). A pressure of 1 bar is equivalent to 760 mm of mercury; the average reading on a standard mercury barometer. Measurements of atmospheric pressure are usually given in millibars (mb), since the meteorologist prefers whole numbers in recording pressures and their changes to the nearest millibar. It is common meteorological practice to refer to heights in the atmosphere by their average pressure in mb, e.g. the top of the stratosphere the stratopause is at the 1 mb level (Fig. 1.2). The air density (ρ) may be obtained from the expression $\rho = p/RT$, where R is the specific gas constant for dry air, (0.29 kJ kg^{-1}K^{-1}) and T K is the air temperature. At sea level, the average $T = 288$ K and thus $\rho = 1.2$ kg m^{-3} (or 1.2×10^{-3} g cm^{-3}) on average at sea level. Air density falls off rapidly with height (Fig. 1.2). Unenclosed air, a compressible fluid, can expand freely, and as pressure and density decrease with height indefinitely, the limit of the atmosphere becomes indeterminate. Within the troposphere however, the lower pressure limit is about 100 mb. At sea level, pressure variations range from about 940 to 1050 mb; the average sea level pressure around the British Isles is 1013 mb. Pressure records form the basis of the meteorologist's synoptic charts with the patterns formed by the isobars (lines of equal pressure) defining areas of high and low pressure (anticyclones and depressions, respectively). Interpretation of the charts plotted from observations made at successive specified times enables the changes in weather systems to be identified and to be forecast ahead. In addition to the sea level measurements, upper air data are plotted and analysed for different levels in the atmosphere.

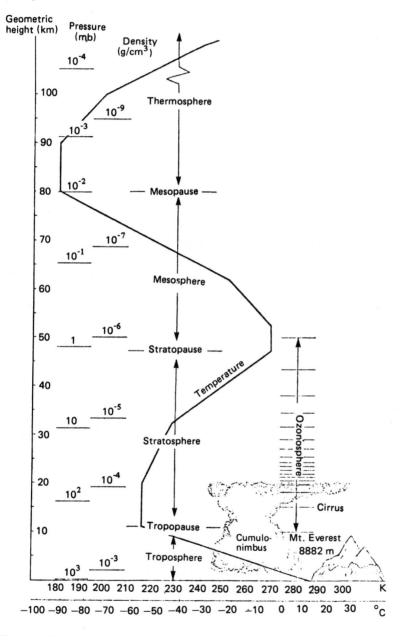

Fig. 1.2   Structure of the atmosphere. (Reproduced from W.L. Donn (1975) *Meteorology, 4th edn.*, by permission of McGraw Hill.)

Chemical composition.   Dry air has a very consistent chemical composition throughout the atmosphere up to the mesopause at 80 km. The proportions of the major constituents are as shown in Table 1.2. The last category contains small proportions of other inert gases and, of particular importance, the stratospheric layer of ozone which filters the sun's radiation. Small quantities of hydrocarbons, ammonia and nitrates may also exist temporarily in the atmosphere. Man-made gaseous pollutants are confined to relatively limited

Table 1.2   Major Constituents of Air

	Percentage (by mass)
Nitrogen	75.51
Oxygen	23.15
Argon	1.28
Carbon dioxide, etc.	0.06

areas of heavy industry, but can have considerable effects on local weather conditions. Traces of radioactive isotopes from nuclear fission also contaminate the atmosphere, particularly following nuclear explosions. Although there is no evidence that isotopes cause weather disturbances, their presence has been found useful in tracing the movement of water through the hydrological cycle.

Water vapour.   The amount of water vapour in the atmosphere (Table 1.3) is directly related to the temperature and thus, although lighter than air, water vapour is restricted to the lower layers of the troposphere because temperature decreases with altitude. The distribution of water vapour also varies over the Earth's surface according to temperature, and is lowest at the Poles and highest in Equatorial regions. The water vapour content or *humidity* of air is usually measured as a vapour pressure, and the units used are millibars (mb).

Several well recognized physical properties concerned with water in the

Table 1.3   Average Water Vapour Values for Latitudes with Temperate Climates (Volume %)

Height (km)	Water vapour	Height (km)	Water vapour
0	1.3	5	0.27
1	1.0	6	0.15
2	0.69	7	0.09
3	0.49	8	0.05
4	0.37		

atmosphere are defined to assist understanding of the complex changes that occur in the meteorological phases of the hydrological cycle.

(a) *Saturation*. Air is said to be saturated when it contains the maximum amount of water vapour it can hold at its prevailing temperature. The relationship between saturation vapour pressure (e) and air temperature is shown in Fig. 1.3. At typical temperatures near the ground, e ranges from 5 to 50 mb. At any temperature $T = T_a$, saturation occurs at corresponding vapour pressure $e = e_a$.

Meteorologists acknowledge that saturated air may take up even more water vapour and become *supersaturated* if it is in contact with liquid water in a sufficiently finely divided state (for example, very small water droplets in clouds). At sub-zero temperatures, there are two saturation vapour pressure

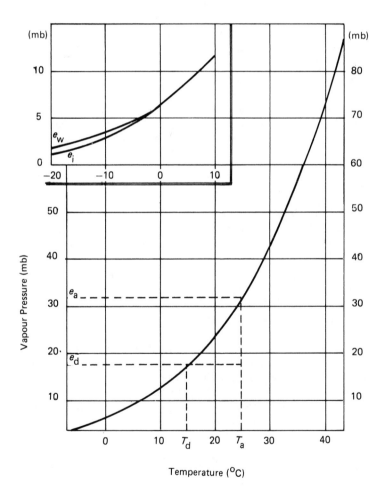

Fig. 1.3 Saturation pressure and air temperature. $e_a - e_d$ = saturation deficit. T_d = dew point temperature.

curves, one with respect to water (e_w) and one with respect to ice (e_i) (Fig. 1.3 inset). In the zone between the curves, the air is unsaturated with respect to water but supersaturated with respect to ice. This is a common condition in the atmosphere as will be seen later.

(b) *Dew point* is the temperature, T_d, at which a mass of unsaturated air becomes saturated when cooled, with the pressure remaining constant. In Fig. 1.3, if the air at temperature T_a is cooled to T_d, the corresponding saturation vapour pressure, e_d, represents the amount of water vapour in the air.

(c) *Saturation deficit* is the difference between the saturation vapour pressure at air temperature, T_a, and the actual vapour pressure represented by the saturation vapour pressure at T_d, the dew point. The saturation deficit, ($e_a - e_d$), represents the further amount of water vapour that the air can hold at the temperature, T_a, before becoming saturated.

(d) *Relative humidity* is the relative measure of the amount of moisture in the air to the amount needed to saturate the air at the same temperature, i.e. e_d/e_a, represented as a percentage. Thus, if $T_a = 30\ °C$ and $T_d = 20\ °C$,

$$\text{relative humidity} = \frac{e_d}{e_a} \times 100 = \frac{23\ \text{mb}}{42.5\ \text{mb}} \times 100 = 54\%$$

(e) *Absolute humidity* (ρ_w) is generally expressed as the mass of water vapour per unit volume of air at a given temperature and is equivalent to the water vapour density. Thus, if a volume V m³ of air contains m_w g of water vapour,

$$\rho_w = \frac{\text{mass of water vapour (g)}}{\text{volume of air (m}^3)} = \frac{m_w}{V}\ (\text{g m}^{-3})$$

(f) *Specific humidity* (q) relates the mass of water vapour (m_w g) to the mass of moist air (in kg) in a given volume; this is the same as relating the absolute humidity (g m⁻³) to the density of the same volume of unsaturated air (ρ kg m⁻³):

$$q = \frac{m_w\ (\text{g})}{(m_w + m_d)\ (\text{kg})} = \frac{\rho_w}{\rho}\ (\text{g kg}^{-1})$$

where m_d is the mass (kg) of the dry air.

(g) *Precipitable water* is the total amount of water vapour in a column of air expressed as the depth of liquid water in millimetres over the base area of the column. Assessing this amount is a specialised task for the meteorologist. The precipitable water gives an estimate of maximum possible rainfall under the unreal assumption of total condensation.

In a column of unit cross-sectional area, a small thickness, dz, of moist air contains a mass of water given by:

$$dm_w = \rho_w dz$$

Thus, in a column of air from heights z_1 to z_2, corresponding to pressures p_1 and p_2:

$$\text{the total mass of water } m_w = \int_{z_1}^{z_2} \rho_w dz$$

Also, $dp = -\rho g\, dz$ and, by rearrangement, $dz = -dp/\rho g$. Thus:

$$m_w = -\int_{p_1}^{p_2} \frac{\rho_w}{\rho g}\, dp$$

$$= \frac{1}{g}\int_{p_2}^{p_1} q\, dp$$

Allowing for the conversion of the mass of water (m_w) to equivalent depth over a unit cross-sectional area, the precipitable water is given by:

$$W\,(\text{mm}) = \frac{0.1}{g}\int_{p_2}^{p_1} q\, dp$$

where p is in mb, q in g kg^{-1} and $g = 9.81$ m s^{-2}

In practice, the integration cannot be performed since q is not known as a function of p. A value of W is obtained by *summing* the contributions for a sequence of layers in the troposphere from a series of measurements of the specific humidity q at different heights and using the average specific humidity $\bar{q}$ over each layer with the appropriate pressure difference:

$$W\,(\text{mm}) = \frac{0.1}{g} \sum_{p_2}^{p_1} \bar{q}\,\Delta p$$

Example.   From a radiosonde (balloon) ascent, the pairs of measurements of pressure and specific humidity shown in Table 1.4 were obtained. The precipitable water in a column of air up to the 250 mb level is calculated ($g = 9.81$ m s^{-2}).

Table 1.4

Pressure (mb)	1005	850	750	700	620	600	500	400	250
Specific humidity q (g kg^{-1})	14.2	12.4	9.5	7.0	6.3	5.6	3.8	1.7	0.2
$P_n - P_{n+1} = \Delta p$		155	100	50	80	20	100	100	150
Mean $q = \bar{q}$		13.30	10.95	8.25	6.65	5.95	4.70	2.75	0.95
$\bar{q}\Delta p$		2061.5	1095.	412.5	532.	119.	470.	275.	142.5

$\Sigma \bar{q}\Delta p = 5107.5$

$\therefore$ The precipitable water up to the 250 mb level:

$$W = \frac{0.1}{g} \sum_{p_2}^{p_1} (\bar{q}\Delta p) = \frac{510.75}{g}\ \text{mm} = 52.1\ \text{mm}$$

1.2.2 Solar Radiation

The main source of energy at the Earth's surface is radiant energy from the sun, termed solar radiation or insolation. It is the solar radiation impinging on the Earth that fuels the heat engine driving the hydrological cycle. The amount of radiant energy received at any point on the Earth's surface (assuming no atmosphere) is governed by the following well defined factors.

(a) *The solar output.* The sun, a globe of incandescent matter, has a gaseous outer layer about 320 km thick and transmits light and other radiations towards the Earth from a distance of 145 million km. The rate of emission of energy is shown in Fig. 1.4 but only a small fraction of this is intercepted by the Earth. Half the total energy emitted by the sun is in the visible light range, with wavelengths from 0.4 to 0.7 μm. The rest arrives as ultraviolet or infrared waves, from 0.25 μm up to 3.0 μm.

The maximum rate of the sun's emission (10 500 kW m^{-2}) occurs at 0.5 μm wavelength in the visible light range. Although there are changes in the solar output associated with the occurrence of sunspots and solar flares, these are disregarded in assessing the amount of energy received by the Earth. The total solar radiation received in unit time on unit area of a surface placed at right angles to the sun's rays at the Earth's mean distance from the sun is known as the *solar constant*. The average value of the solar constant is 1.39 kW m^{-2} (1.99 cal cm^{-2} min^{-1}).

(b) *Distance from the sun.* The distance of any point on the Earth's surface from the sun is changing continuously owing to the Earth's eccentric orbit. The Earth is nearest the sun in January at perihelion and furthest from the sun in July at aphelion. The solar constant varies accordingly.

(c) *Altitude of the sun.* The sun's altitude above the horizon has a marked influence on the rate of solar radiation received at any point on the Earth. The factors determining the sun's altitude are latitude, season and time of day.

(d) *Length of day.* The total amount of radiation falling on a point of the Earth's surface is governed by the length of the day, which itself depends on latitude and season.

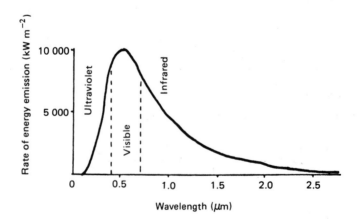

Fig. 1.4   Solar radiation.

Atmospheric effects on solar radiation.   The atmosphere has a marked effect on the energy balance at the surface of the Earth. In one respect it acts as a shield protecting the Earth from extreme external influences, but it also prevents immediate direct loss of heat. Thus it operates as an energy filter in both directions. The interchanges of heat between the incoming solar radiation and the Earth's surface are many and complex. There is a loss of energy from the solar radiation as it passes through the atmosphere known as *attenuation*. Attenuation is brought about in three principal ways.

(a) *Scattering*. About 9% of incoming radiation is scattered back into space through collisions with molecules of air or water vapour. A further 16% are also scattered, but reach the Earth as diffuse radiation, especially in the shorter wavelengths, giving the sky a blue appearance.

(b) *Absorption*. 15% of solar radiation is absorbed by the gases of the atmosphere, particularly by the ozone, water vapour and carbon dioxide. These gases absorb wavelengths of less than 0.3 μm only, and so very little of this radiation penetrates below an altitude of 40 km.

(c) *Reflection*. On average, 33% of solar radiation is reflected from clouds and the ground back into space. The amount depends on the *albedo* (r) of the reflecting surfaces. White clouds and fresh white snow reflect about 90% of the radiation ($r = 0.9$), but a dark tropical ocean under a high sun absorbs nearly all of it ($r \rightarrow 0$). Between these two extremes is a range of surface conditions depending on roughness, soil type and water content of the soil. The albedo of the water surface of a reservoir is usually assumed to be 0.05, and of a short grass surface, 0.25.

Net radiation.   As a result of the various atmospheric losses, only 43% of solar radiation reaches the Earth's surface. The short-wave component of that 43% is absorbed and heats the land and oceans. The Earth itself radiates energy in the long-wave range (Fig. 1.5) and this long-wave radiation is readily absorbed by the atmosphere. The Earth's surface emits more than twice as much energy in the infrared range as it receives in short-wave solar radiation.

The balance between incoming and outgoing radiation varies from the Poles to the Equator. There is a net heat gain in equatorial regions and a net heat loss in polar regions. Hence, heat energy travels through circulation of the atmosphere from lower to higher latitudes. Further variations occur because the distribution of the continents and oceans leads to differential heating of land and water.

The amount of energy available at any particular point on the Earth's surface for heating the ground and lower air layers, and for the evaporation of water, is called the *net radiation*.

The net radiation R may be defined by the equation:

$$R = S^{\downarrow} - r(S^{\downarrow}) + L^{\downarrow} - L^{\uparrow}$$

where S and L are short- and long-wave radiation and r is the albedo, and the arrows indicate incoming and outgoing directions.

Incoming long-wave radiation comes from clouds (from absorbed solar radiation), and this has the following effects in the net radiation equation:

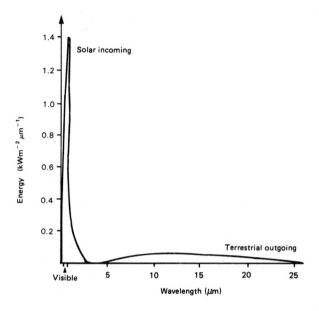

Fig. 1.5   Solar and terrestrial radiation.

In clear conditions, $L^{\downarrow} \simeq (0.6 \text{ to } 0.8)\, L^{\uparrow}$, thus $L^{\downarrow} - L^{\uparrow}$ gives a net loss of long-wave radiation.

For cloudy conditions, $L^{\downarrow} \simeq L^{\uparrow}$ and $L^{\downarrow} - L^{\uparrow}$ becomes 0.

More significant are diurnal variations in net radiation which is the primary energy source for evaporation.

At night, $S = 0$ and $L^{\downarrow}$ is smaller or negligible so that $R \simeq L^{\uparrow}$. In other words, net radiation is negative, and there is a marked heat loss, which is particularly noticeable when the sky is clear.

Some average values of solar (S), terrestrial (L) and net (N) radiation for points on the earth's surface are given in Table 1.5.

Table 1.5   Average Radiation Values for Selected Latitudes (Wm^{-2})

	July season			January season		
	S	L	N	S	L	N
50°N	250	210	40	70	190	− 120
Equator	280	240	40	310	240	70
30°S	170	220	− 50	320	230	90

1.3   Evaporation

Evaporation is the primary process of water transfer in the hydrological cycle. The oceans contain 95 % of the Earth's water and constitute a vast reservoir that remains comparatively undisturbed. From the surface of the seas and oceans, water is evaporated and transferred to temporary storage in the atmosphere, the first stage in the hydrological cycle.

1.3.1   Factors Affecting Evaporation

To convert liquid water into gaseous water vapour at the same temperature a supply of energy is required (in addition to that possibly needed to raise the liquid water to that temperature). The *latent heat of vaporization* must be added to the liquid molecules to bring about the change of state. The energy available for evaporation is the net radiation obtaining at the water surface and is governed by local conditions of solar and terrestrial radiation.

The rate of evaporation is dependent on the temperature at the evaporating surface and that of the ambient air. It also depends on the vapour pressure of the existing water vapour in the air, since this determines the amount of additional water vapour that the air can absorb. From the saturation vapour pressure and air temperature relationship shown in Fig. 1.3, it is clear that the rate of evaporation is dependent on the saturation deficit. If the water surface temperature, T_s, is equal to the air temperature, T_a, then the saturated vapour pressure at the surface, e_s, is equal to e_a. The saturation deficit of the air is given by $(e_s - e_d)$ where e_d is the measure of the actual vapour pressure of the air at T_a.

As evaporation proceeds, the air above the water gradually becomes saturated and, when it is unable to take up any more moisture, evaporation ceases. The replacement of saturated air by drier air would enable evaporation to continue. Thus, wind speed is an important factor in controlling the rate of evaporation. The roughness of the evaporating surface is a subsidiary factor in controlling the evaporation rate because it affects the turbulence of the air flow.

In summary, evaporation from an open water surface is a function of available energy, the net radiation, the temperatures of surface and air, the saturation deficit and the wind speed. The evaporation from a vegetated surface is a function of the same meteorological variables, but it is also dependent on soil moisture. From a land surface it is a combination of the evaporation of liquid water from precipitation collected on the surface, rainfall intercepted by vegetation (mainly trees) and the transpiration of water by plants. A useful concept is potential evapotranspiration, the amount of evaporation and transpiration that would take place given an unlimited supply of moisture. Potential evapotranspiration is the maximum possible evaporation under given meteorological conditions and, as it is easily computed, gives a ready assessment of possible loss of water. Methods for the measurement and assessment of evaporation quantities are presented in detail in Chapter 4.

1.4   Precipitation

The moisture in the atmosphere, although forming one of the smallest storages of the Earth's water, is the most vital source of fresh water for mankind. Water is present in the air in its gaseous, liquid, and solid states as water vapour, cloud droplets and ice crystals, respectively.

The formation of precipitation from the water as it exists in the air is a complex and delicately balanced process. If the air was pure, condensation of the water vapour to form liquid water droplets would occur only when the air became greatly supersaturated. However, the presence of small airborne particles called *aerosols* provides nuclei around which water vapour in normal saturated air can condense. Many experiments, both in the laboratory and in the open air, have been carried out to investigate the requisite conditions for the change of state. Aitken (Mason, 1975) distinguished two main types of condensation nuclei: *hygroscopic particles* having an affinity for water vapour, on which condensation begins before the air becomes saturated (mainly salt particles from the oceans); and *non-hygroscopic particles* needing some degree of supersaturation, depending on their size, before attracting condensation. This latter group derives from natural dust and grit from land surfaces and from man-made smoke, soot and ash particles.

Condensation nuclei range in size from a radius 10^{-3} μm for small ions to 10 μm for large salt particles. The concentration of aerosols in time and space varies considerably. A typical number for the smallest particles is 40 000 per cm^3, whereas for giant nuclei of more than 1 μm radius there might be only 1 per cm^3. Large hygroscopic salt nuclei are normally confined to maritime regions, but the tiny particles called *Aitken nuclei* can travel across continents and even circumnavigate the earth. Although condensation nuclei are essential for widespread condensation of water vapour, only a small fraction of the nuclei present in the air take part in cloud droplet formation at any one time.

Other conditions must be fulfilled before precipitation occurs. First, moist air must be cooled to near its dew point. This can be brought about in several ways:

(a) by an adiabatic expansion of rising air. A volume of air may be forced to rise by an impeding mountain range. The reduction in pressure causes a lowering of temperature without any transference of heat;

(b) by a meeting of two very different air masses. For example, when a warm, moist mass of air converges with a cold mass of air, the warm air is forced to rise and may cool to the dew point. Any mixing of the contrasting masses of air would also lower the overall temperature; or

(c) by contact between a moist air mass and a cold object such as the ground.

Once cloud droplets are formed, their growth depends on hygroscopic and surface tension forces, the humidity of the air, rates of transfer of vapour to the water droplets and the latent heat of condensation released. A large population of droplets competes for the available water vapour and so their growth rate depends on their origins and on the cooling rate of air providing the supply of moisture (Fig. 1.6).

The mechanism becomes complicated when the temperature reaches freezing point. Pure water can be supercooled to about -40 °C (233 K) before freezing spontaneously. Cloud droplets are unlikely to freeze in normal air conditions until cooled below -10 °C (263 K) and commonly

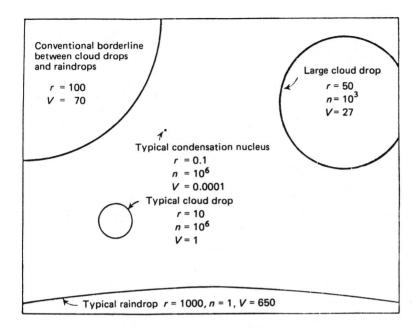

Fig. 1.6   Comparative sizes, concentrations and terminal falling velocities of some particles involved in condensation and precipitation processes. r = radius (μm); n = number per dm^3 (10^3 cm^3); V = terminal velocity (cm s^{-1}). (Reproduced from B.J. Mason (1975) *Clouds, Rain and Rainmaking, 2nd edn.*, by permission of Cambridge University Press.)

exist down to -20 °C (253 K). They freeze only in the presence of small particles called *ice nuclei*, retaining their spherical shape and becoming solid ice crystals. Water vapour may then be deposited directly on to the ice surfaces. The crystals grow into various shapes depending on temperature and the degree of supersaturation of the air with respect to the ice.

Condensed water vapour appears in the atmosphere as clouds in various characteristic forms; a standard classification of clouds is shown in Fig. 1.7. The high clouds are composed of ice crystals, the middle clouds of either water droplets or ice crystals, and the low clouds mainly of water droplets, many of them supercooled. Clouds with vigorous upwards vertical development, such as cumulonimbus, consist of cloud droplets in their lower layers and ice crystals at the top.

1.4.1   Theories of Raindrop Growth

Considerable research has been carried out by cloud physicists on the various stages involved in the transference of atmospheric water vapour into precipitable raindrops or snowflakes. A cloud droplet is not able to grow to raindrop size by the simple addition of water vapour condensing from the air. It is worth bearing in mind that one million droplets of radius 10 μm are equivalent to a single small raindrop of radius 1 mm. Fig. 1.6 shows the principal characteristics of nuclei, cloud droplets and raindrops.

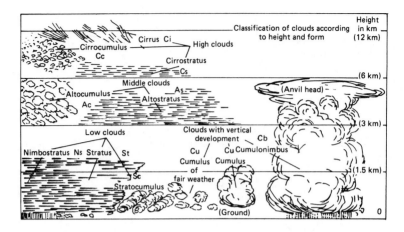

Fig. 1.7   Classification of clouds. (Reproduced from A.N. Strahler (1973) *Introduction to Physical Geography, 3rd edn.*, by permission of John Wiley & Sons Inc.)

Cloud droplets can grow naturally to about 100 μm in radius, and although tiny drops from 100 μm to 500 μm may, under very calm conditions, reach the ground, other factors are at work in forming raindrops large enough to fall to the ground in appreciable quantities. There are several theories of how cloud droplets grow to become raindrops, and investigations into the details of several proposed methods continue to claim the attention of research workers.

The Bergeron process, named after the famous Norwegian meteorologist, requires the coexistence in a cloud of supercooled droplets and ice particles and a temperature less than 0 °C (273 K). The air is saturated with respect to water but supersaturated with respect to ice. Hence water vapour is deposited on the ice particles to form ice crystals. The air then becomes unsaturated with respect to water so droplets evaporate. This process continues until either all the droplets have evaporated or the ice crystals have become large enough to drop out of the cloud to melt and fall as rain as they reach lower levels. Thus the crystals grow at the expense of the droplets. This mechanism operates best in clouds with temperatures in the range – 10 °C to – 30 °C (263–243 K) with a small liquid water content.

Growth by collision.   In clouds where the temperature is above 0 °C (273 K), there are no ice particles present and cloud droplets collide with each other and grow by coalescence. The sizes of these droplets vary enormously and depend on the size of the initial condensation nuclei. Larger droplets fall with greater speeds through the smaller droplets with which they collide and coalesce. As larger droplets are more often formed from large sea-salt nuclei, growth by coalescence operates more frequently in maritime than in continental clouds. In addition, as a result of the dual requirements of a relatively high temperature and generous liquid water content, the growth of raindrops by coalescence operates largely in summer months in low-level clouds.

When cloud temperatures are below 0 °C (273 K) and the cloud is composed of ice particles, their collision causes growth by *aggregation* to form snowflakes. The most favourable clouds are those in the 0 °C to – 4 °C (269 K) range and the size of snowflakes decreases with the cloud temperature and water content.

Growth by accretion occurs in clouds containing a mixture of droplets and ice particles. Snow grains, ice pellets, or hail are formed as cloud droplets fuse on to ice particles. Accretion takes place most readily in the same type of cloud that favours the Bergeron process, except that a large liquid water content is necessary for the water droplets to collide with the ice particles.

Even when raindrops and snowflakes have grown large enough for their gravity weight to overcome updraughts of air and fall steadily towards the ground, their progress is impeded by changing air conditions below the clouds. The temperature may rise considerably near the Earth's surface and the air become unsaturated. As a result snowflakes usually melt to raindrops and the raindrops may evaporate in the drier air. On a summer's day it is not uncommon to see cumulus clouds trailing streams of rain which disappear before they reach the ground. With dry air below a high cloud base of about 3 km, all precipitation will evaporate. Hence it is rare to see rainfall from altocumulus, altrostratus and higher clouds (see Fig. 1.7). Snowflakes rarely reach the ground if the surface air temperature is above 4 °C, but showers of fine snow can occur with the temperature as high as 7 °C if the air is very dry.

1.5   Weather Patterns Producing Precipitation

The main concern of the meteorologist is an understanding of the general circulation of the atmosphere with the aim of forecasting the movements of pressure patterns and their associated winds and weather. It is sufficient for the hydrologist to be able to identify the situations that provide the precipitation, and for the practising civil engineer to keep a 'weather eye' open for adverse conditions that may affect his site work.

The average distribution and seasonal changes of areas of high atmospheric pressure (*anticyclones*) and of low pressure areas (*depressions*) can be found in most good atlases. Associated with the location of anticyclones is the development of homogeneous air masses. An *homogeneous air mass* is a large volume of air, generally covering an area greater than 1000 km in diameter, that shows little horizontal variation in temperature or humidity. It develops in the stagnant conditions of a high-pressure area and takes on the properties of its location (known as a *source region*). In general, homogeneous air masses are either cold and stable, taking on the characteristics of the polar regions from where they originate, or they are warm and unstable, revealing their tropical source of origin. Their humidity depends on whether they are centred over a large continent or over the ocean. The principal air masses are summarized in Table 1.6. Differences in atmospheric pressure cause air masses to move from high- to low-pressure regions and they become modified by the environments over which they pass. Although they remain homogeneous, they may travel so far and become so modified that they warrant reclassification. For example, when polar maritime air reaches the British

Isles from a south-westerly direction, having circled well to the south over warm subtropical seas, its character will have changed dramatically.

Precipitation can come directly from a maritime air mass that cools when obliged to rise over mountains in its path. Such precipitation is known as *orographic rainfall* (or snowfall, if the temperature is sufficiently low), and is an important feature of the western mountains of the British Isles, which lie across the track of the prevailing winds bringing moisture from the Atlantic Ocean. Orographic rain falls similarly on most hills and mountains in the world, with similar locational characteristics, though it may occur only in particular seasons.

When air is cooled as a result of the converging of two contrasting air masses it can produce more widespread rainfall independent of surface land features. The boundary between two air masses is called a *frontal zone*. It intersects the ground at the *front*, a band of about 200 km across. The character of the front depends on the difference between the air masses. A steep temperature gradient results in a strong or *active front* and much rain, but a small temperature difference produces only a *weak front* with less or even no rain. The juxtaposition of air masses across a frontal zone gives rise to two principal types of front according to the direction of movement.

Fig. 1.8 illustrates cloud patterns and weather associated firstly with a *warm front* in which warm air is replacing cold air, and secondly with a *cold front* in which cold air is pushing under a warm air mass. In both cases, the warm air is made to rise and hence cool, and the condensation of water vapour forms characteristic clouds and rainfall. The precipitation at a warm front is usually prolonged with gradually increasing intensity. At a cold front, however, it is heavy and short-lived. Naturally, these are average conditions; sometimes no rain is produced at all.

Over the world as a whole there are distinctive regions between areas of high pressure where differing air masses confront each other. These are principally in the mid-latitudes between 30° and 60° in both hemispheres, where the main boundary, the *polar front*, separates air masses having their origins in polar regions from the tropical air masses.

In addition, there is a varying boundary between air masses originating in the northern and southern hemispheres known as the intertropical convergence zone (ITCZ). The seasonal migration of the ITCZ plays a large part in the formation of the monsoon rains in south-east Asia and in the islands of Indonesia.

Four major weather patterns producing precipitation have been selected for more detailed explanation.

1.5.1   Mid-latitute Cyclones or Depressions

Depressions are the major weather pattern for producing precipitation in the temperate regions. More than 60% of the annual rainfall in the British Isles comes from such disturbances and their associated features. They develop along the zone of the polar front between the polar and tropical air masses. Knowledge of the growth of depressions, the recognition of air masses, and the definition of fronts all owe much to the work of the Norwegian meteorologists in the 1920s, and scholars from that country continue to make valuable contributions to this subject (e.g. Petterssen & Smebye, 1971).

Table 1.6   Classification of Air Masses

Air mass	Source region	Properties at source
Polar maritime (Pm)	oceans; 50° latitude	cool, rather moist, unstable
Polar continental (Pc)	continents in vicinity of Arctic Circle; Antarctica	cool, dry, stable
Arctic or Antarctic (A)	Arctic Basin and Central Antarctica in winter	very cold, dry, stable
Tropical maritime (Tm)	sub-tropical oceans	warm and moist; unstable inversion common feature
Tropical continental (Tc)	deserts in low latitude; primarily the Sahara and Australian deserts	hot and dry

The main features in the development and life of a mid-latitude cyclone are shown in Fig. 1.9. The first diagram illustrates in plan view the isobars of a steady-state condition at the polar front between contrasting air masses. The succeeding diagrams show the sequential stages in the average life of a depression. A slight perturbation caused by irregular surface conditions, or perhaps a disturbance in the lower stratosphere, results in a shallow wave developing in the frontal zone. The initial wave, moving along the line of the front at 15–20 m s⁻¹ (30–40 knots), may travel up to 1000 km without further development. If the wavelength is more than 500 km, the wave usually increases in amplitude, warm air pushes into the cold air mass and active fronts are formed. As a result, the air pressure falls and a 'cell' of low pressure becomes trapped within the cold air mass. Gradually the cold front overtakes the warm front, the warm air is forced aloft, and the depression becomes *occluded*. The low-pressure centre then begins to fill and the depression dies as the pressure rises. On average, the sequence of growth from the first perturbation of the frontal zone to the occlusion takes 3 to 4 days. Precipitation usually occurs along the fronts and, in a very active depression, large amounts can be produced by the occlusion, especially if its speed of passage is retarded by increased friction at the Earth's surface. At all stages, orographic influences can increase the rainfall as the depression crosses land areas. A range of mountains can delay the passage of a front and cause longer periods of rainfall. In addition, if mountains delay the passage of a warm front, the occlusion of the depression may be speeded up.

Considerable research into the distribution of rainfall from depressions has been carried out in the UK in recent years, and meteorologists are progressing rapidly towards an explanation of the cells of intense rainfall that have caused serious floods in medium-sized rural catchments. However, the

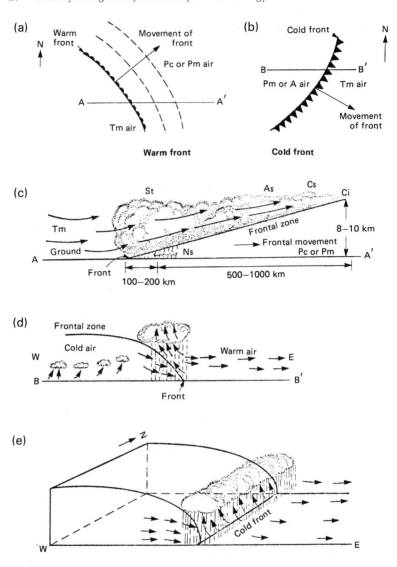

Fig. 1.8 Frontal weather conditions. (a) Map of a warm front. The broken lines show successive positions of the front. (b) Map of a cold front. (c) Cross-section of a warm front showing a typical cloud deck forming in the rising warm air. (d) Cross-section of a cold front showing typical weather characteristics. (e) Three-dimensional view of a cold front showing cold-front cloud and showers. (Reproduced from W.L. Donn (1975) *Meteorology, 4th edn.*, by permission of McGraw-Hill.)

forecasting of such troublesome phenomena is still difficult and awaits success from further work (Browning & Harrold, 1969; Atkinson & Smithson, 1974).

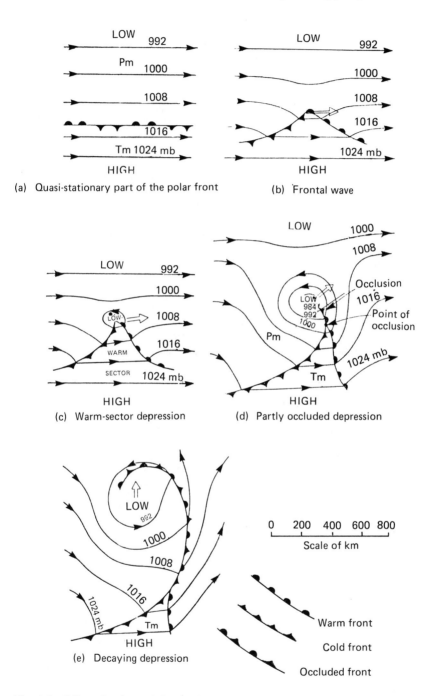

(a) Quasi-stationary part of the polar front

(b) Frontal wave

(c) Warm-sector depression

(d) Partly occluded depression

(e) Decaying depression

Fig. 1.9   Life cycle of a model occluding depression. (Reproduced from *A Course in Elementary Meteorology* (1962), Meteorological Office by permission of the Controller, Her Majesty's Stationery Office. © Crown copyright.)

1.5.2   Waves in the Easterlies and Tropical Cyclones

Small disturbances are generated in the trade wind belts in latitudes 5–25° both north and south of the Equator. Irregular wind patterns showing as isobaric waves on a weather map develop in the tropical maritime air masses on the equatorial side of the subtropical high pressure areas. They have been studied most in the Atlantic Ocean to the north of the South American continent. A typical easterly wave is shown in Fig. 1.10. A trough of low pressure is shown moving westwards on the southern flanks of the Azores anticyclone. The length of the wave extends over 15°–20° longitude (1500 km) and, moving with an average speed of 6.7 m s⁻¹ (13 knots), takes 3 to 4 days to pass. The weather sequence associated with the wave is indicated beneath the diagram. In the tropics, the cloud forming activity from such disturbances is vigorous and subsequent rainfall can be very heavy: up to 300 mm may fall in 24 h.

As in mid-latitudes, the wave may simply pass by and gradually die away, but the low pressure may *deepen* with the formation of a closed circulation with encircling winds. The cyclonic circulation may simply continue as a shallow depression giving increased precipitation but nothing much else. However, rapidly deepening pressure below 1000 mb usually generates hurricane-force winds blowing round a small centre of 30–50 km radius — known as the *eye*. At its mature stage, a *hurricane* centre may have a pressure of less than 950 mb. Eventually the circulation spreads to a radius of about 300 km and the winds decline. Copious rainfall can occur with the passage of

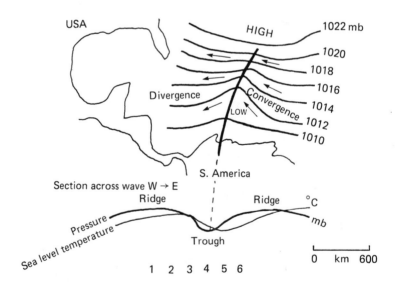

Fig. 1.10   A wave in the easterlies, Weather sequence: 1 — small Cu, no pp.; 2 — Cu, a few build-ups, haze, no pp.; 3 — larger Cu, Ci & Ac, better visibility, pr … pr … pr (showers); 4 — very large Cu, overcast Ci Ac, prpr or rr (continuous rain); 5 — Cu & Cb, Sc, As, Ac, Ci, pRpR (heavy showers), (thunderstorm); 6 — large Cu, occasional Cb, some Sc, Ac, Ci, pRpr – prpr.

a hurricane; record amounts have been measured in the region of southeast Asia, where the effects of the storms have been accentuated by orography. However, the rainfall is difficult to measure in such high winds. In fact, slower moving storms usually give the higher records. Hurricanes in the region of Central America often turn northwards over the United States and die out over land as they lose their moisture. On rare occasions disturbances moving along the eastern coastal areas of the United States are carried into westerly air-streams and become vigorous mid-latitude depressions.

Hurricanes tend to be seasonal events occurring in late summer when the sea temperatures in the areas where they form are at a maximum. They are called *typhoons* in the China Seas and *cyclones* in the Indian Ocean and off the coasts of Australasia. These tropical disturbances develop in well defined areas and usually follow regular tracks; an important fact when assessing extreme rainfalls in tropical regions (Riehl, 1979).

1.5.3   Convectional Precipitation

A great deal of the precipitation in the tropics is caused by local conditions that cannot be plotted on the world's weather maps. When a tropical maritime air mass moves over land at a higher temperature, the air is heated and forced to rise by convection. Very deep cumulus clouds form, becoming cumulonimbus extending up to the tropopause. Fig. 1.11 shows the stages in the life cycle of a typical cumulonimbus. Sometimes these occur in isolation, but more usually several such convective cells grow together and the sky is completely overcast.

The development of convective cells is a regular daily feature of the weather throughout the year in many parts of the tropics, although they do not always provide rain. Cumulus clouds may be produced but evaporate again when the air ceases to rise. With greater vertical air velocities, a large

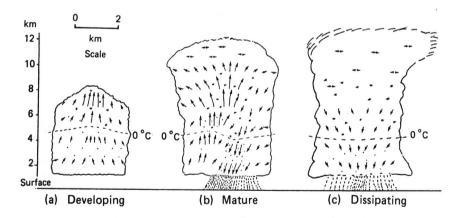

Fig. 1.11   Convective cells — stages in the life cycle. Time scales: (a) approximately 20 min; (b) approximately 20 min, heavy rain and hail, thunder may develop; (c) 30 min to 2 h, rainfall intensity decreasing. Total life cycle 1 - 2 h. ↔ Ice * Snow · Rain and hail ↑↓ Winds.

supply of moisture is carried upwards. As it cools to condensation temperatures, rainfall of great intensity occurs. In extreme conditions, hail is formed by the sequential movement of particles up and down in the cloud, freezing in the upper layers and increasing in size by gathering up further moisture. As the rain and hail fall, they cause vigorous down draughts, and when these exceed the vertical movements, the supply of moisture is reduced, condensation diminishes and precipitation gradually dies away. Thunder and lightning are common features of convectional storms with the interaction of opposing electrical charges in the clouds. The atmospheric pressure typically is irregular during the course of a storm.

Convectional activity is not confined to the tropics; it is a common local rain-forming phenomenon in higher latitudes, particularly in the summer. Recent studies have shown that convection takes place along frontal zones thus adding to rainfall intensities. Wherever strong convectional forces act on warm moist air, rain is likely to form and it is usually of high intensity over a limited area (Harrold, 1973).

1.5.4   Monsoons

Monsoons are weather patterns of a seasonal nature caused by widespread changes in atmospheric pressure. The most familiar example is the monsoon of southeast Asia where the dry, cool or cold winter winds blowing outwards from the Eurasian anticyclone are replaced in summer by warm or hot winds carrying moist air from the surrounding oceans being drawn into a low pressure area over northern India. The seasonal movements of the ITCZ play a large part in the development and characteristics of the weather conditions in the monsoon areas. The circulation of the whole atmosphere has a direct bearing on the migration of the ITCZ, but in general the regularity of the onset of the rainy seasons is a marked feature of the monsoon. Precipitation, governed by the changing seasonal winds, can be caused by confrontation of differing air masses, low pressure disturbances, convection, and orographic effects. A map of monsoon areas is shown in Fig. 1.12. Actual quantities of rain vary but as in most tropical and semi-tropical countries, intensities are high (Riehl, 1979).

References

Atkinson, B.W. and Smithson, P.A. (1974). 'Meso-scale circulations and rainfall patterns in an occluding depression.' *Quart. Jl. Roy. met. Soc.*, **100**, 3–22.

Biswas, A.K. (1970). *History of Hydrology*. North-Holland Pub. Co., 336 pp.

Browning, K.A. and Harrold, T.W. (1969). 'Air motion and precipitation growth in a wave depression.' *Quart. Jl. Roy. met. Soc.*, **95**, 288–309.

Donn, W.L. (1975). *Meteorology*. McGraw Hill, 4th ed., 512 pp.

Dooge, J.C.I. (1959). 'Un bilan hydrologique au XVIIe siecle.' *Houille Blanche*, 14ᵉ annee No. 6, 799–807.

Harrold, T.W. (1973). 'Mechanisms influencing the distribution of precipitation within baroclinic disturbances.' *Quart. Jl. Roy. met. Soc.*, **99**, 232–251.

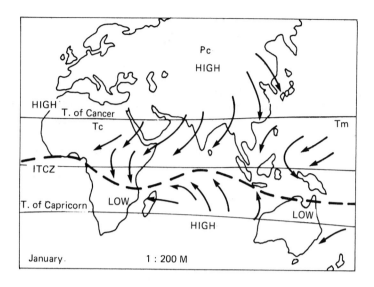

- - - - Inter-tropical convergence zone

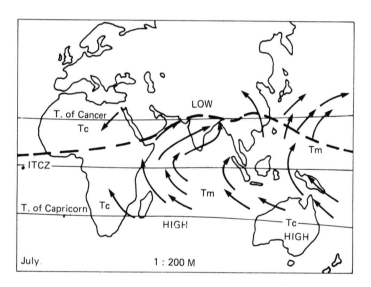

Fig. 1.12   Monsoon lands — pressure systems and winds.

Kirkby, M.J. (ed.) (1978). *Hillslope Hydrology.* John Wiley, 389 pp.

L'vovich, M.I. (1979). *World Water Resources and their Future.* Trans. by R.L. Nace, AGU, 415 pp.

Mason, B.J. (1975). *Clouds, Rain and Rainmaking.* Cambridge University Press, 2nd ed., 189 pp.

McIntosh, D.H. and Thom, A.S. (1969). Essentials of Meteorology. Wykeham Pub. (London) Ltd., 239 pp.

Meteorological Office (1978). *A Course in Elementary Meteorology*. HMSO, 2nd ed., 208 pp.

Petterssen, S. and Smebye, S.J. (1971). 'On the development of extra-tropical cyclones.' *Quart Jl. Roy. met. Soc.*, **97**, 457–482.

Riehl, H. (1979). *Climate and Weather in the Tropics*. Academic Press, 611 pp.

Strahler, A.N. (1973). *Introduction to Physical Geography*. John Wiley, 3rd ed., 468 pp.

Ward, R. (1975). *Principles of Hydrology*. McGraw Hill, 2nd ed., 367 pp.

Part I
Hydrological Measurements

2

Hydrometric Networks

The concept of the hydrological cycle forms the basis for the engineering hydrologist's understanding of the sources of water at or under the Earth's surface and its consequent movement by various pathways back to the principal storage in the oceans. Two of the greatest problems for the hydrologist are quantifying the amount of water in the different phases in the cycle and evaluating the rate of transfer of water from one phase to another within the cycle. Thus measurement within the components of the cycle is a major function; this has not been subject to coordinated planning until the last two decades.

Most hydrological variables such as rainfall, streamflow or groundwater have been measured for many years by separate official bodies, private organizations or even individual amateurs, but there has been very little logical design in the pattern of measurements. The installation of gauges for rainfall or streamflow has usually been made to serve a single, simple purpose, e.g. the determination of the yield of a small mountain catchment for a town's water supply. Nowadays, with the growth of populations and the improvement in communication to serve modern needs, hydrometric schemes are tending to become multipurpose. Nationwide schemes to measure hydrological variables are now considered essential for the development and management of the water resources of a country. As a result, responsibility for measurement stations is becoming focused much more on central or regional government agencies and more precise considerations are being afforded to the planning of hydrological measurements. Cost – benefit assessments are also being made on the effectiveness of data gathering, and hence scientific planning is being recommended to ensure optimum networks to provide the required information (WMO, 1972).

2.1 Gauging Networks

One of the main activities stimulated by the International Hydrological Decade (IHD, 1965–74) was the consideration of hydrological network design, a subject that, it was felt, had been previously neglected (Rodda, 1969). It was recognized that most networks, even in developed countries, were inadequate to provide the data required for the increasing need of hydrologists charged with the task of evaluating water resources for expanding populations.

Before approaching the problem, it is pertinent to define a network. Langbein (1965) gave a broad definition: 'A network is an organized system for the collection of information of a specific kind: that is, each station, point or region of observation must fill one or more definite niches in either space or time'. Expressed more simply, Kohler (1958) infers that a hydrological network is one that provides data commonly used by the hydrologist.

The broad objectives in hydrological network planning and design have been outlined more recently by Hofmann (WMO, 1976). Three main uses for the data are proposed; for planning, which requires long-term records; for management, which requires real-time measurements for daily opera-tions and forecasting; and for research, which generally requires high-quality intensive data. The design of the optimum hydrometric network must be based on quantified objectives wherever possible, with costs and benefits included in the design procedure. One approach is the evaluation of the worth of the data collected which sometimes means realising the benefits lost through lack of data. Closely connected with network design and data collection is an appreciation of the quality of the data. (Aspects of data quality control are considered in Chapter 9.)

2.2   Design Considerations

There are several well defined stages in the design of a network of gauging stations for the measurement of hydrological variables. The first comprises initial background research on the location and known characteristics of the area to be studied. The size of the area and whether it is a political entity or a natural drainage basin are of prime importance. Many of the guide lines on minimum networks have been related to individual countries or states and their populations (Langbein, 1960). However, when assessing the design problem, it is advisable to think in terms of natural catchment areas even if the total area is defined by political boundaries. The physical features of the area should be studied. These include the drainage pattern, the surface relief (altitudinal differences), the geological structure and the vegetation. The general features of the climate should be noted; seasonal differences in tem-perature and precipitation can be identified from good atlases or standard climatological texts (Kendrew, 1961). The characteristics of the precipita-tion also affect network design and the principal meteorological causes of the rainfall or snowfall should be investigated.

The second stage in network design involves the practical planning. Exist-ing measuring stations should be identified, visited for site inspection and to determine observational practice, and all available data assembled. The station sites should be plotted on a topographical map of the area or, if the area is too large, one overall locational map should be made and separate topographical maps compiled for individual catchment areas. The distribu-tion of the measuring stations should be studied with regard to physical fea-tures and data requirements, and new sites chosen to fill in any gaps or provide more detailed information for special purposes. The number of new gauging sites required depends on the density of stations considered to be an optimum for the area. (Indications of desirable station densities are given in following sections.) Any new sites in the network are chosen on the map, but then they must be identified on the ground. Visits to proposed locations are essential for detailed planning and selected sites may have to be adjusted to accord with ground conditions.

The third stage involves the detailed planning and design of required installations on the new sites. These vary in complexity according to the hydrological variable to be measured, ranging from the simple siting of a single storage rain gauge to the detailed designing of a compound weir for stream measurements or the drilling of an observational borehole for monitoring groundwater levels. The costing of the hydrometric scheme is usually done at this stage and when this is approved and the finance is available, steps can be taken to execute the designed scheme.

A procedure that may be carried out at any stage is the testing of the validity of the data produced by the network with or without any new stations, provided that there are enough measurements available from the existing measurement stations to allow significant statistical analyses. These may take various forms depending on the variability of the measurements being tested. The worth of data produced is now an important factor in network design, but such cost/benefit evaluation is complicated by the many uses made of the data and by the unknown applications that may arise in the future.

The ideal hydrometric scheme includes plans for the measurement of all the many different hydrological variables. Some contributions to the considerable literature on the subject include such items as sediment transport and deposition, water quality and flood damages. These variables are becoming increasingly important in contemporary considerations of the quality of the environment. Designed networks of water quality monitoring stations are now being established in conjunction with arrangements for measurements of quantity. In the following sections, further particulars of network design for the more usual variables, precipitation, evaporation, surface flow and groundwater will be given.

2.3   Precipitation Networks

The design of a network of precipitation gauging stations is of major importance to the hydrologist since it is to provide a measure of the input to the river catchment system. The rainfall input is irregularly distributed both over the catchment area and in time. Another consideration in precipitation network design must be the rainfall type as demonstrated in the areal rainfall errors obtained over a catchment of 500 km² having 10 gauges (Table 2.1). This also shows that a higher density of gauges is necessary to give acceptable areal values on a daily basis.

Table 2.1   Areal Rainfall Errors (%)

Reproduced from J.C. Rodda (1969) *Hydrological Network Design — Needs, Problems and Approaches*. WMO/IHD Report No. 12.

Type of rain	Day	10 days	Month	Season
Frontal	19	8	4	2
Convective	46	17	10	4

As a general guide to the density of precipitation stations required, Table 2.2 gives the absolute minimum density for different parts of the world. The more variable the areal distribution of precipitation, as in mountainous areas, the more gauges are needed to give an adequate sample. In regions of low rainfall totals, the occurrence is variable but the infrequent rainfall events tend to be of higher intensity and thus network designers should ensure adequate sampling over areas that would be prone to serious flooding.

In the UK, recommended minimum numbers of rain gauges for reservoired moorland areas were laid down by water engineers many years ago (IWE, 1937) (Table 2.3). For real time operation of an upland impounding reservoir at least one autographic or digital recording gauge would now be recommended to record heavy falls over short periods. To estimate monthly areal rainfalls over the catchment areas of the major river gauging

Table 2.2   Minimum Density of Precipitation Stations

(Reproduced from World Meteorological Organization (1965) *Guide to Hydrometeorological Practices.*)

Region	Minimum density range (km²/gauge)
Temperate, mediterranean & tropical zones	
Flat areas	600 – 900
Mountainous areas	100 – 250
Small mountainous islands (< 20 000 km²)	25
Arid & polar zones	1500 – 10 000

Table 2.3   Rain Gauge Networks for the UK: Minimum Numbers of Rain Gauges Required in Reservoired Moorland Areas

(Reproduced from Institute of Water Engineers (1937) Trans. Inst. Water Eng. **XLII**, 231–259.)

Area		Rain gauges		
Square miles	Square kilometres	Daily	Monthly	Total
0.8	2	1	2	3
1.6	4	2	4	6
7.8	20	3	7	10
15.6	41	4	11	15
31.3	81	5	15	20
46.2	122	6	19	25
62.5	162	8	22	30

Table 2.4   Minimum Numbers of Rain Gauges for Monthly Percentage of Average Rainfall Estimates

(Reproduced from A. Bleasdale (1965) *Proc. WMO/IASH Symp. on the Design of Hydrological Networks*, IASH Pub. No. 67, pp. 46–54.)

Square miles	Square kilometres (Approx.)	Number of rain gauges
10	26	2
100	260	6
500	1300	12
1000	2600	15
2000	5200	20
3000	7600	24

stations, the Meteorological Office uses the minimum gauge numbers over the range of areas shown in Table 2.4 (Bleasdale, 1965). In designing a rain gauge network as part of the hydrometric scheme for the former Devon River Board area, an overall density of 25 km² per gauge was taken as a guide line for sampling daily rainfalls (Shaw, 1965). A higher density than this was needed in mountainous areas where higher relief causes greater rainfall variability. Over the whole River Board area, there were already several established gauges, especially in the reservoired mountain valleys, but there were many gaps in the networks of the drainage basins of the main rivers. For the total area of 6252 km², there were 174 existing gauges; 56 proposed new gauges brought the total number up to 230, which made an overall density of 27.2 km² per gauge. To improve the observations of heavy rainfalls, 23 new autographic gauges were recommended to be sited in the upland valleys; these formed part of a flood warning scheme.

The cost of maintaining the rain gauge networks in the UK has come under review, and a study of their effectiveness in producing the necessary data has been carried out by the Institute of Hydrology (1977). The applications of rainfall information were assembled together with the related margins of error that could be tolerated. The precision of point rainfalls and areal rainfall estimates derived from selected area networks was then tested by several statistical and hydrological techniques. Table 2.5 shows some of the results with the recommended spacing of gauges to meet the data requirements at point locations or over designated areas. It will be noted that for the specified daily data requirements, a greater density of gauges than the 1 per 25 km² guideline used previously, is now recommended.

2.4   Evaporation Networks

The assessment of evaporation loss over a drainage basin by means of point sample measurements is the next to be considered. The various recommended methods of measurement are outlined in Chapter 4. Since evapora-

Table 2.5   UK Rain Gauge Spacings to Meet Data Requirements

Date use	Time period	Area	Gauge spacing (km)
Water balance	Month	100 km²	7.5 – 9
Long range forecast	½ Month	10 000 km²	20 – 25
Soil moisture deficit	1 day	400 km²	3.5 – 4
Agriculture			
(seed germination)	Month	Point	1.5 – 4
Flood design	Day 10 mm	Point	0.9 – 1.2

tion and transpiration over an area are relatively conservative quantities in the hydrological cycle, fewer gauging stations are required to give areal evaporation estimates than for areal rainfalls. Evaporation and transpiration are dependent on altitude, and thus a network of measuring stations should sample different altitudinal zones within a catchment area. To give some idea of numbers, for experimental catchments covering 18 km² in Wales with a range in altitude of 460 m, whereas 25 rain gauges are needed for evaluations of areal rainfall with a 2% error only three or four hydrometeorological stations would be necessary for areal evaporation estimates (McCulloch, 1965). A reliable single station would provide adequate information over a flat plain or plateau.

2.5   Surface Water Networks

The establishment of river gauging stations is the most costly item in a hydrometric shceme and as such, river gauging is usually the responsibility of a Regional Water Authority (UK) or a national authority as in the USA. The density of gauging station depends on the nature of the terrain and the population creating a water demand. In England and Wales, it was proposed that there should be 400 primary gauging stations, equivalent to a density of 1 in 375 km² (Boulton, 1965). When the Water Resources Act, 1963, came into force, the number of gauging stations producing records for publication was approaching this figure, and coordination of further planning of the nationwide network was undertaken by the central authority, the Water Resources Board. The individual River Authorities were advised on the status of the gauging stations required:

(a) primary or principal stations defined as permanent stations to measure all ranges of discharges and observations and records to be accurate and complete;

(b) secondary or subsidiary stations to operate for as long as necessary to obtain a satisfactory correlation with the record of a primary station; their function is to provide hydrological knowledge of streams likely to be used for water supply abstractions; the range of a secondary station should be as comprehensive as possible and the observations and records should be of primary station standard;

(c) special stations are those serving particular needs, such as reservoir levels and dry weather flow stations for controlling abstractions; these may be permanent or temporary stations according to requirements and they can be related to primary and secondary stations.

For water resources evaluation, 20 years of records from a secondary station would suffice to give an acceptable correlation coefficient between the monthly discharges of the secondary station and a primary station. Then the secondary station could be discontinued. However, for other purposes such as recording flood flows, the continued running of the secondary station may be viable. Extension of discharge information for a short term secondary station can also be made by relating the flow duration curves (Chapter 12).

The ultimate design and establishment of a river gauging network depends on the data requirements, the hydrological characteristics of the area and the achievement of an acceptable cost – benefit relationship for the scheme.

2.6   Groundwater Networks

The main purposes of groundwater investigations are to identify productive aquifers, to determine their hydraulic properties, and to make arrangements for monitoring the water levels within the aquifers. Siting of observation wells must take into account differences in aquifer properties within an aquifer in addition to variations between aquifers. For example, in the UK a fairly homogeneous Triassic sandstone may require a basic network of one observation well per 260 km^2. However, Chalk aquifers can be very variable in permeability and consequently a denser network of observation wells, say 1 in 5 km^2, may be needed to record water level fluctuations in the more permeable areas (Ineson, 1965).

In planning a network of observation wells, the sites of existing wells must be noted, as these may have reliable consistent water level records which give the long-term fluctuations of the water table. The importance of ground-water networks has been enhanced by the development of groundwater resources and the recharge of aquifers depleted by over-pumping.

Groundwater measurement networks are well established in the aquifers of the sedimentary rocks of the lowlands of England, but groundwater developments have been of less importance in Scotland.

The following chapters of Part I dealing with Hydrological Measurements describe the methods of measurement of the different hydrological variables and the instruments in most common use in the UK.

References

Bleasdale, A. (1965). 'Rain gauge networks development and design with special reference to the United Kingdom,' in *Proc. WMO/IASH Symp. on the Design of Hydrological Networks*. IASH Pub. No. 67, pp. 46-54.

Boulton, A.G. (1965). *Surface Water, Basic Principles Related to Network Design*. IASH Pub. No. 67, 234-244.

Ineson, J. (1965). *Ground Water Principles of Network Design*. IASH Pub. No. 68, 476–483.

Institution of Hydrology (1977). *Methods for Evaluating the UK Rain Gauge Network*.

Institute of Water Engineers (IWE) (1937). 'Report of Joint Committee to consider methods of determining general rainfall over any area.' *Trans. Inst. Water Eng.*, **XLII**, 231–259.

Kendrew, W.G. (1961). *Climates of the Continents*. Oxford University Press, 5th ed., 608 pp.

Kohler, M. (1958). *Design of Hydrological Networks*. WMO Tech. Note No. 25, 16 pp.

Langbein, W.B. (1960). *Hydrologic Data Networks and Methods of Extrapolating or Extending Hydrologic Data*. UN, ECAFE/WMO Flood Control Series No. 15, 157 pp.

Langbein, W.B. (1965). *National Networks of Hydrological Data*. IASH Pub. No. 67, pp. 5-11.

McCulloch, J.S.G. (1965). *Hydrological Networks for Measurement of Evaporation and Soil Moisture*. IASH Pub. No. 68, pp. 579–584.

Rodda, J.C. (1969). *Hydrological Network Design—Needs, Problems and Approaches*. WMO/IHD Report No. 12, 57 pp.

Shaw, E.M. (1965). *A Rain Gauge Network Prepared for the Devon River Board Area of the United Kingdom*. IASH Pub. No. 67, 63-70.

World Meteorological Organization (WMO) (1965). *Guide to Hydrometeorological Practices*.

World Meteorological Organization (1972). *Casebook on Hydrological Network Design*. WMO Pub. No. 324.

World Meteorological Organization (1976). *Hydrological Network Design and Information Transfer*. WMO Operational Hydrology Report No. 8, 185 pp.

3
Precipitation

Of all the components of the hydrological cycle, the elements of precipitation, particularly rain and snow, are the most commonly measured. It would appear to be a straightforward procedure to catch rain as it falls and the depth of snow lying can be determined easily by readings on a graduated rod. Men have been making these simple measurements for more than 2000 years; indeed the first recorded mention of rainfall measurement came from India as early as 400 BC. The first rain gauges were used in Korea in the 1400s AD, and 200 years later, in England, Sir Christopher Wren invented the self-recording rain gauge.

However, climatologists and water engineers appreciate that making an acceptable precipitation measurement is not as easy as it may first appear. It is not physically possible to catch all the rainfall or snowfall over a drainage basin; the precipitation over the area can only be *sampled* by rain gauges. The measurements are made at several selected points representative of the area and values of the total volume (m³) or equivalent areal depth (mm) over the catchment are calculated later. Such are the problems in obtaining representative samples of the precipitation reaching the ground that, over the years, a comprehensive set of rules has evolved. The principal aim of these rules is to ensure that all measurements are comparable and consistent. All observers are recommended to use standard instruments installed uniformly in representative locations and to adopt regular observational procedures.

Many investigations carried out in England into the problems of rainfall measurement owe their origin to the enthusiasm of one man, G.J. Symons. Symons, a civil servant in the Meteorological Department of the Board of Trade in the 1850s, instigated and encouraged formal scientific experiments by such volunteers as retired army officers or clergymen whose spare time interests included observations of the weather and measurements of meteorological variables (Mill, 1901). The results of this work were incorporated by Symons into his *Rules for Rainfall Observers*, a publication that is revised from time to time and can still be obtained from the UK Meteorological Office (Meteorological Office, 1982). Symons' rules continue to form the basis of the practice of precipitation measurement in the UK today, although considerable fundamental re-thinking has taken place in the past 20 years.

The Meteorological Office, which in 1919 inherited the advisory functions of Symons and his successors in the British Rainfall Organization, has approved instruments of several designs having the salient features recommended as a result of the early experiments. These include various types of storage rain gauge and rainfall recorders. For the assessment of water resources, monthly totals may suffice; for evaluating flood peaks in urban areas, rainfall intensities over an hour or even minutes could be required, and recording rain gauges are used.

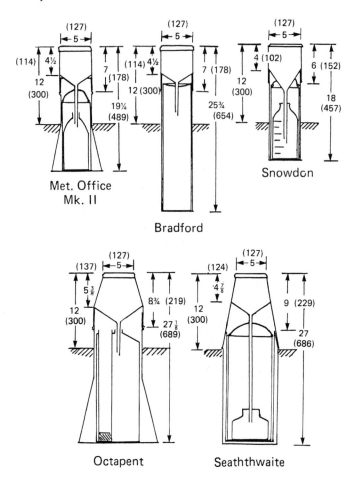

Fig. 3.1   Standard rain gauges. Dimensions in inches and (millimetres).

3.1   Storage Gauges

The rain gauges shown in Fig. 3.1 vary in capacity depending on whether they are to be read daily or monthly. The period most generally sampled is the day, and most precipitation measurements are the accumulated depths of water caught in simple storage gauges over 24 h.

For many years, the UK's recognized standard daily rain gauge has been the Meteorological Office Mark II instrument (Meteorological Office, 1981). The gauge has a sampling orifice of diameter 5" (127 mm). The $\frac{1}{2}$" rim (12.7 mm) is made of brass, the traditional material for precision instruments, and the sharply tooled knife edge defines a permanent accurate orifice. The Snowdon funnel forming the top part of the gauge has a special design. A straight-sided drop of 4" (102 mm) above the funnel prevents losses from outsplash in heavy rain. Sleet and light snowfall also collect

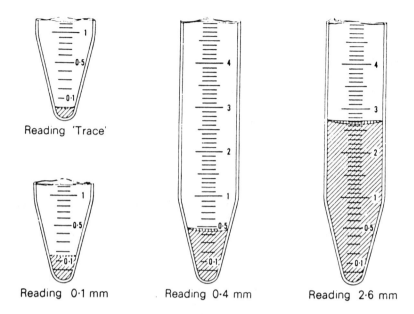

Reading 'Trace'

Reading 0·1 mm

Reading 0·4 mm

Reading 2·6 mm

Fig. 3.2   Reading the rain measure. Millimetre graduations. (Reproduced from *Handbook of Meteorological Instruments, Vol. 5* (1981) Meteorological Office, by permission of the Controller, Her Majesty's Stationery Office. © Crown copyright.)

readily in the deep funnel and except in very low temperatures, the melted water runs down to join the rain in the collector. The Snowdon funnel, the main outer casing of the gauge and an inner can are all made of copper, a material that has a smooth surface, wets easily and whose surface once oxidized, does not change. The inside of the collecting orifice funnel should never be painted, since the paint soon cracks, water adheres to the resulting rough surface and there are subsequent losses by evaporation. The main collector of the rain water is a glass bottle with a narrow neck to limit evaporation losses. The gauge is set into the ground with its rim level and 12" (300 mm) above the ground surface, which should ideally be covered with short grass, chippings, or gravel to prevent insplash in heavy rain.

During very wet weather, the rain collected in the bottle may overflow into the inner can. Bottle and can together hold the equivalent of 150 mm (6") rainfall depth. The inner can is easily removed from the outer casing and its contents can be emptied and measured without disturbing the installation.

The gauge is inspected each day at 0900 h GMT, even if it is thought that no precipitation has occurred. Any water in the bottle and inner can is poured into a glass measure (Fig. 3.2) and the reading taken at the lowest point of the meniscus. The glass measure is graduated in relation to the orifice area of the rain gauge and so gives a direct reading of the depth of rain that fell on the area contained by the brass rim. The glass measure has a capacity of 10 mm; if more than 10 mm of rain has fallen, the water in the gauge must be measured in two or more operations. The glass measure is tapered at the bottom so that small quantities can be measured accurately. If no water is found in the gauge and precipitation is known to have fallen, this should be noted as a 'trace' in the records. The glass bottle and inner can should be quite empty before they are returned to the outer case. It is advisable to check the instrument regularly for any signs of external damage or general wear and tear. Severe sharp frosts can sometimes loosen the joints of the casing and if this is suspected, testing for leaks should be carried out.

The *Bradford* rain gauge (Fig. 3.1) was first made for the engineers of the Bradford Waterworks who required a larger capacity instrument for the higher rainfalls in the Pennines in the north of England. The greater depth of the container below the ground surface reduces the risk of the collected water freezing. The *Snowdon* gauge, a Meteorological Office Mark I instrument, remains in favour among private observers since without the splayed base it is easily maintained in a garden lawn. It is however more difficult to keep rigid with the rim level.

Monthly rain gauges hold larger quantities of precipitation than daily gauges (Fig. 3.1). The catch is measured using an appropriately graduated glass measure holding 50 mm. Monthly gauges are designed for remote mountain areas and are invaluable on the higher parts of reservoired catchments. Measurements are made on the first day of each month to give the previous month's total, and corrections may need to be made to readings obtained from remote gauges recorded late in the day in wet weather.

The *Octapent* gauge, a hybrid of the 5″ and old 8″ diameter gauges, is made in two sizes with capacities of 685 mm and 1270 mm. The large inner can contains a lead-weighted hose. The hose absorbs increased pressure if the water freezes and so prevents damage to the copper container. The *Seathwaite* gauge, developed for use in the Lake District mountains of north-west England, is insulated to protect the catch against extremes of temperature.

During recent years the rapid increase in the price of copper and the development of new types of synthetic materials have encouraged the design of new gauges. At the same time there occurred a rapid growth of interest in hydrological measurements and the need for more observations of all kinds. In the later 1950s and early 1960s, the Meteorological Office, in collaboration with a few manufacturers, made extensive investigations with numerous trial gauges and finally decided on a system of rainfall measurement using two funnel sizes and two interchangeable bases (Maidens, 1965). The gauge orifices have areas of 150 cm^2 and 750 cm^2. The new gauges are made of a green glass-fibre laminate. However, the material does not wear well and the sharp edges of the orifices are easily damaged; the gauge can also be adversely affected by condensation on the underside of the funnel which can drip into the container.

The operation of the new system of measurement was made more flexible. Both orifices may be used with a base as simple storage gauges to be read daily or monthly as required, but there may also be a recording mechanism housed in the large base or a telemetering device in the small base. In all installations the rim of the orifice is set level at 300 mm above the ground surface, which should be made similar to that specified for the installation of the Mark II instrument.

In general, copper rain gauges remain in use for standard rainfall measurements, whilst the synthetic materials are used for raingauges of the tipping bucket type.

3.2   Rainfall Recorders

The need for the continuous recording of precipitation arose from the need to know not just how much rain has fallen but when it fell and over what period. Numerous instruments have been invented, usually built by enthusiasts of mechanical devices. Two main types have endured, the tilting-syphon rain recorder developed by Dines, and the tipping-bucket gauge which had its origins with Sir Christopher Wren. Both of these types of instrument are used by the UK Meteorological Office, although with reservations about models marketed by some manufacturers.

The *Dines tilting-syphon* rain recorder, now Mk2 (Fig. 3.3) is installed with its rim 500 mm above ground level. The rain falling into the 287 mm diameter funnel is led down to a collecting chamber containing a float. A pen attached to the top of the plastic float marks a chart on a revolving drum driven by clockwork. The collecting chamber is balanced on a knife edge. When there is no rain falling, the pen draws a continuous horizontal line on the chart; during rainfall, the float rises and the pen trace on the chart slopes upwards according to the intensity of the rainfall. When the chamber is full, the pen arm lifts off the top of the chart and the rising float releases a trigger disturbing the balance of the chamber which tips over and activates the syphon. A counter-weight brings the empty chamber back into the upright position and the pen returns to the bottom of the chart. With double syphon tubes, syphoning should be completed within 8 s, but the rain trap reduces the loss during heavy rainfall. It is recommended, however, that a standard daily storage gauge is installed nearby and that quantities recorded are amended to match the daily total. Each filling of the float chamber is equivalent to 5 mm (0.2″) of precipitation.

Charts normally record by the day, but modifications to the instrument can allow a strip chart to be used which gives continuous measurements for as long as a month and which has an extended time scale for intense falls over very short periods. In cold weather, the contents of the float chamber may freeze and special insulation with thermostatically controlled heating equipment, the simplest being a low wattage lamp bulb, can be installed. The provision of heating assists in the melting of snow, but in very cold weather or during heavy snow existing low-powered heating devices will not be adequate and there will be a time lag in the melted water being recorded on the chart. Care must be taken to avoid too much heating since evaporation of the melted snow would result in low measurements. Adequate drainage below the gauge should be provided during installation, especially in heavy clay soils and in areas liable to heavy storms, for the syphon system will fail if the delivery pipe enters flood water in the soakaway. A model for use in the tropics has a 128 mm diameter receiving aperture and the filling of the float chamber represents 25 mm on the chart.

The principle of the *tipping-bucket* rain gauge is shown in Fig. 3.4. Rain is led down a funnel into a wedge-shaped bucket of fixed capacity. When full,

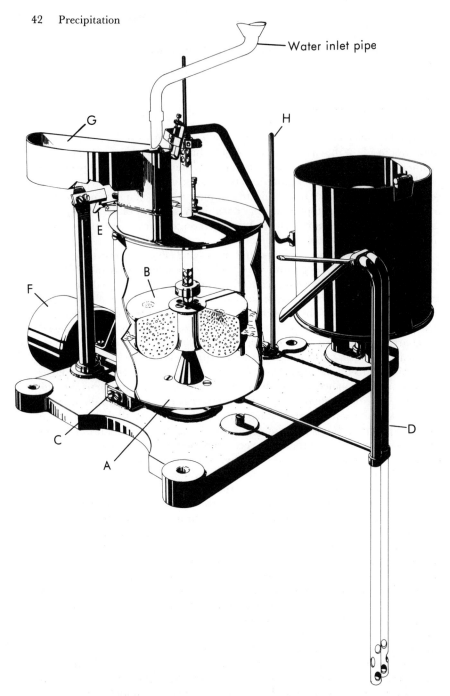

Fig. 3.3 Meteorological Office tilting-syphon rain recorder. A = Collecting chamber; B = Plastic float; C = Knife-edges; D = Double siphon tubes; E = Trigger; F = Counterweight; G = Rain trap; H = Pen-Lifting rod. (Reproduced from *Observer's Handbook*, 4th edition, 1982, Meteorological Office, by permission of the Controller, Her Majesty's Stationery Office © Crown copyright.)

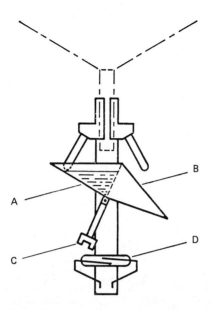

Fig. 3.4   Principle of the tipping-bucket mechanism. A, B: buckets. C: magnet. D: switch. (Reproduced from *Observer's Handbook, 3rd edn.* (1969) Meteorological Office, by permission of the Controller, Her Majesty's Stationery Office. © Crown copyright.)

the bucket tips to empty and a twin adjoining bucket begins to fill. At each tip, a magnet attached to the connecting pivot closes a circuit and the ensuing pulse is recorded on a counter. The mechanism can be used in a variety of gauges. The 15 g of water in one bucketful represents 1 mm of rain caught in a 150 cm^2 gauge, and 0.2 mm in a 750 cm^2 gauge. A small adjustment allows the tipping buckets to be used with the 5″ gauge, so that each tip represents 1 mm of precipitation. It is advisable to arrange for the water to be collected below the buckets so that a day's or month's total can be measured if the recording mechanism fails.

Improvements to tipping-bucket gauges continue to be made and the present Meteorological Office instrument is the Mk 5 tipping bucket rain gauge shown in Fig. 3.5. At many stations, particularly in remote locations, the measurements are recorded on magnetic tape or solid state event recorders and the cassette is usually changed at monthly intervals.

3.3   Siting the Rain Gauge

Choosing a suitable site for a rain gauge is not easy. The amount measured by the gauge should be representative of the rainfall on the surrounding area. What is actually caught as a sample is the amount that falls over the orifice area of a standard gauge, that is, 150 cm^2. Compared with the area of even a small river catchment of 15 km^2 for example, this 'point' measurement

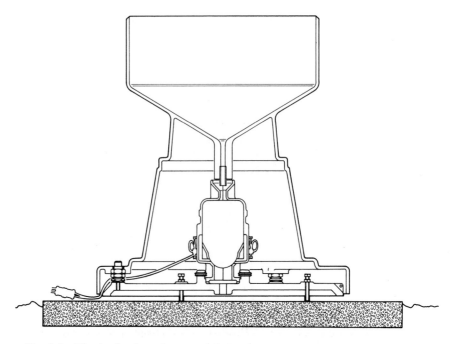

Fig. 3.5   Tipping bucket rain gauge Mk 5. (Reproduced from *Observer's Handbook*, 4th edition, 1982, Meteorological Office, by permission of the Controller, Her Majesty's Stationery Office © Crown copyright.)

represents only a 1 in 10^9 fraction of the total catchment area. Thus even a small error in the gauged measurement due to poor siting represents a very substantial volume of water over a catchment.

It is best to find some level ground if possible, definitely avoiding steep hillsides, especially those sloping down towards the prevailing wind. In the UK, the wind comes mainly from westerly directions. A sheltered, but not over-sheltered, site is the ideal (Fig. 3.6). It is advisable to measure the height of sheltering objects in determining the best site, taking into account anticipated growth of surrounding vegetation.

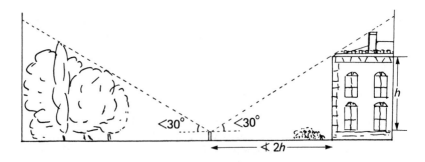

Fig. 3.6   A rain gauge site: maximum shelter allowed.

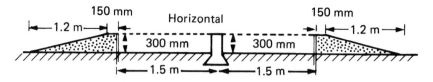

Fig. 3.7   A turf wall installation.

In overexposed locations on moorlands, plateaus, and extensive plains where natural shelter may be scarce, a turf wall of the kind designed by Hudleston is recommended (Fig. 3.7) (Hudleston, 1934). The surrounding small embankment prevents wind eddies which can inhibit rain drops from falling into an unprotected gauge. The disadvantages of this enclosure are that drifting snow may engulf the gauge and very heavy rain may flood it if there is no drainage channel beneath the wall.

Rain gauge sites should be examined occasionally to note any possible changes in the exposure of the instrument. Removal of neighbouring trees or the growth of adjacent plants are modifications of the natural surroundings that could affect the rain gauge record. Observers should be encouraged to report any major structural changes to buildings near the gauge because they could result in changing wind patterns in the vicinity of the instrument which could also affect the homogeneity of the catch record. When inconsistancies in a record caused by such changes in the exposure of a gauge are reported or discovered, the data processors make suitable amendments to the measurements. (Chapter 9).

3.4   Snowfall

There are various solid forms of precipitation, and all except hail require the surface air temperature to be lower than about 4 °C if they are to reach the ground.

Small quantities of snow, sleet or ice particles fall into a rain gauge and eventually melt to yield their water equivalent. If the snow remains in the collecting funnel, it must be melted to combine the catch with any liquid in the gauge. If practicable, the gauge may be taken indoors to aid melting but any loss by evaporation should be avoided. Alternatively, a quantity of warm water measured in the graduated rain measure for the rain gauge type can be added to the snow in the funnel and this amount subtracted from the measured total.

When snow has accumulated on the ground, its depth is also measured. A representative smooth cover of the ground is selected where there has been no drifting and sample depths are taken with a metre stick held vertically. For a rough estimate of water equivalent, the average snow depth is converted taking 300 mm of *fresh* snow equal to 25 mm of rain. However, the density of fresh snow may range between 50 and 200 g l^{-1} according to the character of the snow flakes.

Deep snow provides problems especially if it is wind blown and the rain gauge has been buried. Samples of snow may be taken from a level area by inverting a rain gauge funnel of the Snowdon pattern and pressing it down through the snow to the ground. The compacted core is then melted to obtain the water equivalent. In very deep snow several samples of the core may be necessary. The method is valid only if all the snow fell after the previous observation. When a series of snowy days occurs, a clean board should be placed on the snow surface, or part of the ground could be cleared, and each day's accumulation measured down to the cleared surface. The position of the board or cleared ground should be marked.

When compacted snow lies for several days and there are subsequent accumulations, the observer is advised to take density measurements at selected points over the higher parts of important catchment areas. The density of snow increases with compaction to around 300 g l^{-1}. Sample volumes of the snow are taken at different depths, are weighed and the density calculated. Thence the water equivalent of the total depth of snow can be obtained. If the snow is not too deep, a total sample core can be weighed to give an overall density or melted to give the water equivalent directly. A continuous monitoring of the water equivalent of lying snow is essential to promote warnings against flooding if a sudden thaw occurs.

In countries that experience annually a snow cover lasting throughout a whole winter season, for example, Canada, northern USA, USSR, Norway and Switzerland, special arrangements are made for a regular survey of the snow pack. At the end of the accumulating season, it is known how much snow (and its water equivalent), is lying 'in store' on the land. Hydrologists are then very much concerned as the spring season progresses with the rate of snow melt and the resulting river discharges. The rate at which the snow melts is entirely dependent on meteorological conditions. Solar radiation, air temperature, humidity, wind speed and falls of rain all affect melting of a snow pack. When there is snow to melt, the most favourable conditions are provided by a combination of low cloud preventing reflected radiation loss, a warm humid air mass and heavy rain on the snow cover.

Calculations of snow melt are made in the USSR from formulae based on the energy budget of the pack. These require careful measurements of short and long wave radiation, snow and air temperatures, vapour pressures, wind speed and degree of cloudiness. In the USA, for melting with rain falling, the Corps of Engineers use a simple formula for forested areas:

$$M = (0.3 + 0.012R)T + 1.0$$

where M = daily melt water (mm), R = daily rain (mm), and T = mean daily temperature (°C).

For grassland or partly forested areas, wind speed measurements (v, m s^{-1}, at 10 m height) and a basin parameter (k) ranging from 0.3 for moderately dense forest to 1.0 for open plains, are introduced:

$$M = (0.1 + 0.12R + 0.8kv)T + 2.0$$

Both these formulae have been developed for conditions of complete cloud cover and saturated air (WMO, 1965).

For more detailed considerations of snow melt assessment for hydrological purposes in the United Kingdom, the reader is referred to the *Flood Studies Report* (NERC, 1975).

3.5   International Practice

Although the UK standard instruments and observation practices are found in many countries, particularly those of the Commonwealth, many other types of instruments and siting characteristics have been developed to suit different climatic conditions.

In Europe, each country has its own standard instruments recommended by the national hydrological or meteorological services. In those regions where snowfall is a regular feature, it has been found more satisfactory to set a gauge with its rim 1 m above ground level. This is also an advantage in areas with very heavy rainfall because drops can splash up for considerable heights above the ground, especially if the surrounding surface has become hard and compacted and there is no short grass cover. However, the greater the height of the gauge above ground the greater the losses in windy weather. To overcome these deficiencies, shields have been designed to fix round the orifice of the gauge. Investigations into the effectiveness of various types of shields have been undertaken in the USA and the Nipher and Alter shields have both proved satisfactory (Kurtyka, 1953).

In order to have comparable rainfall measurements from one country to another, the member nations of the World Meteorological Organization agreed on an international gauge. It is known as the interim reference precipitation gauge (IRPG) and it consists of a 5″ (127 mm) diameter Snowdon gauge with the rim 1 m above the ground, protected by a US Alter shield 36″ (0.914 m) in diameter. Each country is expected to have at least one of these instruments to which their national measurements may be related, whence international records may be compiled.

3.6   Recent Developments

It has always been appreciated in the UK that the compromise setting of the gauge rim 300 mm (1 ft) above the ground surface is not altogether satisfactory. Hydrologists have led the move to require gauges to be set with the rim at ground level and various methods to prevent insplash have been developed. The most acceptable installation is one in which the gauge is set in the centre of a pit about 1 m square, which is then covered with a plastic or metal grid with a hole in the middle to accommodate the funnel. The square grid slats should be less than one millimetre thick at the top edge and be 50 mm deep with 50 mm spacing. An early antisplash device was a Venetian blind arrangement around the gauge with the slats directing splash away from the orifice.

These provisions add considerably to the cost of a rain gauge station and make the task of taking daily readings more difficult for amateur observers. However, if wind speeds at the level of the rim can be kept below 5 m s^{-1} by

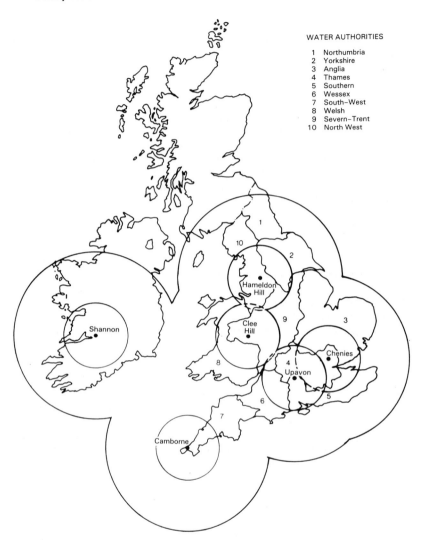

WATER AUTHORITIES

1   Northumbria
2   Yorkshire
3   Anglia
4   Thames
5   Southern
6   Wessex
7   South–West
8   Welsh
9   Severn–Trent
10 North West

Fig. 3.8   Weather radars 1984. Circles at 75 km and 200 km radius. (Reproduced from Collinge, V.K. and Kirby, C., eds, *Weather Radar and Flood Forecasting*. John Wiley & Sons, 1977.)

finding a suitably sheltered site for the instrument, the standard installation set at 300 mm above the ground remains acceptable and the results are comparable with a ground-level gauge. Nevertheless, it has been shown by several researchers that a standard daily gauge in its conventional setting catches 6–8% less rain than a properly installed ground-level gauge (Rodda, 1967; Green, 1969).

For many years, pilot studies in the use of radar for measuring precipitation have been in progress in the UK. The former Water Resources Board, the Meteorological Office, Plessey Radar Ltd., and the former Dee and Clwyd River Authority combined resources to fund and operate special radar equipment and a network of monitoring rain gauges in the River Dee catchment of North Wales. The advantage of the method to the hydrologist is that it produces a measure of the rainfall over the whole of a catchment area as it is falling. However, ground catches in rain gauges are required to calibrate the radar scans. Promising results were reported (Harrold, *et al.*, 1974), but widespread use of radar is being inhibited by cost, though one or two permanent installations have been installed in the UK by Water Authorities in conjunction with the Meteorological Office. The present operational network of radar stations measuring rainfall is shown in Fig. 3.8; additional radars are planned to complete countrywide coverage, and two are presently being built, one near Lincoln (Ingham) and the other in Northern Ireland (Castor Bay). The development of modern techniques in areal rainfall measurement is being coordinated by a Meteorological Office team at the Radar Research Laboratory at Malvern, and now rainfall calculations from satellite imagery are linked to the measurements at the radar stations. The two sets of data can be displayed on the same television screen, an overall coverage being provided by the satellite data and greater details of the rainfall inserted by the ground-based radar (Browning, 1979).

References

Browning, K.A. (1979). 'The FRONTIERS plan: a strategy for using radar and satellite imagery for very-short-range precipitation forecasting.' *Meteorol. Mag.*, **108**, 1283, 161–184.

Collinge, V.K., and Kirby, C. (ed.) (1987). *Weather Radar and Flood Forecasting.* John Wiley, 296 pp.

Green, M.J. (1969). 'Effects of exposure on the catch of rain gauges.' *Water Res. Assoc. TP.* **67**, 28 pp.

Harrold, T.W., English, E.J. and Nicholass, C.A. (1974). 'The accuracy of radar-derived rainfall measurements in hilly terrain.' *Quart. Jl Roy. met. Soc.*, **100**, 331–350.

Hudleston, F. (1934). A summary of seven years' experiments with rain gauge shields in exposed positions, 1926–1932 at Hutton John, Penrith. *British Rainfall*, 1933, 274–293.

Kurtyka, J.C. (1953). *Precipitation Measurements Study.* Investigation Report No. 20, Illinois State Water Survey Division, 178 pp.

Maidens, A.L. (1965). 'New Meteorological Office rain gauges.' *Meteorol. Mag.* **94**, 1114, 142–144.

Meteorological Office (1969 and 1982) *Observer's Handbook*, 3rd and 4th eds. HMSO, London.

Meteorological Office (1982). *Rules for Rainfall Observers.* Meteorological Office Leaflet No. 6,12 pp.

Meteorological Office (1981). *Handbook of Meteorological Instruments*, Vol. 5. HMSO, London. 2nd ed., 34 pp.

Mill, H.R. (1901). 'The development of rainfall measurement in the last 40 years.' *British Rainfall*, 1900, 23–41.

Natural Environment Research Council (NERC). (1975). *Flood Studies Report*, Vols. I and II.

Rodda, J.C. (1967). The systematic error in rainfall measurement. *J. Inst. Water Eng.*, **21**, 173–177.

World Meteorological Organization (WMO). (1975). *Guide to Hydrometeorological Practices*. WMO No. 168 TP. 82, p.A.24.

4

Evaporation

Of the several phases in the hydrological cycle, that of evaporation is one of the most difficult to quantify. Certainly, it is difficult to define the unseen amounts of water stored or moving underground, but above the ground surface, the great complexities of evaporation make it an even more elusive quantity to define; yet evaporation can account for the large differences that occur between incoming precipitation and water available in the rivers. In the UK, if annual totals are considered, evaporation would appear to deprive much of south-east England of all of its rainfall; in actual practice, evaporation amounts vary seasonally and surplus surface water feeds the rivers in winter. In hotter climates with seasonal rainfall, evaporation losses cause rivers to dry up and river flows are dependent on excessive, heavy rainfall in the wet season.

For the engineering hydrologist, the loss of water by evaporation must be considered from two main aspects. The first, *evaporation* from an open water surface, E_o, is the direct transfer of water from lakes, reservoirs and rivers to the atmosphere. This can be relatively easily assessed if the water body has known capacity and does not leak. The second form of evaporation loss occurs from the *transpiration* from vegetation, E_t. This is sometimes called *evapotranspiration*, since loss by direct evaporation of intercepted precipitation and transpired water on plant surfaces is also included. Thus E_t is usually thought of as the total loss by both evaporation and transpiration from a land surface and its vegetation. The value of E_t varies according to the type of vegetation, its ability to transpire and to the availability of water in the soil. It is much more difficult to quantify E_t than E_o since transpiration rates can vary considerably over an area and the source of water from the ground for the plants requires careful definition.

Both forms of evaporation, E_o and E_t, are influenced by the general climatic conditions. Although the instrumental measurements are not so simple and straightforward as for rainfall, it is a compensating factor that evaporation quantities are less variable from one season to another, and therefore more easily predicted than rainfall amounts. With unlimited supplies of water, evaporation is one of the more consistent elements in the hydrological cycle.

4.1   Factors Affecting Evaporation

The physical process in the change of state from liquid to vapour operates in both E_o and E_t and thus the general physical conditions influencing evaporation rates are common to both.

(a) Latent heat is required to change a liquid into its gaseous form and in nature this is provided by energy from the sun. The latent heat of vaporization comes from solar (short-wave) and terrestrial (long-wave) radiation. The incoming *solar radiation* is the dominant source of heat and affects evaporation amounts over the surface of the earth according to latitude and season.

(b) *Temperature* of both *air* and the *evaporating surface* is important and is also dependent on the major energy source, the sun. The higher the air temperature, the more water vapour it can hold, and similarly, if the temperature of the evaporating water is high, it can more readily vaporize. Thus evaporation amounts are high in tropical climates and tend to be low in polar regions. Similar contrasts are found between summer and winter evaporation quantities in mid-latitudes.

(c) Directly related to temperature is the water vapour capacity of the air. A measure of the amount of water vapour in the air is given by the vapour pressure, and a unique relationship exists between the saturated vapour pressure and the air temperature (Chapter 1, Fig. 1.3). Evaporation is dependent on the *saturation deficit* of the air, which is the amount of water vapour that can be taken up by the air before it becomes saturated. The saturation deficit is given by the difference between the saturation vapour pressure at the surface temperature and the actual vapour pressure of the air. Hence more evaporation occurs in inland areas where the air tends to be drier than in coastal regions with damp air from the sea.

(d) As water evaporates, the air above the evaporating surface gradually becomes more humid until finally it is saturated and can hold no more vapour. If the air is moving, however, the amount of evaporation is increased as drier air replaces the humid air. Thus *wind speed* at the surface is an important factor. Evaporation is greater in exposed areas that enjoy plenty of air movement than in sheltered localities where air tends to stagnate.

It will be noted that the temperature and wind speed factors may be in conflict in affecting evaporation since windy areas tend to be cooler and sheltered areas are often warmer. Over a large catchment area, it is the general characteristics of the prevailing air mass that will have the major affect on evaporation (apart from the direct solar radiation).

The principle influences on the physical process of evaporation enumerated above are in their turn affected by wider considerations. The following factors outline more generally larger-scale influences.

(e) The prevailing weather pattern indicated by the *atmospheric pressure* affects evaporation. The edge of an anticyclone provides ideal conditions for evaporation as long as some air movement is operating in conjunction with the high air pressure. Low atmospheric pressure usually has associated with it damp unsettled weather in which the air is already well charged with water vapour and conditions are not conducive to aid evaporation.

(f) The nature of the *evaporating surface* affects evaporation by modifying the wind pattern. Over a rough, irregular surface, friction reduces wind speed but has a tendency to cause turbulence so, with an induced vertical component in the wind, evaporation is enhanced. Over an open water surface, strong winds cause waves which provide an increased surface for evaporation in addition to causing turbulence. As wind passes over smooth, even surfaces there is little friction and turbulence and the evaporation is affected predominantly by the horizontal velocity.

Variations in some of the dominant factors operating over different surfaces can result in noticeable changes in evaporation rates over small adjacent areas in short time periods. Diurnal fluctuations are considerable since during the night there is no solar radiation. However, evaporation totals over neighbouring areas show relatively smaller differences over periods of a week or a month.

Evaporation is necessarily dependent on a supply of water and thus the availability of moisture is a crucial factor. With all the other factors acting favourably, once the body of water disappears then open water evaporation E_o ceases. For E_t, the availability of water is not so easily observed. Plants draw their supply from the soil where the moisture is held under tension, and their rate of transpiration is governed by the stomata in the leaves which act like valves to regulate the passage of water through the pores according to the incidence of light. The pores are closed in darkness and hence transpiration ceases at night. When there is a shortage of water in the soil, the stomata regulate the pores and reduce transpiration. Thus E_t is controlled by soil moisture content and the capacity of the plants to transpire, which are conditioned by the meteorological factors.

If there is a continuous supply and the rate of evaporation is unaffected by lack of water, then both E_o and E_t are regulated by the meteorological variables, *viz* radiation, temperature, vapour pressure and wind speed. The evaporation plus transpiration from a vegetated surface with unlimited water supply is known as *potential evaporation* (PE) and it constitutes the maximum possible loss rate due to the prevailing meteorological conditions. Thus although E_t is the actual evaporation, PE is the maximum value of E_t. PE $= E_t$ when water supply is unlimited.

4.2 Measurement of E_o

The direct measurement of the rate of evaporation from open water or from a vegetated surface continuously and at selected sampling sites over a catchment area is an ideal which has not as yet been realized. An instrument, the 'evapotron', has been developed (Dyer and Maher, 1965) to monitor the continuous upward flux of moisture from an evaporating surface in the field, but the complex variability of atmospheric conditions makes the development of comprehensive sensors for humidity and wind movement, to accommodate all eventualities, a very difficult problem. A more recent instrument is the Hydra, an apparatus developed at the Institute of Hydrology (see Chapter 11).

Currently, an indirect measurement of evaporation from open water is made by taking the difference in storage of a body of water measured at two known times, which gives a measure of the evaporated water over the time interval. If rain has fallen during the time period then the rainfall quantity must be taken into account. In practice, this *water budget* method is used on two widely differing spatial scales, by measurements at reservoirs and by measurements with specially designed instruments maintained at meteorological stations.

4.2.1   Water Budget of Reservoirs

The evaporation from a reservoir over a time period is given by:

$$E_o = I - O - \Delta S$$

where I = Inflow into the reservoir plus precipitation on to the reservoir surface, O = outflow from the reservoir plus subsurface seepage and ΔS = increase in storage. Although water engineers are anxious to assess evaporation losses from their surface reservoir sources of supply, not many impounding reservoirs are instrumented to give the measurements required for the water budget equation. In the UK it is rare to find river flow gauging stations on the streams flowing into reservoirs. There are usually several feeder tributaries, which add to the complexity and cost of total inflow measurements. The evaluation of outflow from a reservoir is, however, usually made regularly. Measurements of supply or consumption water led off by pipeline together with the statutory compensation water and surplus releases to the stream are made regularly and the water levels in the reservoir give changes in storage. Difficulties sometimes arise in the assessment of flood flows over spillways while the amount of leakage beneath the dam and through the sides and bottom of a reservoir can only be roughly estimated. The measurement of evaporation from an operating reservoir using the water budget method can only give a broad approximation to water loss unless a thorough knowledge of the different components is available.

A valuable study of reservoir evaporation was made by Lapworth for the Kempton Park Reservoir from 1956–62 (Lapworth, 1965). During these years, there was no inflow and no outflow from this storage reservoir on the Thames flood plain. Hence, it was expected that E_o would be equal to ΔS, the change in storage with rainfall deducted. In addition to the necessary rainfall measurements, meteorological stations were set up to make the observations required for calculating evaporation (to be described later) so that comparisons of the different methods of evaporation evaluation could be made. The results showed that there were marked seasonal differences between the measured changes in storage E_{Res} and the calculated E_o. The average annual evaporation total over the seven years was 663 mm (26.1 in) and the monthly means are given in Table 4.1. In explanation, the values for E_{Res}, the evaporation from the reservoir, have two components, E_o, evaporation due to surface water conditions, plus E_{St}, evaporation caused by the changing heat storage of the water in the reservoir. As seen in Table 4.1, during the autumn months, evaporation from the reservoir due to heat storage effects is enhanced. This results from heat diffusing and convecting to the surface from the lower water layers which had absorbed the energy of the summer sun. In the spring, the temperature of the water body is low following the

Table 4.1   Kempton Park Reservoir Evaporation 1956–62 (mm)

	J	F	M	A	M	J	J	A	S	O	N	D
E_{Res}	15	18	28	48	76	94	107	94	74	58	33	18
E_{St}	+ 3	− 3	− 13	− 20	− 18	− 23	− 5	+ 5	+ 15	+ 23	+ 20	+ 10
E_o	12	21	41	68	94	117	112	89	59	35	13	8
E_o tank	5	13	30	56	89	109	103	84	58	33	15	8

colder winter months and the evaporation is reduced as some of the available incoming energy is absorbed by the cold lower layers. The overall effects on reservoir loss by these seasonal fluctuations of stored energy are dependent on the dimensions of the reservoir. Wide shallow bodies of water are more readily affected by marked seasonal temperature changes, whereas in deep narrow reservoirs, the smaller seasonal fluctuations of stored energy will have less effect on water loss. These considerations assume major proportions in tropical countries where annual evaporation exceeds precipitation. For an operational impounding reservoir, heat storage is also affected by water temperatures of inflows and outflows. To obtain monthly estimates of E_{Res} in practice, measured values of E_o can be obtained by other methods, to which are added estimated values of E_{St}, either positive or negative according to season.

4.2.2   Tanks and Pans

Although there may be difficulties in relating the measurements of evaporation from small bodies of water to the real losses from a large reservoir, the advantages in using tanks and pans are numerous. These relatively small specially designed instruments, with either circular or square plan sections, are easily managed and can be transported to any required location for simple installation. Originally designed to be kept at meteorological stations where readings are made regularly at a fixed time each day, their operation has been improved by the attachment of self-filling devices and by the continuous measurement of the water level, automatically recorded on charts or magnetic tape. However, the general opinion in the UK is that this method of evaporation measurement is unreliable and the data collected are incapable of being adequately quality controlled. The tendency is to use calculated estimates of evaporation (Chapter 11).

There are many different tanks and pans, since each country or organization seems to have designed its own instrument to suit particular needs and conditions. Only a selection of the most frequently used instruments will be described; for a more comprehensive list the reader is referred to the World Meteorological Organization Technical Note No. 83 (WMO, 1966).

Of the many evaporimeters used experimentally in the 1860s, the *tank* ascribed to *Symons* became the British standard instrument (Fig. 4.1a). It is a galvanized iron tank, 1.83 m (6 ft) square and 0.61 m (2 ft) deep and set in

Table 4.2

Date		Observations made at 0900 h GMT (mm)		Daily records (mm) (from 0900 h)	
		Rain	Hook gauge	Evaporation	Rainfall
June	1	0.3	21.1	1.0	—
	2	—	20.1	1.6	8.9
	3	8.9	27.4	0.7	12.7
	4	12.7	39.4	1.8	—
	5	—	37.6	2.3	—
	6	—	35.3	2.0	—
	7	—	33.3	—	—

the ground with the rim 100 mm (4 in) above ground level. The tank holds about 400 gallons (1.8 m³), the water level being kept at near ground level and never allowed to fall more than 100 mm below the rim. Measurements of the water level are made daily using a hook gauge attached to a vernier scale and any rainfall measured in the previous 24 h must be added. The depth of evaporation is evaluated as shown in the example in Table 4.2. Records compiled in this way from a British standard tank kept at Kempton Park are given in the last row of Table 4.1. (It should be noted that rainfall observations made at 0900 h are normally allocated to the previous day).

The most widely used instrument nowadays is the *American Class A pan* (Fig. 4.1b). This is circular with a diameter of 1.21 m (47.5 in) and is 255 mm (10 in) deep. It is set with the base 150 mm above the ground surface on an open wooden frame so that the air circulates freely round and under the pan. The water level in the pan is kept to about 50 mm (2 in) below the rim. The level is measured daily with a hook gauge and the difference between two readings gives a daily value of evaporation. Alternatively, evaporation can be obtained by bringing the water level in the pan back to a fixed level with a measured amount of water. Again any rainfall must be allowed for. Since the sides of the pan are exposed to the sun, the contained water tends to attain a higher temperature than in pans set in the ground and thus the measured evaporation is higher than otherwise. For example, in the Kempton Park study, a US Class A pan was installed in 1959, and for the four years of records, 1959–62, the average annual total evaporation measured by the pan was 963 mm (37.9 in) compared with 673 mm (26.5 in) from the reservoir and 625 mm (24.6 in) from a Symons tank. On an annual basis, the reservoir evaporation was 0.7 times the pan measurement. This factor, 0.7, is known as the *pan coefficient* and its value varies slightly over different climatic regions. (A range of 0.67 – 0.81 is noted (WMO, 1966)). If seasonal evaporation values are required, the heat storage effects cause greater differences between pan and reservoir. Monthly pan coefficients must be obtained and used to give monthly estimates of reservoir evaporation from pan measurements. For example, in the classical evaporation studies at Lake Hefner (USGS, 1954), for monthly observations in 1950 the

Class A pan coefficient ranged from 0.34 in May to 1.31 in November. Thus the pan evaporation greatly exceeded the lake evaporation in the early summer, but extra heat storage gave higher lake evaporation in the autumn.

Many experiments with modified installations of the US Class A pan have

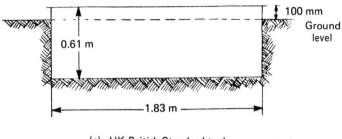

(a) UK British Standard tank

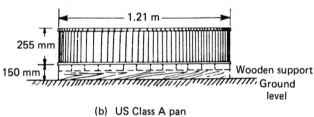

(b) US Class A pan

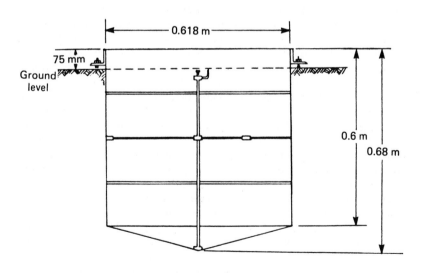

(c) USSR GGI-3000 tank

Fig. 4.1   Evaporimeters.

been made in attempts to inhibit the exaggerated evaporation due to the overheating of the water. In India, the outside has been painted white to increase radiation reflection and some studies have recommended setting the pan in the ground. In arid regions, the pan is an attraction to birds and animals and is usually covered with a wire mesh. Such a screen of 25 mm (1 in) chicken wire mesh, while preventing the bulk removal of water gave an average reduction of 14% in measurements of monthly mean evaporation over two years of measurements in Kenya due to reduction in radiation. In using evaporation measurements from a US Class A pan careful note must be made of its siting and installation.

Another instrument that has been accepted by many countries is the Russian tank (Fig. 4.1c). The *USSR GGI-3000 tank* has a smaller surface area (0.3 m², 0.618 m diameter) than the other instruments, but has the depth of the British tank (0.60 – 0.685 m). It is cylindrical with a conical base and is made of galvanized iron. The tank is installed in the ground with the rim about 75 mm above the surface. A comparison between the GGI-3000 tank and the Class A pan was made at Valday (USSR) over 11 summer seasons and an average ratio (tank/pan) for the seasonal evaporation totals was 0.78. Thus like the Symons tank, the GGI-3000 tank gives a measure of E_o of the correct order of magnitude, and a measure of heat storage effects is required before reservoir loss can be evaluated.

4.2.3 Atmometers

These are devices that can give direct measurement of evaporation. A water supply is connected to a porous surface and the amount of evaporation over a designated time period is given by a measure of the change in water stored. Thus $E_o = \Delta S$. (This evaporation mechanism has been likened to transpiration from leaves, but as the biological control is not simulated, atmometer data are considered as measures of E_o.) It is essential to have a constant instrument exposure to ensure consistent observations and it has been found satisfactory to have atmometers set in a well ventilated screen as is used for exposing thermometers to register air temperature. Atmometers are simple, inexpensive and easy to operate, but care must be taken to see that the porous surfaces from which the evaporation takes place are kept clean. Two types are described here.

The *Piche* evaporimeter consists of a glass tube 14 mm in diameter and 225 mm long with one end closed. A circular disc 32 mm diameter of absorbent blotting paper is held against the open end by a small circular metal disc with a spring collar. The evaporating surface area is 1300 mm² and this is fed constantly by the water in the tube hung up by its closed end. The tube is graduated to give a direct reading of evaporation (E_o) over a chosen time period, usually a day. The measurement in millimetres is related to the evaporating surface of both sides of the paper. The tube holds an equivalent of 20 mm of evaporation; the water is replenished when necessary.

When the Piche evaporimeter is exposed in a standard temperature screen, the annual values have been found to be approximately equivalent to the open water evaporation from a US Class A pan. This type of instrument is used widely in the developing countries of Africa and the Near East.

In the *Bellani* atmometer, the porous surface is provided by a thin ceramic disc, 85 mm in diameter. This is attached to a graduated burette holding the water supply. As with the Piche evaporimeter, the difference in burette reading over a specified time gives the measure of evaporation.

4.3   Measurement of E_t

The measurement of evaporation loss from a vegetated land surface is even more complex than the measurement of loss from an open water body. The extra mechanism of plant transpiration must be added to considerations of water availability and the ability of the atmosphere to absorb and carry away the water vapour. However, similar approaches for the measurement of E_t may be adopted.

4.3.1   Water Budget Method

To establish the E_t loss from a catchment area draining to a gauging station on a river, the water balance over a selected time period can be evaluated:

$$E_t = P - Q - G - \Delta S$$

where P = precipitation, Q = river discharge, G = groundwater discharge and ΔS = increase in storage. For a natural catchment, measurements of the precipitation and river discharge may be made satisfactorily with some degree of precision, but the measurement of groundwater movements into or out of the drainage area cannot be made easily. In water balance studies, it is usually assumed that the catchment is watertight and that no subsurface movement of water across the defined watershed is occurring. However, if there are aquifers noted in the area, groundwater movements should be investigated. The evaluation of change in storage depends on the time period over which the water balance is being made. On an annual basis, the time at which the balance is effected is chosen so that the water stored in the ground and in surface storage is approximately the same each year and thus in the equation, $\Delta S = 0$. Nevertheless, significant differences in the amount stored may occur from one year to another. In the UK, the end of September marks the end of the Water Year when most of the transpiration of the summer season is over and the groundwater replenishment of the winter months is about to begin. If monthly losses are required then values of ΔS must be obtained. Measurements of soil moisture content can be made regularly each month or can be budgeted from potential evaporation calculations (see Chapter 11) and rainfall measurements.

Measurements of the components of the water balance equation for a drainage basin are even less reliable than for a reservoir and hence its use for evaluating E_t is recommended only for annual values. For shorter time periods, changes in storage should be measured and in all instances a thorough knowledge of the catchment area is essential.

4.3.2   Percolation Gauges

These are instruments specially designed for measuring evaporation and transpiration from a vegetated surface, E_t, and are comparable with the tanks and pans used for measuring E_o. Similarly, there are very many different designs and, in general, these are regarded as research tools rather than standard instruments to be installed at every climatological station. A cylindrical or rectangular tank about 1 m deep is filled with a representative soil sample supporting a vegetated surface and is then set in the ground. A pipe from the bottom of the tank leads surplus percolating water to a collecting container. The surface of the gauge should be indistinguishable from the surrounding grass or crop covered ground. A rain gauge is sited nearby and the evaporation plus transpiration is given by the following equation:

$$E_t = \text{Rainfall} - \text{Percolation}$$

Percolation gauges do not take into account changes in the soil moisture storage and thus measurements should be made over a time period defined by instances when the gauge is saturated so that any difference in the soil moisture storage is small. Records are generally compiled on a monthly basis in climates with rainfall all the year round. A recommended installation is shown in Fig. 4.2.

4.3.3   Lysimeters

By taking into account change in water storage in the ground, lysimeters improve on the E_t measurements of percolation gauges. Compared with the latter, lysimeters are much more complex, more expensive to construct and maintain, and therefore even more associated with research installations or specially funded studies.

A large block of undisturbed soil covered by representative vegetation is surrounded by a watertight container driven into the ground. A sealing base with a drain pipe is secured to the bottom of the block and a weighing device established underneath. Then:

$$E_t = \text{Rainfall} - \text{Percolation} \pm \text{Weight change}$$

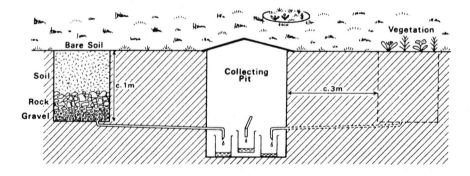

Fig. 4.2   A percolation gauge installation. (Reproduced from K.J. Gregory & D.E. Walling (1973) *Drainage Basin Form and Process*, by permission of Edward Arnold.)

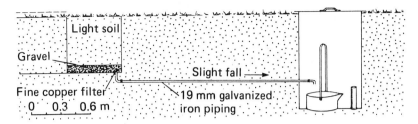

Fig. 4.3 Installation for measuring potential evapotranspiration (PE). (Reproduced from F.H.W. Green (1979) *Rep. Inst. Geol. Sci.* **79/6**, 4–6, by permission.)

All units of measurements are referred to the area of the lysimeter orifice at ground level. The accuracy in the measurement of actual evaporation by lysimetry is dependent on the sensitivity of the weighing mechanism and, to detect small changes in soil moisture storage, large block samples are required. However, once the complications of the elaborate installations have been overcome with a suitable balance, the lysimeters can be easy and inexpensive to run (McIlroy, 1966).

4.4   Measurement of Potential Evaporation

Potential evaporation (PE), the evaporation plus transpiration from a vegetated surface when the water supply is unlimited, is measured regularly at several climatological stations in the UK using an irrigated 'lysimeter'. One of the earlier installations, shown in Fig. 4.3, closely resembles a percolation gauge. The principal difference is in the operation of the apparatus, with the contained soil being kept at field capacity (Chapter 5) by sprinkling a known quantity of water on the tank when rainfall is deficient. Field capacity is assured by maintaining continuous percolation from the bottom of the tank. Thus the vegetation cover is allowed to transpire freely, and the total evaporation loss is dependent entirely on the ability of the air to absorb the water vapour (Guerrini, 1953). Then:

PE = Rainfall + Irrigation – Percolation

Later installations using standard galvanized iron dust bins with the drainage pipe added proved to be most economical to establish. One of the disadvantages of these gauges is that the soil sample has been disturbed, but with careful filling of the bin and after the establishment of the vegetation cover, the gauge gradually becomes representative of the surrounding terrain. Certain difficulties in operating the gauges are encountered in winter with snow cover and freezing temperatures, but discrepancies are not of great importance since evaporation losses are low and often negligible under such conditions. Measured values of PE using these irrigated gauges can be exaggerated in very dry periods and hot climates when an oasis effect might be experienced. Surrounding parched ground heating and drying the air above tends to cause increased evaporation from the continuously watered and transpiring vegetation of the gauge.

Comparisons of evaporation measurements with calculated values will be made elsewhere (Chapter 11). Since this chapter is devoted to practical measurement, it is concluded with descriptions of the various instruments (and their recommended exposures plus observational practice), used in the assemblage of meteorological data required for the calculation of evaporation.

4.5   Measurement of Meteorological Variables

The factors affecting evaporation have already been described. The principal source of energy, the sun, transmits its radiation through the atmosphere. This is measured by solarimeters maintained by the Meteorological Office at observatories and major climatological stations, and an engineering hydrologist can obtain solar radiation data from the appropriate authorities. However, the measurement of net radiation, the difference between incoming radiation (short and long wave) and outgoing radiation (reflected short wave and ground-emitted long wave) is of particular concern in the calculation of evaporation. The setting up of a net radiometer for hydrological studies is fairly straightforward and will be mentioned further in connection with automatic weather stations. The more conventional measurements of air temperature, humidity, wind speed and direction, and sunshine are made at many types of installations for meteorological, climatological and agricultural purposes. The hydrologist can use all these observations for his own purpose, but increasingly it is found that there are shortages of data for particular areas and it may be necessary to set up special stations to provide the data required for evaporation estimations. Wherever possible it is recommended that the rules and guide lines developed over the years by the expert organization should be followed by the engineer establishing a new meteorological station.

4.5.1   Siting a Meteorological Station

Level ground about 10 m by 7 m in extent covered with short grass is selected and enclosed by open fencing or railings. The site should not have any steep slopes in the immediate vicinity and should not be located near trees or buildings. Although a certain amount of shelter is required for rain gauges (see Chapter 3), a very open site is advisable for sunshine recorders and anemometers. If the enclosure is not satisfactory for particular instruments, they may be installed elsewhere according to their special requirements. (For details, see Meteorological Office, 1982.)

A recommended plan for the instrument enclosure is shown in Fig. 4.4. The geographical coordinates of the station latitude and longitude should be determined from the relevant topographical map if available, and the height above sea level established from the nearest Ordnance Survey bench mark. Particular attention should be paid to noting the exact orientation of the enclosure since the setting up of the anemometer and sunshine recorder is dependent on direction.

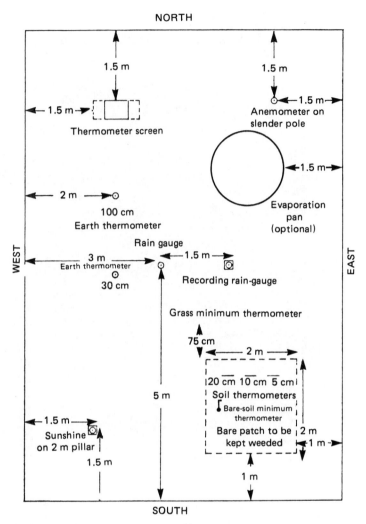

NORTH

1.5 m

1.5 m

←1.5 m→

←1.5 m→
Anemometer on
slender pole

Thermometer screen

←1.5 m→

2 m
100 cm
Earth thermometer

Evaporation
pan
(optional)

Rain gauge
←1.5 m→
3 m
Earth thermometer
30 cm

Recording rain-gauge

Grass minimum thermometer

75 cm

←2 m→

5 m

20 cm 10 cm 5 cm
Soil thermometers
Bare-soil minimum
thermometer

←1.5 m→
Sunshine
on 2 m pillar

Bare patch to be
kept weeded

2 m
←1 m→

1.5 m

1 m

WEST   **EAST**

SOUTH

Fig. 4.4   Plan of a meteorological station for the northern hemisphere (dimensions in metres). (Reproduced from *Observer's Handbook, 4th edn.* (1982) Meteorological Office, by permission of the Controller, Her Majesty's Stationery Office. © Crown copyright.)

4.5.2   Instruments

The most prominent feature of a meteorological station is the Stevenson screen which houses the air thermometers. The ordinary screen (Fig. 4.5) provides the standard exposure for the *air thermometers* with their bulbs 1.25 m above the ground surface. The double-louvered screen painted white is set firmly in the ground so that it opens away from the direction of the midday sun (i.e. to the north in the northern hemisphere). The two vertically hung

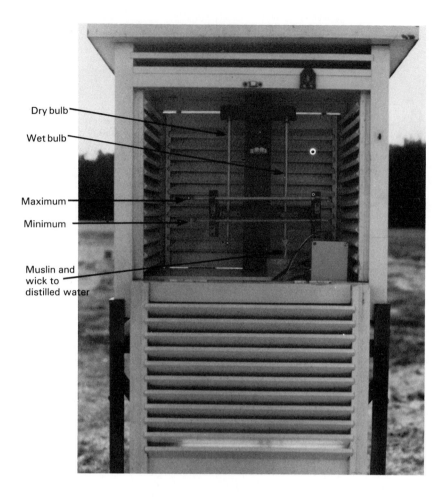

Dry bulb

Wet bulb

Maximum

Minimum

Muslin and
wick to
distilled water

Fig. 4.5 Ordinary Stevenson screen. (Reproduced from Meteorological Office, 1986, *Making Weather Observations*, Met. O. Leaflet No. 5, by permission of the Controller, Her Majesty's Stationery Office © Crown copyright.)

thermometers are for the direct reading of the air temperature (dry bulb) and the reading of the wet bulb, covered with muslin kept moist by a wick leading from a small reservoir of distilled water. With these two temperature readings, the dew point, vapour pressure and relative humidity of the air are obtained from hygrometric tables or a humidity slide rule. Supported horizontally are maximum and minimum thermometers. The four thermometers are read at 0900 h GMT each day in the UK and at this time the maximum and minimum thermometers are reset. At some climatological stations, the dry and wet bulb thermometers are read again at 1500 h, but where more detailed observations are required, further readings may be taken. For continuous air temperature and humidity records, a bimetallic

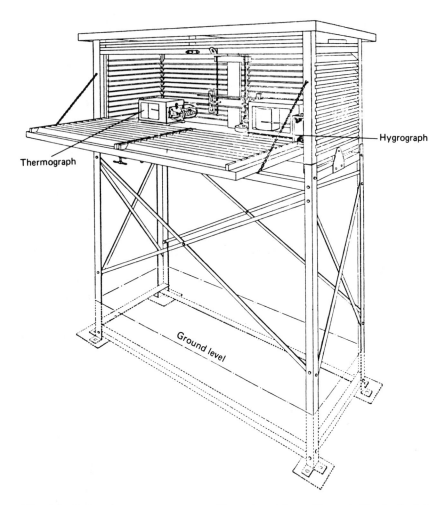

Thermograph

Hygrograph

Ground level

Fig. 4.6   Large thermometer screen. (Reproduced from *Observer's Handbook, 3rd edn.* (1969) Meteorological Office, by permission of the Controller, Her Majesty's Stationery Office. © Crown copyright.)

thermograph and a hair hygrograph may be installed in a large Stevenson screen (Fig. 4.6) which also contains the more accurate thermometers for cross-checking with the autographic instruments. Both the thermograph and hygrograph have weekly charts rotating on clockwork drums upon which ink pens draw the temperature and humidity traces.

Although not directly connected with the computation of evaporation, there is provision in the meteorological station (Fig. 4.4) for temperature measurements of the soil and ground. Special thermometers of the types shown in Fig. 4.7 are set at selected depths in the soil of a bare plot and in ground covered with short grass. These measurements may be of value to the hydrologist concerned with irrigation or with flood forecasting. The grass minimum thermometer is the scientific indicator of ground frost which is not always visible as a hoar frost deposit.

The rain gauges and the evaporation pan have been discussed earlier (Sections 3.1 and 4.2.2). The remaining instruments of significance in the

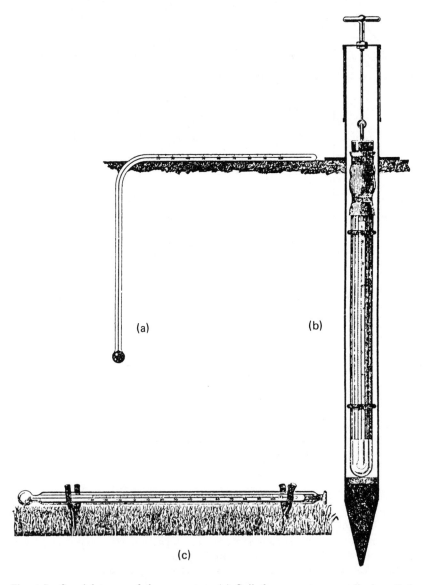

Fig. 4.7   Special types of thermometer. (a) Soil thermometer, usually installed under bare soil. (b) Earth thermometer and steel tube, usually installed under a grass-covered surface. (c) Grass minimum thermometer. (Reproduced from *Observer's Handbook, 3rd edn.* (1969) Meteorological Office, by permission of the Controller, Her Majesty's Stationery Office. © Crown copyright.)

enclosure and shown in Fig. 4.4 are the sunshine recorder and anemometer. A standard Campbell – Stokes *sunshine recorder* is shown in Fig. 4.8. The glass sphere focuses the sun's rays on to a specially treated calibrated card where they burn a trace. The accumulated lengths of burnt trace give a measure of the total length of bright sunshine in hours. Three sizes of cards are used with the recorder according to the season, i.e. over the winter or summer solstice or the equinoxes. A later pattern Mark III has the glass sphere held between two screws. A single card records a day's sunshine and is therefore changed each day at the normal observational time 0900 h GMT with the sunshine before and after 0900 h being credited to the correct days.

The direction and speed of the wind are most important features of the weather. Although the hydrologist concerned with evaporation may not be unduly worried by wind direction, his other duties with regard to real time hydrological events should encourage him to instal an instrument to measure both characteristics. The most sophisticated instrument is a Dines pressure tube anemograph, which continuously records direction and velocity with two traces on a graduated chart revolving daily on a clockwork drum. However, these are usually installed on buildings and the mechanism is housed safely inside. For measurements in the meteorological station enclosure, a *cup anemometer* is recommended and the most recent instruments are cup generator anemometers incorporating a remote indicating wind vane (Fig. 4.9). The instrument is fixed on a pole 2 m from the ground and the electrical recording apparatus is housed conveniently away from the installation. The cup anemometer can give instantaneous readings of wind velocity (knots or m s^{-1}) or provide a run-of-the-wind, a collective distance in kilometres when the counter is read each day.

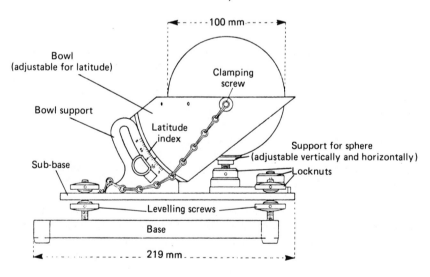

Fig. 4.8 Sunshine recorder Mk. 2 (Campbell – Stokes). (Reproduced from *Observer's Handbook, 3rd edn.* (1969) Meteorological Office, by permission of the Controller, Her Majesty's Stationery Office. © Crown copyright.)

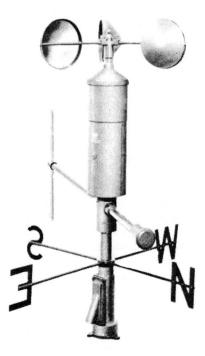

Fig. 4.9   Cup generator anemometer and wind vane. (Reproduced by permission of R.W. Munro Ltd.)

4.5.3   Recent Developments

The modern developments in instrumentation of all kinds have led to marked improvements in the recording of measurements of meteorological variables. The standard installation of a meteorological station described in the foregoing section has been adapted and modified as new devices became available. The major improvements are new sensors for temperature and humidity and the translation of the various measurements into signals for recording on magnetic tape ready to be processed by a digital computer. Thus all the 'thermometers', the humidity sensor and the anemometer can be connected by cable to a tape recorder buried below the Stevenson screen or housed safely off the site. Where required, the data may be transmitted to distant offices.

At all stages in the introduction of new instruments, the Meteorological Office has carried out exhaustive comparability studies with the old standard instruments to ensure homogeneity in the observations.

For remote locations, the *automatic weather station* serves ideally the needs of the hydrologist. Various organizations have developed such installations; the UK Meteorological Office was particularly concerned with obtaining weather observations at sea, and has worked on new transmitting instruments. The Institute of Hydrology has produced, after many years of careful development, a set of instruments with appropriate recording devices which

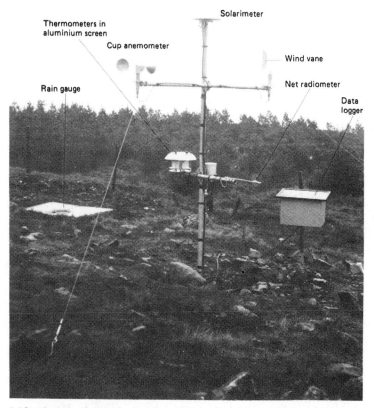

Fig. 4.10   Automatic weather station. (Reproduced by permission of the Institute of Hydrology.)

is finding favour with hydrologists in the UK and overseas (Strangeways & McCulloch, 1965). An annotated photograph of the Institute of Hydrology's automatic weather station is given in Fig. 4.10. The instruments, excluding the rain gauge, are mounted on an aluminium mast with two cross arms. The Kipp solarimeter for incoming short wave radiation and the eight inch Rimco tipping bucket rain gauge are both well tried and recognized instruments. The net radiometer, anemometer and wind vane and the temperature sensors have all been designed at the Institute of Hydrology. The air temperature and wet bulb depression (difference between dry and wet bulb readings) are measured by two dry and one wet bulb platinum-resistance thermometers. These are housed in a small specially designed thermal radiation screen instead of a conventional type Stevenson screen. All the instruments are wired to an interface unit and thence to a data logger. (Such an automatic weather station can now be purchased directly from the Didcot Instrument Company Ltd.) The data are fed from the logger tape into a computer which can be programmed to check the observations and print out hourly or daily values for all the meteorological variables (Strangeways, 1972). The data logger can also be used for recording other meteorological measurements such as barometric pressure and hydrological measurements, river or well levels.

References

Dyer, A.J. and Maher, E.J. (1965). *The Evapotron: an Instrument for the Measurement of Eddy Fluxes in the Lower Atmosphere*. Div. of Meteorological Physics Tech. Paper 15, CSIRO, Australia.

Green, F.H.W. (1979). 'Potential evaporation determined from lysimeters.' *Rep. Inst. Geol. Sci.*, No. 79/6, 4–6.

Guerrini, V.H. (1953). *Evaporation and Transpiration in the Irish Climate*. Tech. Note No. 14, Meteorological Service, Dublin.

Lapworth, C.F. (1965). 'Evaporation from a reservoir near London.' *Jl Inst. Water Eng.*, **19**, (2), 163–181.

McIlroy, I.C. (1966). 'Evaporation and its measurement' in *Agricultural Meteorology*. Proc. WMO Seminar, Melbourne, pp. 243–263.

Meteorological Office (1969 and 1982). *Observer's Handbook*. 3rd and 4th eds. HMSO.

Meteorological Office (1986). *Making Weather Observations*. Met. O. Leaflet No. 5.

Strangeways, I.C. (1972). 'Automatic weather stations for network operation'. *Weather*, **27**, (10), 403–408.

Strangeways, I.C. and McCulloch, J.S.G. (1965). 'A low priced automatic hydrometeorological station'. *Bull. IASH*, **10**, (4), 57–62.

US Geological Survey (USGS). (1954). *Water Loss Investigations, Lake Hefner*. US Geol. Surv. Prof. Papers 269 and 270.

World Meteorological Organization (WMO), (1966). *Measurement and Estimation of Evaporation and Evapotranspiration*.' Tech. Note 83, WMO No. 201 TP 105, Reprinted 1968.

5
Soil Moisture

In the general progression of the hydrological cycle beginning with atmospheric water vapour and ensuing precipitation, the top layers of material near the land surface provide the first of the sub-surface water storages. Water on the Earth's surface either in solid or liquid form occurs in a fairly homogeneous body and can be seen and measured with varying degrees of accuracy. In the ground however, water is contained as a mixture among a heterogeneous collection of solids; quantities are much more difficult to assess. The study of soil moisture is of vital interest to the agriculturalist, especially in those countries where irrigation can improve the yield of food crops. More recently, the role played by soil moisture content in the management of water yields and flood control is being more fully appreciated.

In this chapter, consideration is given to water in those layers of the soil near the surface that contain the rooting zones of vegetation. The larger scale occurrence of water generally at greater depths below the land surface is dealt with in Chapter 7 on Groundwater.

5.1 Soil Structure and Composition

The original parent material of soil is the solid rock of the Earth's outer skin. Weathering and erosion have broken down the surface layers of the solid geological strata and in many areas large deposits of unconsolidated material, possibly moved considerable distances by wind or water, have been deposited. Thus a soil may be a direct product of underlying weathered rocks or may be formed from loose deposits unrelated to the solid rocks below. Soil depths and their composition can therefore be very variable. Another most important constituent of a soil, especially in its upper layers, is the organic material derived from living plants and other organisms. An idealized section through a typical sequence of soil layers is shown in Fig. 5.1. Layers of vegetation litter and partly decomposed debris lie on the surface above what is termed the A horizon, a layer that is generally friable and rich in humus. The B horizon is mainly composed of well weathered parent material with its structure modified by roots and living creatures such as earthworms. The C horizon is unconsolidated rock material containing a wide range of particle and stone sizes. The thickness of soil layers depends upon their location relative to the geological structure and the geomorphology of surface features.

Soil composition is studied in detail by pedologists and soil physicists. Dried samples of soil are analysed by sequential sieving and the different particle sizes separated and weighed. It is generally agreed that particles of less than 2 μm in diameter form the clay fraction and consist of chemically and

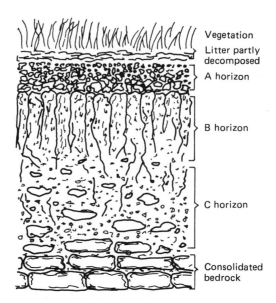

Fig. 5.1   Idealized soil section. (Reproduced from D. Hillel (1980) *Fundamentals of Soil Physics*, by permission of Academic Press.)

physically reactive minerals (colloids). Particles or pebbles greater than 2 mm in diameter are universally classed as gravels. The range in between clay and gravel is variously subdivided into silts and sands. Three standard classifications are given in Fig. 5.2. Soils are labelled according to the proportions of the different fractions they contain. Fig. 5.3 shows the definition of soils using the US Department of Agriculture particle-size classification. A loam is a well mixed soil and the diagram demonstrates that it can be formed from 8 – 28% clay, 28 – 50% silt and 22 – 52% sand. Other soils are named relative to their proportions of the three major constituents. A guide to their identification in the field is given in Table 5.1.

ISSS	clay	silt		fine sand			coarse sand			gravel
USDA	clay	silt		v.f. sand	fine sand	m. sand	co. sand	v. co. sand		gravel
BSI	clay	fine silt	medium silt	coarse silt	fine sand		medium sand	coarse sand		gravel

```
       1      2     5    10  20      50    100   200        500  1000 2000
```
Particle diameter (μm)

Fig. 5.2   Particle size classification. ISSS: International Society of Soil Science. USDA: US Department of Agriculture. BSI: British Standards Institution.

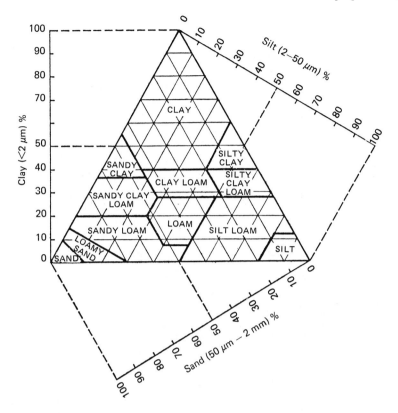

Fig. 5.3   Soil definition with USDA classification. (Reproduced from C.J. Wiesner (1970) *Climate, Irrigation and Agriculture*, by permission of Angus & Robertson (UK) Ltd.)

5.2   Soil Properties

Before considering water in the soil, it is necessary to appreciate in more detail the composite mixture of a soil. The functioning of a soil as a water store depends on the packing of the clay or sand particles and the amount of space available between the solids. This *pore space* may contain gas and/or liquid, which are usually air, water vapour and/or liquid water. The composition of the soil mixture or matrix is described by the relationships of some of the basic properties of the different constituents.

Thus, if V_s, V_w, V_g, V_t are volumes of solids, liquids, gases and the total soil content and m_s, m_w, m_g, m_t, are the corresponding masses, then the following properties are defined:

Density of solids (mean particle density):
$$\rho_s = m_s / V_s$$
(average value for mineral soils 2.65 g cm^{-3})

Table 5.1 Classification of Soil by Observation and Feel

(Reproduced from C. J. Wiesner (1970) *Climate, Irrigation and Agriculture*, by permission of Angus & Robertson (UK) Ltd.)

Type	Texture	Feel	Squeezed when dry	Squeezed when moist
Sand	Single grained and loose	Individual grains clearly seen or felt	Falls apart when pressure released	Forms cast but will crumble when touched
Sandy loam	Contains much sand (50% or more) but enough silt, clay to give coherence	Individual grains can be seen or felt	Forms cast which falls apart readily	Forms cast which can be carefully handled without breaking
Loam	Sand, silt, and clay, in such proportions that no one predominates	A mellow to gritty feel slightly plastic when moist	Forms cast which bears careful handling	Forms cast which can be handled freely
Silt loam	Contains much silt (50% or more) moderate amount of sand, and a small quantity of clay	When dry appears in clods which are readily broken. Pulverised, it feels smooth, soft, and floury. Runs together when wet	Forms cast which can be freely handled without breaking	Forms cast which can be freely handled without breaking. Will not ribbon if squeezed between finger and thumb but gives broken appearance
Clay loam	Contains each component	A fine textured soil which breaks into clods or lumps when dry hard	Generally hard and cloddy lumps when dry	Plastic forms cast which will bear much handling. Will form thin ribbon which breaks easily if squeezed between finger and

Table 5.1 (cont'd)

(Reproduced from C.J. Wiesner (1970) *Climate, Irrigation and Agriculture*, by permission of Angus & Robertson (UK) Ltd.)

Type	Texture	Feel	Squeezed when dry	Squeezed when moist
				thumb. Kneeded in hand, works into compact mass; does not crumble
Clay	Contains much clay	Fine textured forms very hard lumps or clods when dry. Plastic and sticky when wet	Very hard and cloddy when dry	Plastic and sticky. Will form long flexible ribbon when pinched out

Bulk density of soil:

$$\rho_b = m_s / V_t$$

(sandy soils 1.6 g cm^{-3}; clays 1.1 g cm^{-3})

Porosity

$$f = \frac{V_w + V_g}{V_t}$$

(range 0.2 – 0.6 from coarse to fine soils)

Since V_t may change with moisture content and compaction, an alternative measure of pore space is given by:

Void ratio:

$$e = \frac{V_w + V_g}{V_s}$$

(range 0.25 – 2.0)

The porosity and void ratio, which are interrelated, can also be expressed in terms of the densities. These simple measures of soil characteristics are important in assessing a soil's water-storage capacity.

5.3   Water in the Soil

Most of the water content of a soil comes from rainfall or melting snow, and is

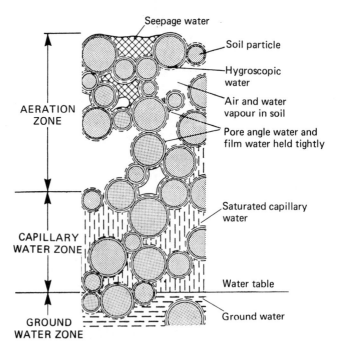

Fig. 5.4   Water in the soil. (Reproduced from C.J. Wiesner (1970) *Climate, Irrigation and Agriculture*, by permission of Angus & Robertson (UK) Ltd.)

shown in Fig. 5.4, infiltrating as seepage water moving by gravity and surface tension through the pore spaces. Its pathways are smoothed by a thin film of hygroscopic water on each of the soil particles. The hygroscopic moisture is held tightly by electrostatic forces and is not readily moved by other forces, including plant roots. Below the percolating flow, the voids in the soil are filled with air and/or water vapour. This layer is a zone of aeration where there is a complex mixture of solid particles, liquids and gases. With increase in depth, the aeration zone gives way to a layer of saturated soil where all the pore space is occupied by water. In the saturated capillary zone, the water is held by capillary forces between the soil particles and is at less than atmospheric pressure. At greater depths in this saturated zone the water pressure exceeds atmospheric pressure. The surface over which the pressure equals atmospheric pressure is defined as the *water table*. The extent of the capillary zone is dependent on the soil composition and packing of the soil particles. It ranges from a few centimetres in a coarse sandy soil to a few metres in a clay soil.

Using the same basic properties defined previously, the wetness of the soil can be assessed in the following terms.

Water content by mass

$$\theta_m = m_w/m_s$$

The mass of the dry soil, m_s, is obtained by heating the soil in an oven to 105 °C with repeated weighings until the mass is constant.

Water content, volume fraction

$$\theta = V_w/V_t$$

The volume fraction, θ, is equivalent to the depth ratio of soil water, i.e. the equivalent depth of free water relative to the depth of soil for a unit plan area. In this way, the soil moisture can be related easily to precipitation and evaporation depths.

The water content (by volume) is shown for three different soil types for three soil water conditions in Table 5.2. These data show that there is much less water available at field capacity in a sandy soil that drains quickly. The higher the clay content of a soil, the greater is its retention capability. At the permanent wilting point, a clay soil still contains a significant amount of water, but it is in a finely divided state held under high tension in the minute pores and adsorbed to the surface of the clay particles.

Degree of saturation

$$S = \frac{V_w}{V_w + V_g}$$

(relating the volume of water to volume of pore space)

In relating plant water use to available water in the soil, there are two further terms to be defined.

Table 5.2   Soil Water Content (Volume Fraction θ)

(Reproduced from T.J. Marshall & J.W. Holmes (1979) *Soil Physics*, Cambridge University Press.)

Soil	Clay content (%)	Saturation	Field Capacity	Permanent wilting point
Sand	3	0.40	0.06	0.02
Loam	22	0.50	0.29	0.05
Clay	47	0.60	0.41	0.20

Field capacity is the water content of the soil (volume fraction) after the saturated soil has drained under gravity to equilibrium (usually for about two days).

Permanent wilting point is the volume water content of the drying soil beyond which wilting plants will not recover when provided again with humid conditions. (These terms are met again in Chapter 11).

5.4   Soil Water Retention

In addition to thinking of soil water in terms of relative masses or volumes, it can be considered from the amount of energy needed for its movement (kinetic) or for its retention in the soil (potential). Since the movement of water through a soil is very slow, the kinetic energy is extremely small and may safely be neglected. Potential energy is thus the dominant influence and results from gravity, capillary and adsorptive forces. Hence the *soil water potential* represents the work (energy) required to overcome the forces acting on the soil water referred from a given datum to the point of interest. If this potential energy in joules is expressed in terms of energy per unit volume $(J\ m^{-3})$, then

$$\begin{aligned} \text{Total potential} &= \text{Gravity potential} + \text{Capillary potential} \\ \phi_t &= \phi_g + \phi_p \end{aligned}$$

The gravity potential is given by $\phi_g = z \rho_w g$, where z is the height of the point of interest above some arbitrary datum, ρ_w is the density of water, and g the gravitational acceleration.

The capillary potential is exemplified in Fig. 5.5 showing a rise of water in a capillary tube to a height ψ above the surrounding water level. Above the meniscus at A the pressure is atmospheric, but just below the meniscus the pressure in the water is $-\psi \rho_w g$; this is the energy per unit volume at A due to capillarity, i.e. the capillary potential, ϕ_p. Thus the total potential at A is:

$$\begin{aligned} \phi_t &= z \rho_w g - \psi \rho_w g \\ &= \rho_w g (z - \psi) \\ &= \rho_w g h \end{aligned}$$

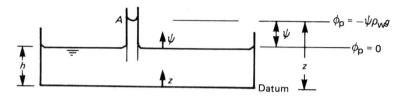

Fig. 5.5   Soil water potential.

where h, in Fig. 5.5, is the height of the water surface above the datum for z. In the static water situation of Fig. 5.5, $(z - \psi)$ is constant (at h) for all points in the water; there are no water movements if there are no differences in total potential. Water will move through a soil between points of different total potential. The latter is often written as $\phi_t / \rho_w \, g$, i.e. $(z - \psi) \, m$, or the equivalent height of water (energy per unit weight). The negative capillary pressure potential, $-\psi \rho_w g$, is often referred to as a *suction* pressure or *tension* pressure.

This capillary pressure potential, ϕ_p, is of dominant importance in the assessment of soil moisture and this can be demonstrated in reference to Fig. 5.6. A sample of non-shrinking (constant-volume) soil is saturated and at Stage 1 covered by a layer of water. The hydrostatic pressure at the soil surface is given by $\psi \rho_w g$ and therefore ϕ_p at and near the soil surface is greater than atmospheric pressure. With a thin soil, increases in ψ are small to points within the soil and thus it is said that the soil water is at the same positive pressure potential ϕ_p. When water is withdrawn from the soil, the water level falls until it reaches the soil surface at Stage 2. Then the head of water has disappeared and the soil water pressure at and near the soil surface is zero but the soil is still saturated. Further withdrawal of water is only possible by applying suction pressure and the air – water interface takes on the form of the line of Stage 3 with the solid particles in contact with the air. The

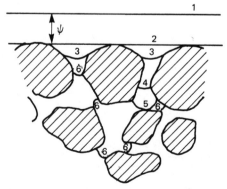

Fig. 5.6   Stages in the drying of a soil. (From E.C. Childs (1969) *An Introduction to the Physical Basis of Soil Water Phenomena*, copyright 1969. Reprinted by permission of John Wiley & Sons Ltd.)

Table 5.3   Soil Water Pressure Units

Head of water (cm)	bar ($100 \text{ kPa} = 10^5 \text{ N m}^{-2}$)	pF
− 1	− 0.001	0
− 10	− 0.01	1
− 100	− 0.1	2
− 1000	− 1.0	3
− 10,000	− 10.0	4

soil water is then under surface tension forces seeking to prevent its removal and the soil water pressures are thus negative in relation to atmospheric pressure. With increased suction, the surface tension in the larger pores is overcome, further water will be withdrawn from the soil (Stage 4, 5, 6) and air enters the emptied pores. The magnitude of these negative pressures or tensions can be quite considerable, and a logarithmic scale of suction pressure heads (pF) is sometimes used. Comparative values are given in Table 5.3. In general, at pF = 0 (1 mb suction), soil is saturated, field capacity is about pF = 2, and permanent wilting point about pF 4.2 (Marshall, 1963).

The relationship between soil water content and water pressure in the pores is shown for contrasting soils in Fig. 5.7. The finely structured clay soil has a higher initial water content owing to a higher porosity, but with small or moderate soil water tensions, the sandy soil will release more water from its larger pores. The withdrawal of water from a soil due to increasing tension

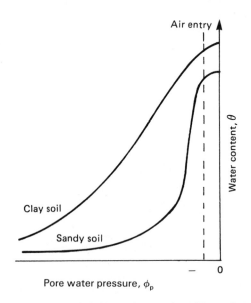

Fig. 5.7   Soil water retention with increasing suction. (Reproduced from D. Hillel (1980) *Fundamentals of Soil Physics*, by permission of Academic Press.)

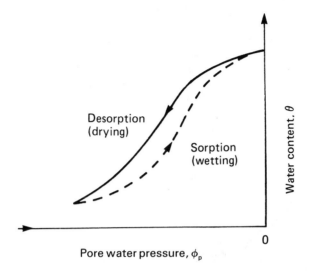

Fig. 5.8   Wetting and drying curves.

(the drying process) is called *desorption*. The reverse wetting process, the addition of water to an unsaturated soil is called *sorption*, Fig. 5.8. The wetting curve does not follow the same relationship as the drying curve. The water content – pore water pressure relationship can be very complex and a family of curves may be obtained for varying initial states of the soil. This general hysteresis effect is caused by the interaction of the soil pores but it is also influenced by entrapped air and the changing of soil volume due to shrinking and swelling. Thus in determining the state of water in the soil, the soil-water content θ and the soil-water tension ϕ_p must be found, since there is no single unique relationship between these two variables.

5.5   Methods of Measurement

The complexities of the physical composition of a soil and the great variety of soils that are to be found over the size of areas for which hydrological measurements are often required make the assessment of soil moisture quantities another sampling procedure. A knowledge of the soils of a catchment area is a first prerequisite; in the UK, such information can be obtained from the maps of the national Soil Survey. In developing countries, the hydrologist should engage the services of a specialist soil scientist to make professional soil surveys for any area where detailed hydrological measurements are required. Then sampling points to cover the main soil types and vegetational zones can be selected over the area.

There are many techniques used for soil moisture measurements and the two most common methods for each of the soil water content θ and soil water potential ϕ_p will be described.

5.5.1   Soil Water Content θ (Volumetric Fraction)

Gravimetric determination.   A soil sample of known volume (V_t) is removed from the ground with a soil auger. The sample is weighed (m_t) and then dried in a special oven with temperature controlled to between 100 and 110 °C until the final weight (m_s) is constant. Then m_w is given by $m_t - m_s$ whence the gravimetric water content is obtained:

$$\theta_m = \frac{m_w}{m_s} = \frac{m_t - m_s}{m_s}$$

Calculating the bulk density $\rho_b = m_s/V_t$ and knowing ρ_w (1 g cm^{-3}), the volumetric water content can be found from:

$$\theta = \theta_m \frac{\rho_b}{\rho_w}$$

This can be a reliable and accurate method (if the measurements are made precisely) and it is used to calibrate other techniques. However, it is destructive of the soil and therefore a 'one-off' determination that cannot be repeated. A new sample must be taken for another measurement and all subsequent samples should be similar in composition to the original sample. Thus a large enough sampling area with homogeneous soil must be selected when siting the sampling points. The main disadvantages of this direct method are that it is time consuming and laboratory facilities are needed.

Neutron scattering.   This technique developed in the 1950s gives a direct measurement of volumetric soil moisture content in the field. A radioactive source is lowered into an augered hole in the soil; the fast neutrons emitted are impeded by the hydrogen nuclei of soil water. The collisions with the hydrogen nuclei cause a scatter of slowed nuclei and the density of these, dependent on the number of hydrogen nuclei present, is sensed by a detector. Thus a reading of the detector of the slow neutrons gives a measure of the amount of water present in the soil. A compact portable instrument has been developed by a special research team of the Institute of Hydrology and the operation of the *neutron probe* is fully explained in the instrument manual (Institute of Hydrology, 1979). The neutron source and detector are mounted in the probe which is lowered down an access tube previously set in the ground at a selected sampling point. Measurements are made at various depths in the soil so that a water-content profile is obtained. A diagram of the installation and the probe together with a graph of a profile are shown in Fig. 5.9. The instrument must be calibrated for different soil types and each site should have its own probe reading to moisture content relationship. The application of the neutron probe technique is ideal for catchment areas where soil moisture measurements are required on a regular basis. A network of access tubes for sampling different soil types can be established on catchments for which water-balance studies at weekly or monthly intervals are to be made.

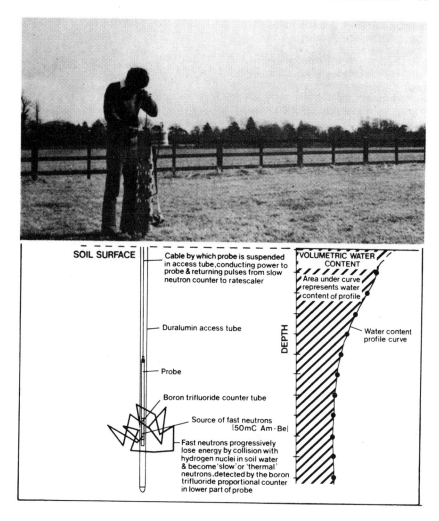

Fig. 5.9   The neutron probe. The probe in use in the field, showing diagrammatically the scattering of fast neutrons and return of slow neutrons in soil around the centre of sensitivity of the probe; the graph shows a typical soil moisture profile. (Reproduced from Natural Environment Research Council (1979) *Neutron Probe System IH II: Instruction Manual*, by permission of the Institute of Hydrology.)

5.5.2   Soil Water Potential ϕ_p

Although changes is soil moisture content over fixed time periods would suffice for computing water balances, soil water potential is necessary for determining moisture fluxes. These are required particularly by hydrologists concerned with plant water use and with the advising of agriculturalists on irrigation practices.

The tensiometer.   As its name implies, this is an instrument to measure the negative pressures or suctions in the soil. The principle of the tensiometer is shown in Fig. 5.10. A porous ceramic cup filled with water is set in the soil and connected to a pressure-measuring device. It is important that the porous cup is firmly contained and in good contact with the soil matrix. If the water is initially at atmospheric pressure and the soil moisture pressure is negative, then the water will move through the porous cup into the soil until equilibrium is reached. Then $-\psi$ gives the soil moisture tension in centimetres of water. The pressure-measuring device is more usually a mercury manometer (Fig. 5.11) or in commercial instruments a dial vacuum gauge. In Fig. 5.11, the pressure potential at B:

$$\phi_B = \phi_P - \rho_w gh_w + \rho_m gh_m$$

But $\phi_B = 0$ (atmospheric pressure) so:

$$\phi_P = \rho_w gh_w - \rho_m gh_m$$

where ρ_m is the density of mercury. There are certain limitations in the use of tensiometers. At suctions of about 1 bar (pF 3), air dissolved in the water comes out of solution and the water column in the tensiometer breaks. In practice, tensiometers are only useful up to suctions of about 0.85 bar, which is a comparatively low tension for soils with a high clay content. Thus as an irrigation guide, tensiometers are helpful for crops needing nearly saturated soil and crops grown in sandy or light loamy soils in which there is little water left when the pore pressures approach this limit (see Tables 5.2 and 5.3). A commercial instrument is available for automatically controlling an irrigation system according to a preset limit of soil moisture tension measured in the rooting zone.

Electrical resistance blocks.   These simple, inexpensive instruments have been widely used since 1940 for measuring soil water potential. A porous block of gypsum with a pair of electrodes embedded is buried in the soil. Water from the soil seeps into the gypsum until the pore pressures in the soil and block reach equilibrium. Then the water content of the block is measured by the electrical resistance between the electrodes. A direct relationship is obtained between the electrical resistance and the water content, which is dependent on the soil water potential. One of the disadvantages of the method is that the block has a long response time and is dependent on continuous close contact

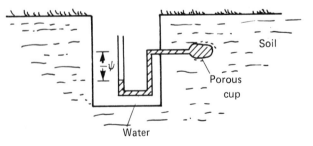

Fig. 5.10   A tensiometer.

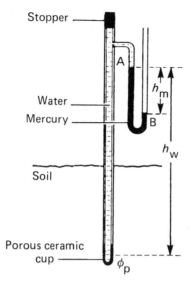

Fig. 5.11 Tensiometer with a mercury manometer. (Reproduced from T.J. Marshall & J.W. Holmes (1979) *Soil Physics*, by permission of Cambridge University Press.)

with the soil. However, the method can be used to measure higher potentials than the tensiometer and indeed they are generally unsatisfactory when the soil is wetter than pF 2.6 (Marshall, 1959). Electrical resistance blocks are made commercially and are a useful indicator of irrigation need in crop production. There is a hysteresis effect between the wetting and drying curve relationships between electrical resistance and soil moisture tension, but calibration data are usually given for the drying curve. Gypsum blocks are not long lasting since there are solution effects and they soon deteriorate in wet soils.

Field instruments developed for the measurement of moisture tension can also be connected to recorders and thus make valuable monitoring devices for continuous assessment of soil conditions. Many of them are, however, not very accurate and are more adapted to agricultural and engineering requirements rather than to scientific research of hydrological processes.

Considerations of water movement in the complex earth storage will be covered in Chapter 7.

References

Childs, E.C. (1969). *An Introduction to the Physical Basis of Soil Water Phenomena*. John Wiley, 493 pp.

Hillel, D. (1980). *Fundamentals of Soil Physics*. Academic Press, 413 pp.

Institute of Hydrology (1979). *Neutron Probe System IH II. Instruction Manual*. Natural Environment Research Council, 32 pp.

Marshall, T.J. (1959). *Relations Between Water and Soil*. Commonwealth Agricultural Bureau, Tech, Comm. No. 50, 91 pp.

Marshall, T.J. and Holmes, J.W. (1979). *Soil Physics*. Cambridge University Press, 345 pp.

Wiesner, C.J. (1970). *Climate, Irrigation and Agriculture*. Angus & Robertson, 246 pp.

6

River Flow

For intending hydrologists having a parallel training in civil engineering, the mechanics of open channel flow is a familiar subject. A full detailed treatment is given by standard texts listed in the References. However, to introduce the measurement of river discharges, a summary of the salient features of open channel flow is given here. The main topic of this chapter is hydrometry, which in its restricted sense means river flow measurement. A more mathematical treatment of unsteady open channel flow is contained in Chapter 16 on Flood Routing.

6.1 Open-Channel Flow

Water in an open channel is effectively an incompressible fluid that is contained but can change its form according to the shape of the container. In nature, the bulk of fresh surface water either occupies hollows in the ground, as lakes, or flows in well defined channels. Open channel flow also occurs in more regular man-made sewers and pipes as long as there is a free water surface and gravity flow.

The hydrologist is interested primarily in flow rate or discharge of a river in terms of cubic metres per second ($m^3 s^{-1}$), but in the study of open channel flow, although the complexity of the cross-sectional area of the channel may be readily determined, the velocity of the water in metres per second ($m s^{-1}$) is also a characteristic of prime importance. The variations of velocity both in space and in time provide bases for the standard classifications of flow.

Uniform flow. In practice, uniform flow usually means that the velocity pattern within a constant cross-section does not change in the direction of the flow. Thus in Fig. 6.1, the flow shown is uniform from A to B in which the depth of flow, y_0, called the *normal depth*, is constant. The values of velocity, v, remain the same at equivalent depths. Between B and C, the flow shown is non-uniform; both the depth of flow and the velocity pattern have changed. In Fig. 6.1, the depth is shown as decreasing in the direction of flow ($y_1 < y_0$). A flow with depth increasing ($y_1 > y_0$) with distance would also be non-uniform.

Velocity distributions. Over the cross-section of an open channel, the velocity distribution depends on the character of the river banks and of the bed and on the shape of the channel. The maximum velocities tend to be found just below the water surface and away from the retarding friction of the banks. In Fig. 6.2a, lines of equal velocity show the velocity pattern across a stream

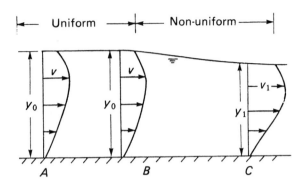

Fig. 6.1   Uniform and non-uniform flow.

with the deepest part and the maximum velocities typical of conditions on the outside bend of a river. A plot of the velocities in the vertical section at depth y is shown in Fig. 6.2b. In practice, the average velocity of such a profile has often been found to occur at or near 0.6 depth.

Laminar and turbulent flow.   When fluid particles move in smooth paths without lateral mixing, the flow is said to be *laminar*. Viscous forces dominate other forces in laminar flow and it occurs only at very small depths and low velocities. It is seen in thin films over paved surfaces. Laminar flow is identified by the Reynolds number $Re = \rho vy/\mu$ where ρ is the density and μ the viscosity. (For laminar flow in open channels, Re is less than about 500.) As the velocity and depth increase, Re increases and the flow becomes *turbulent*, with considerable mixing laterally and vertically in the channel. Nearly all open channel flows are turbulent.

Critical, slow and fast flow.   Flow in an open channel is also classified according to an energy criterion. For a given discharge, the energy of flow is a function of its depth and velocity, and this energy is a minimum at one particular depth, the critical depth, y_c. It can be shown (Francis, 1975) that the flow is characterized by the dimensionless Froude number $Fr = v/\sqrt{(gy)}$. For Fr <

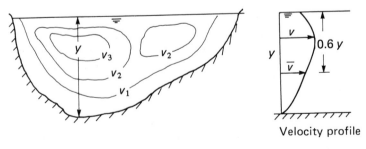

(a) Transverse cross-section v_i — velocity contours   (b) Vertical section

Fig. 6.2   Velocity distributions.

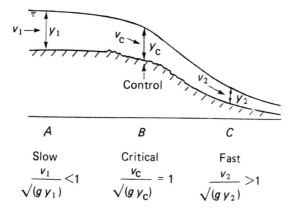

Fig. 6.3   Critical flow.

1, flow is said to be subcritical (slow, gentle or tranquil). For Fr = 1, flow is critical, with depth equal to y_c the critical depth. For Fr > 1, flow is super-critical (fast or shooting). Fig. 6.3 demonstrates these flow conditions. Larger flows have larger values of v_c and y_c.

Steady flow occurs when the velocity at any point does not change with time. Flow is unsteady in surges and flood waves in open channels (although they may sometimes appear steady to a moving observer). The analytical equations of unsteady flow are complex and difficult to solve (Chapter 16) but the hydrologist is most often concerned with these unsteady flow conditions. With the more simple conditions of steady flow, some open channel flow problems can be solved using the principles of continuity, conservation of energy and conservation of momentum.

Natural water courses ranging from small tributary streams to mature rivers flowing into the sea have very varied channel cross-sections and profiles along their length. Rapid changes in river discharges produce the dominant influences on the river banks and bed profile. A sudden increase in flow causes increased erosion and sediment load while the lowering of discharge may result in the sequential deposition of boulders, gravel and silt.

6.2   River Gauging

As in the measurement of precipitation, measurement of river flow is a sampling procedure. For springs and very small streams, accurate volumetric quantities over timed intervals can be measured. For a large stream, a continuous measure of one variable, river level, is related to the discharge calculated from sampled values of other variables, velocity and depth, so that the final result is strictly an 'estimated measurement'.

The discharge of a river, Q, is obtained from the summation of the product of mean velocities in the vertical, $\bar{v}$, and related segments, a, of the total cross-sectional area, A. (Fig. 6.4). Thus:

$$Q = \Sigma(\bar{v}.b.y) = \Sigma(\bar{v}.a)$$

The fixed cross-sectional area is determined with relative ease, but it is much more difficult to ensure consistent measurements of the flow velocities to obtain values of $\bar{v}$.

To obtain a measured estimate of the discharge of a river, it is first necessary to choose a site or short stretch of the channel where variations in discharge will cause the least modification to the cross-section. Ideally, a site where all discharges would be contained within the banks should be used, but almost invariably severe floods exceed the maximum known flow and the river breaks out over an extended flood plain. The second major requisite of a good river gauging site is a well regulated stable bed profile.

A single estimate of river discharge can be made readily on occasions when access to the whole width of the river is feasible and the necessary velocities and depths can be measured. However, such 'one off' values are of limited use to the hydrologist. Continuous monitoring of the river flow is essential for assessing water availability; the continuous recording of velocities across a river is not a practical proposition. It is, however, relatively simple to arrange for the continuous measurement of the river level. A fixed and constant relationship is required between the river level (stage) and the discharge at the gauging site. This occurs along stretches of a regular channel where the flow is slow and uniform and the stage – discharge relationship is under 'channel control'. In reaches where the flow is usually non-uniform, it is important to arrange a unique relationship between water level and discharge. It is therefore necessary either to find a natural 'bed control' as in Fig. 6.3 where critical flow occurs over some rapids with a tranquil pool upstream, or to build a control structure across the bed of the river making the flow pass through critical conditions (Fig. 6.4b). In both cases, the discharge, Q, is a unique function of y_c and hence of the water level just upstream of the control. In establishing a permanent critical section gauging station, care has to be taken to verify that the bed or structural control regulates the upstream flow for all discharges. At very high flows, the section of critical flow may be 'drowned out' as higher levels downstream of the control eliminate the critical depth. Then the flow depths will be greater than y_c throughout the control and the relationship between the upstream water level and discharge reverts to 'channel control'.

At a gauging site, when the flow is contained within the known cross-section and is controlled by a bed structure, then the discharge Q is a function of H (head), the difference in height between the water level upstream and the crest level of the bed control (Fig. 6.4b). The functional stage – discharge relationship is established by estimating Q from sampled measurements of velocity across the channel, when it is convenient, for different values of H. Regularly observed or continuously recorded stages or river levels can then be converted to corresponding discharge estimates. For a structural control, e.g. a weir built to standard specifications, the stage – discharge or $Q \sim H$ relationship is known, and velocity – area measurements are used only as a check on the weir construction and calibration.

After flood flows, cross-sectional dimensions at a gauging station should be checked and if necessary, the river level – discharge relationships

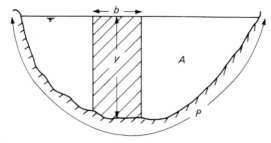

(a) Cross-section

> P = wetted perimeter
> A = total cross-sectional area
> Hydraulic mean depth $R = A/P$
> $b \times y = a$, area of segment

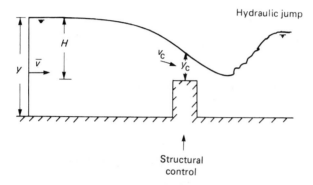

(b) Profile for $y \gg y_c$, H = head

Fig. 6.4   Channel definitions.

amended by a further series of velocity – area measurements.

The type of river gauging station depends very much on the site and character of the river. To a lesser extent, its design is influenced by the data requirements, since most stations established on a permanent basis are made to serve all purposes. Great care must therefore be afforded to initial surveys of the chosen river reach and the behaviour of the flow in both extreme conditions of floods and low flows should be observed if possible. Details of methods used for stage and discharge measurements will be given in the following sections.

6.3   Stage

The water level at a gauging station, the most important measurement in hydrometry, is generally known as the *stage*. It is measured with respect to a datum, either a local bench mark or the crest level of the control, which in turn should be levelled into the geodetic survey datum of the country

(Ordnance Survey datum in the UK). All continuous estimates of the discharge derived from a continuous stage record depend on the accuracy of the stage values. The instruments and installations range from the most primitive to the highly sophisticated, but can be grouped into a few important categories.

6.3.1   The Staff Gauge

This is a permanent graduated staff generally fixed vertically to the river bank at a stable point in the river unaffected by turbulence or wave action. It could be conveniently attached to the upstream side of a bridge buttress but is more likely to be fixed firmly to piles set in concrete at a point upstream of the river flow control. The metre graduations, resembling a survey staff, are shown in Fig. 6.5 and they should extend from the datum or lowest stage to the highest stage expected. The stage is read to an accuracy of $\pm$ 3 mm. Where there is a large range in the stage with a shelving river bank, a series of vertical staff gauges can be stepped up the bank side with appropriate overlaps to give continuity. For regular river banks or smooth man-made channel sides, specially made staff gauges can be attached to the bank slope with their graduations, extended according to the angle of slope, to conform to the vertical scale of heights. All staffs should be made of durable material insensitive to temperature changes and they should be kept clean especially in the range of average water levels.

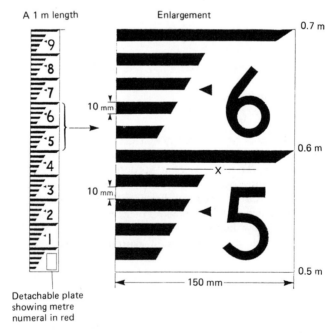

Fig. 6.5   A staff gauge. Stage reading at X = 0.585 m. (Reproduced from BS 3680 (Part 7) 1971 by permission of BSI, 2 Park Street, London W1A 2BS, from whom complete copies can be obtained.)

Depending on the regime of the river and the availability of reliable observers, single readings of the stage at fixed times of the day could provide a useful regular record. Such measurements may be adequate on large mature rivers, but for flashy streams and rivers in times of flood critical peak levels may be missed. Additionally, in these days of increasing modification of river flow by man, the sudden surges due to releases from reservoirs or to effluent discharges could cause unexpected irregular discharges at a gauging station downstream which could be misleading if coincident with a fixed time staff reading. To monitor irregular flows, either natural storm flows or man-made interferences, continuous level recording is essential.

6.3.2   Crest Gauges

Where there is no continuous level recorder, there are several simple devices for marking the peak flood level. These are recommended for minor gauging stations where flood records are particularly important. The standard crest gauge consists of a 50 mm diameter steel tube perforated near the bottom and closed at the top with one or two holes under a lid to allow air to escape. Inside the tube is a removable rod that retains the highest water mark from floating granular cork supplied near the base. The crest gauge is levelled into a normal staff gauge or bench mark on the bank. The rod is cleaned and the crest gauge reset after each reading. The main disadvantage of this device is that a sequence of minor peaks over a short period might be missed, the reading after such a series of events giving only the highest peak level.

6.3.3   Autographic Recorders

The most reliable means of recording water level is provided by a float operated chart recorder. To ensure accurate sensing of small changes in

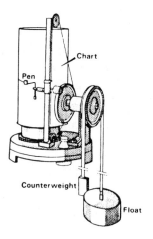

Fig. 6.6   Vertical float-recorder. (Reproduced from K.J. Gregory & D.E. Walling (1973) *Drainage Basin Form and Process*, by permission of Edward Arnold.)

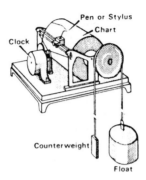

Fig. 6.7   Horizontal float-recorder. (Reproduced from K.J. Gregory & D.E. Walling (1973) *Drainage Basin Form and Process*, by permission of Edward Arnold.)

water level, the float must be installed in a *stilling well* to exclude waves and turbulence from the main river flow. These are two basic mechanisms used by manufacturers:

(a) The moving float looped over a geared pulley with a counterweight activates a pen marking the level on a chart driven round on a vertical clockwork drum (Figs 6.6 and 6.9). The level calibration of the chart should accommodate the whole range of water levels, but extreme peaks are sometimes lost. The time scale of the chart is usually designed to serve a week, but the trace continues round the drum until the chart is changed or the clock stops.

(b) The float with its geared pulley and conterweight turns the charted drum set horizontally and the pen arm is moved across the chart by clockwork (Fig. 6.7). With this instrument all levels are recorded, but the time scale is limited.

More detailed descriptions are given in Grover and Harrington (1966). (Autographic level recorders may be specially calibrated by the makers for a control structure and the stage scale is then replaced by a discharge scale.)

Many modern instruments have been designed to overcome the shortcomings of the two simple mechanisms described. Additional gearing to the pen arm and a trip device can reverse the trace on the vertical chart and thereby record excessive peak levels as a mirror image in the top of the chart. Another mechanism can return the pen arm on the horizontal drum chart to extend the time period. Another added development is the introduction of a more rapidly moving strip chart that can give an improved record by providing greater detail of rapid level variations on a larger time scale as well as by operating continuously for a longer time period. The Stevens recorder in Fig. 6.8 is an example of a later design. The pen traces on autographic charts provide a visible record of the water levels and thus the behaviour of the river can be readily appreciated. On visiting a gauging station, a hydrologist can see at once whether or not a current storm event has peaked. However, chart records require careful analysis and time must be spent in abstracting data for the calculations of discharge. Increasing automation of data processing has led to the development of a new range of level recorders.

Fig. 6.8   Stevens type A35 recorder. A more up-to-date model (A-71) is currently available. (Reproduced by permission of Leupold & Stevens Inc.)

6.3.4   Punched-Tape and Magnetic-Tape Recorders

One of the first punched-tape water level recorders was the Fischer and Porter instrument produced in the US. The float attached to its counter-weight is geared to move two calibrated discs from which the level can be read. At 15 min intervals, the freely moving discs lock and the level measurement is punched in binary code on to a $2\frac{1}{2}''$ wide paper tape. The instrument can operate at remote sites from batteries, and at the 15 min frequency of recording, the tape can last 12 months. With experience, the punched tape can be read up to the take-up spool when the gauging station is visited but an immediate picture of the river behaviour is not so easily obtained as from an autographic chart. A major disadvantage of this recorder is that a special translator is needed for the data since the punched tape is not compatible with the input systems for digital computers. Many of these recorders have now been replaced by computer-compatible punched-paper-tape recorders. These do not display the level record in an immediately recognizable form and thus are usually installed alongside a conventional autographic chart recorder. In their turn, the vulnerable paper tapes are now less favoured than the more robust magnetic tapes easily handled in small cassettes. Solid state data loggers with 'no moving parts' are the latest technical innovation. These robust recorders using microelectronic technology are simple to use, have low maintenance costs, and can be linked directly to computers for processing the data. Technological advances soon make instruments and equipment obsolete, and in setting up a new gauging station, advice should be sought from several reliable specialist manufacturers.

Fig. 6.9   The ALPHA Water Level Recorder (automatically records water level using a float or pressure transducer). (Photograph kindly supplied by Hydrokit Instrumentation, 42 Tewkesbury Terrace, Bounds Green, London N11 2LT.)

6.3.5   Flood-Warning Gauges

The simple crest gauge has also been modernized. At strategic locations on major rivers liable to flood, water level gauges may be installed to give instantaneous water level readings in times of high flows. The instruments, interrogated by telephone, relay the water level in a number of sounds in different tones according to the units of measurement. These 'teletones' are invaluable to hydrologists with the responsibility for forecasting flood levels downstream. Some instruments may be set to initiate an alarm when the river has risen to a prescribed level, and computer-controlled electronic speaking devices are now being developed. At many sites, such installations are part of a river gauging station and as such, the levels relate to known discharge values, but where water levels are the single vital measure, the flood warning station may not be calibrated.

6.3.6   Pressure Gauges

The measurement of stage by pressure transducers, an indirect method converting the hydrostatic pressure at a submerged datum to the water level above, has not previously found favour in hydrometry owing to the power and technical skills required in their operation. With the development of electronic circuitry reducing the power needed to drive the transducers, battery-operated equipment is becoming available. Pressure gauges for water level measurement are particularly useful for gauging stations where it is impractical to build stilling wells for float gauges.

6.4   Discharge

The most direct method of obtaining a value of discharge to correspond with a stage measurement is by the *velocity – area* method in which the flow velocities are measured at selected verticals of known depth across a measured section of the river.

At a river gauging station, the cross-section of the channel is surveyed and considered constant unless major modifications during flood flows are suspected, after which it must be resurveyed. The more difficult component of the discharge computation is the series of velocity measurements across the section. The variability in velocity both across the channel and in the vertical must be considered. To ensure adequate sampling of velocity across the river, the ideal measuring section should have a symmetrical flow distribution about the mid vertical and this requires a straight and uniform approach channel upstream, in length at least twice the maximum river width. Then measurements are made over verticals spaced at intervals no greater than 1/15th of the width across the flow. With any irregularities in the banks or bed, the spacings should be no greater than 1/20th of the width (BS 3680: Part 3). Guidance in the number and location of sampling points is obtained from the form of the cross-section with verticals being sited at peaks or troughs. To assess the sampling requirements in the depth of the stream, the pattern of flow in the vertical is studied. From theoretical considerations of two-dimensional turbulent flow in a rough lined natural channel, it can be shown (Francis, 1975, p. 276) that the average velocity in the vertical occurs at 0.63 depth below the surface, which is in agreement with the empirical 0.6 times depth rule (Fig. 6.2). It has also been found by observation that a good estimate of the average velocity in the vertical is obtained by taking the mean of velocity measurements at 0.2 and 0.8 depths.

6.4.1   Measurement of Velocity

The simplest method for determining a velocity of flow is by timing the movement of a *float* over a known distance. Surface floats comprising any available floating object are often used in rough preliminary surveys; these measurements give only the surface velocity and a correction factor must be applied to give the average velocity over a depth. A factor of 0.7 is recommended for a river of 1 m depth with a factor of 0.8 for 6 m or greater (BS 3680). Specially designed floats can be made to travel at the mean velocity of the stream (Fig. 6.10). The individual timing of a series of floats placed across a stream to determine the cross-sectional mean velocity pattern could become a complex procedure with no control of the float movements. Therefore this method is recommended only for reconnaissance discharge

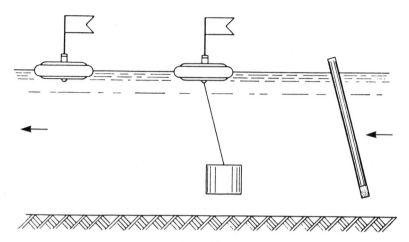

Fig. 6.10   Floats: (a) surface float, (b) canister float for mean velocity, (c) rod float for mean velocity. (Reproduced from R.W. Herschy (Ed.) (1978) *Hydrometry, Principles and Practices*, © 1978, by permission of John Wiley & Sons Ltd.)

estimates. (Freely moving floats are also very useful for estuary surveys of flow patterns.)

The determination of discharge at a permanent river gauging station is best made by measuring the flow velocities with a *current meter*. This is a reasonably precise instrument that can give a nearly instantaneous and consistent response to velocity changes, but is also of simple construction and robust enough to withstand rough treatment in debris-laden flood flows. There are two main types of meter; the cup type has an assembly of six cups revolving round a vertical axis, and the propeller type has a single propeller rotating on a horizontal axis (Fig. 6.11). The cup type is the more robust, but has a high drag. The cup meter registers the actual velocity whatever its direction, rather than the required velocity component normal to the measuring section. The more sensitive propeller type is easily damaged, but has a low drag and records the true normal velocity component with actual velocities up to 15° from the normal direction. However, with an attainable accuracy of the order of 2%, the propeller type current meter is superior to the cup type, with a 5% accuracy, in the range of velocities between 0.15 and 4 m s^{-1}.

Both types of instrument need to be calibrated to obtain the relationship between the rate of revolutions of the cups or propeller and the water velocity. Each individual instrument generally has its own calibration curve or rating table and if in regular use, it should have a calibration check every year. In operating the current meter, the number of revolutions is signalled by a battery-operated buzzer or digital counter.

6.4.2   Operational Methods

The sampling of the velocities across a gauging section depends on the size of the river and its accessibility.

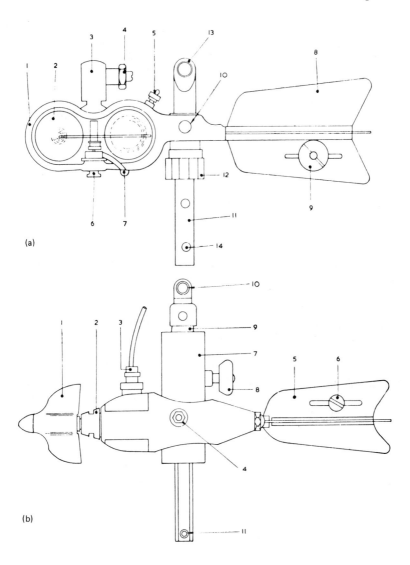

Fig. 6.11   Current meters. (a) A typical cup-type current meter: 1, yoke of main frame; 2, cup rotor; 3, contact box; 4, contact box electric plug adaptor; 5, earth terminal for electric lead; 6, pivot bearing adaptor; 7, rotor lifting lever; 8, tail fin; 9, balance weights; 10, horizontal pivots; 11, hanger bar; 12, hanger bar clamp; 13, hole for suspension cable attachment hook; hole for attachment of sinker weight.
(b) A typical propeller-type current meter: 1, rotor; 2, rotor retaining adaptor; 3, electric cable plug; 4, horizontal pivots; 5, tail fin; 6, balance weights; 7, hanger bar adaptor; 8, hanger bar clamp; 9, hanger bar; 10, hole for suspension cable attachment hook; 11, hole for attachment of sinker weight. (Reproduced from BS 3680 (Part 8A) 1973, by permission of BSI, 2 Park Street, London W1A 2BS, from whom complete copies can be obtained.)

Wading.   The current meter is carried on special rods and held in position by the gauger standing on the stream bed a little to the side and downstream of the instrument. This is the ideal method, since the gauger is in full control of the operation, but it is only practicable in shallow streams with low or moderate velocities.

Bridge.   Where there is a clear span bridge aligned straight across the river near the gauging station, the current meter can be lowered on a line from a gauging reel carried on a trolley. Care must be taken to sample a section with even flow.

Boat.   For very wide rivers, gaugings may have to be made from a boat either held in position along a fixed wire or under power across the section.

Cablecar.   The gauger travels across the river in a specially designed cablecar hung from an overhead cable, and the current meter is suspended on a steel line.

Cableway.   The gauger remains on the river bank, but winds the current meter across the section on a cableway, lowering it to the desired depth by remote mechanical control (Fig. 6.12).

In the UK, most gauging stations, established in rural areas and calibrated by the velocity – area method, adopt the wading or cableway methods for current meter gauging. Normally, the cableway controls are installed in the recorder house alongside the stilling well installed for stage recording.

6.4.3   Gauging Procedure

At the gauging station or selected river cross-section, the mean velocities for small sub-areas of the cross-section ($\bar{v}_i$) obtained from point velocity measurements at selected sampling verticals across the river are multiplied by the corresponding sub-areas (a_i) and the products summed to give the total discharge:

$$Q = \sum_{i=1}^{n} \bar{v}_i \, a_i$$

where n = the number of sub-areas.

(a) The estimate Q is the discharge related to the stage at the time of gauging; therefore, before beginning a series of current-meter measurements the *stage* must be *read* and *recorded*.

(b) The width of the river is divided into about 20 sub-sections so that no sub-section has more than 10% of the flow.

(c) At each of the selected sub-division points, the water depth is measured by sounding and the current meter operated at selected points in the vertical to find the mean velocity in the vertical, e.g. at 0.6 depth (one-point method) or at 0.2 and 0.8 depths (two-point method). At a

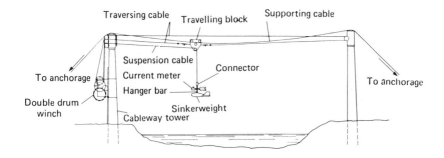

Fig. 6.12   Cableway system. (Reproduced from BS 3680 (Part 8B) 1973, by permission of BSI, 2 Park Street, London W1A 2BS, from whom complete copies can be obtained.)

new gauging station where the vertical velocity distribution is at first unknown, more readings should be taken to establish that the best sampling points to give the mean are those of the usual one or two-point methods.

(d) For each velocity measurement, the number of complete revolutions of the meter over a measured time period (about 60 s) is recorded using a stopwatch. If pulsations are noticed, then a mean of three such counts should be taken.

(e) When velocities at all the sub-division points across the river have been measured, the *stage* is read again.

Should there have been a difference in stage readings over the period of the gaugings, a mean of the two stages is taken to relate to the calculated discharge. Once gaugers have gained experience of a river section at various river stages, the procedure can be speeded up and one velocity reading only at 0.6 depth taken quickly at each point across the stream, with the depths relating to the stage already known. Such an expedited procedure is absolutely essential when gauging flood flows with rapid changes in stage.

6.4.4   Calculating the Discharge

The calculation of the discharge from the velocity and depth measurements can be made in several ways. Two of these are illustrated in Fig. 6.13. In the *mean section* method, averages of the mean velocities in the verticals and of the depths at the boundaries of a section sub-division are taken and multiplied by the width of the sub-division, or segment:

$$Q = \Sigma q_i = \Sigma \bar{v}.a = \sum_{i=1}^{n} \frac{(\bar{v}_{i-1} + \bar{v}_i)}{2} \ \frac{(d_{i-1} + d_i)}{2} \ (b_i - b_{i-1})$$

where b_i is the distance of the measuring point (i) from a bank datum and there are n sub-areas. In the *mid section* method, the mean velocity and depth measured at a subdivision point are multiplied by the segment width measured between the mid-points of neighbouring segments.

Mean section method

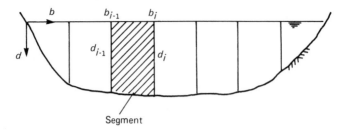

Segment

Mid-section method

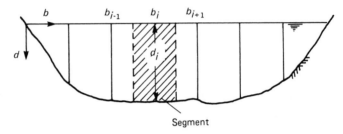

Segment

Fig. 6.13   Calculating discharge.

$$Q = \Sigma q_i = \Sigma \bar{v}.a = \sum_{i=1}^{n} \bar{v}_i.d_i \; \frac{(b_{i+1} - b_{i-1})}{2}$$

with n being the number of measured verticals and sub-areas. In the *mid-section* calculation, some flow is omitted at the edges of the cross-section, and therefore the first and last verticals should be sited as near to the banks as possible.

An example of the discharge calculations using both methods is shown in Table 6.1. For many routine gaugings the calculations are now done by computer.

6.4.5   Problems in Gauging

Small streams.   The depth of flow may be insufficient to cover the ordinary current meter. Smaller instruments known as Pygmy meters, both cup type and propeller type, are used for shallow streams and low-flow gaugings. They are attached to a graduated rod and operated by the gauger wading across the section.

Mountain torrents.   Streams with steep gradients and high velocities cannot be gauged satisfactorily by the velocity – area method and alternative means must be used, e.g. dilution gauging.

Large rivers.   Across wide rivers, there is always difficulty in locating the instruments accurately at the sampling points and inaccuracies invariably occur. Problems in locating the bed of the river may also arise in deep and fast flows and a satisfactory gauging across such a river may take many hours to complete. Check readings of the stage *during* such an operation are advisable. In deep swift-flowing rivers, heavy weights according to the velocity are attached, but the force of the current usually causes a drag downstream from the vertical. Measurements of depth have to be corrected using the measured angle of inclination of the meter cable (Corbett *et al*, 1962).

6.5   Stage – Discharge Relationship

The establishment of a reliable relationship between the monitored variable stage and the corresponding discharge is essential at all river gauging stations when continuous-flow data are required from the continuous stage record. This calibration of the gauging station is dependent on the nature of the channel section and of the length of channel between the site of the staff gauge and discharge measuring cross-section. Conditions in a natural river are rarely stable for any length of time and thus the stage – discharge relationship must be checked regularly and, certainly after flood flows, new discharge measurements should be made throughout the range of stages. In most organizations responsible for hydrometry, maintaining an up-to-date relationship is a continuous function of the hydrologist.

The stage – discharge relationship can be represented in three ways: as a graphical plot of stage versus discharge, (the *rating curve*), in a tabular form (*rating table*; see Table 6.2), and as a mathematical equation, discharge, Q, in terms of stage, H, (*rating equation*).

The rating curve.   All the discharge measurements, Q, are plotted against the corresponding mean stages, H, on suitable arithmetic scales. The array of points usually lies on a curve which is approximately parabolic (Fig. 6.14) and a best fit curve should be drawn through the points by eye. At most gauging stations, the zero stage does not correspond to zero flow. If the points do not describe a single smooth curve, then the channel control governing the Q versus H relationship has some variation in its nature (in Fig. 6.14, there is a change in the slope of the river banks between 0.7 and 0.8 m). For example, the stage height at which a small waterfall acting as a natural control is drowned out at higher flows is usually indicated by a distinct change in slope of the rating curve. Another break in the curve at high stages can often be related to the normal bankful level above which the Q versus H relationship could be markedly different from the within-banks curve owing to the very different hydraulics of flood plain flow.

The rating table.   When a satisfactory rating curve has been established, values of H and Q may be read off the curve at convenient intervals and a rating table is constructed by interpolation for required intervals of stage (Table 6.2). This is the simplest and most convenient form of the stage – discharge relationship for the manual processing of sequential stage records.

The rating equation.   The rating curve can often be represented approximately by an equation of the form:

Table 6.1 Velocity – Area Discharge Calculations

Distance b_i (m)	Depth d_i (m)	Velocity $\bar{v}_i$ (m s^{-1})	Mean section method				Mid section method	
			$\dfrac{\bar{v}_{i-1} + \bar{v}_i}{2}$	$\dfrac{d_{i-1} + d_i}{2}$	$b_i - b_{i-1}$	q_i	$\dfrac{b_{i+1} - b_{i+1}}{2}$	q_i
4.0	0.000	0.000					0	0
9.0	1.131	0.330	0.165	0.565	5.0	0.466	4.0	1.493
12.0	1.740	0.357	0.343	1.435	3.0	1.477	3.0	1.864
15.0	1.993	0.358	0.358	1.867	3.0	2.005	3.0	2.140
18.0	2.057	0.353	0.356	2.025	3.0	2.163	3.0	2.178
21.0	2.057	0.340	0.347	2.057	3.0	2.141	3.0	2.098
24.0	1.905	0.346	0.343	1.981	3.0	2.038	3.0	1.977
27.0	1.753	0.341	0.343	1.829	3.0	1.882	3.0	1.793
30.0	1.753	0.314	0.327	1.753	3.0	1.720	3.0	1.651
33.0	1.600	0.322	0.318	1.676	3.0	1.599	3.0	1.546
36.0	1.295	0.318	0.320	1.447	3.0	1.389	3.0	1.235

39.0	1.436	0.247	0.283	1.365	3.0	1.159	3.0	1.064
42.0	1.308	0.181	0.214	1.372	3.0	0.881	3.0	0.710
45.0	1.640	0.104	0.143	1.474	3.0	0.632	3.0	0.512
48.0	1.512	0.066	0.085	1.576	3.0	0.402	3.5	0.349
52.0	0.000	0.000	0.033	0.756	4.0	0.100	0	0
				$\Sigma q_i =$	20.054	$\Sigma q_i =$	20.610 m³ s⁻¹	

$\Sigma q_i = 20.610 \ \mathrm{m^3 \ s^{-1}}$

$$Q = aH^b$$

If Q is not zero when $H = 0$, then a stage correction, a realistic value of H_0, for $Q = 0$ must be included:

$$Q = a (H - H_0)^b$$

The values of the constants a, b and H_0 can be found by a least-squares fit using the measured data and trial values of H_0 and with b expected to be within limits depending on the shape of the cross section. The equation can then be used for converting stage values into discharge by digital computer.

When the rating curve does not plot as a simple curve on arithmetic scales, plotting the values of Q and $(H - H_0)$ on logarithmic scales helps in identifying the effects of different channel controls. The logarithmic form of the rating equation:

$$\log Q = \log a + b \log (H - H_0)$$

may then plot as a series of straight lines, and changes in slope can be seen more clearly. Two such straight-line plots for Thorverton can be seen in Fig. 6.15. There are two corresponding equations, one for each control range, derived from the appropriate measurements:

$$Q = 70.252 \, (H - 0.034)^{2.18} \text{ for } H \text{ up to } 0.709 \text{ m}$$
$$Q = 65.048 \, (H - 0.139)^{1.39} \text{ for } H \text{ over } 0.709 \text{ m and up to } 2.540 \text{ m}$$

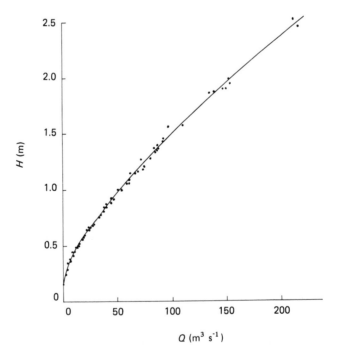

Fig. 6.14   Rating curve for Thorverton (River Exe). (Based on data from the South West Water Authority.)

Table 6.2 A Sample from the Rating Table for Thorverton

Stage (m)	Discharges in (m³ s⁻¹)									
	0.000	0.001	0.002	0.003	0.004	0.005	0.006	0.007	0.008	0.009
0.40	7.846	7.893	7.940	7.987	8.034	8.082	8.130	8.177	8.225	8.273
0.41	8.321	8.370	8.418	8.467	8.516	8.565	8.614	8.663	8.712	8.762
0.42	8.812	8.861	8.911	8.962	9.012	9.062	9.113	9.164	9.215	9.266
0.43	9.317	9.368	9.420	9.472	9.524	9.576	9.628	9.680	9.732	9.785
0.44	9.838	9.891	9.944	9.997	10.050	10.104	10.158	10.212	10.266	10.320
0.45	10.374	10.428	10.483	10.538	10.593	10.648	10.703	10.759	10.814	10.870
0.46	10.926	10.982	11.038	11.094	11.151	11.207	11.264	11.321	11.378	11.435
0.47	11.493	11.550	11.608	11.666	11.724	11.782	11.840	11.899	11.958	12.016
0.48	12.075	12.134	12.194	12.253	12.313	12.373	12.432	12.493	12.553	12.613
0.49	12.674	12.734	12.795	12.856	12.917	12.979	13.040	13.102	13.164	13.226
0.50	13.288	13.350	13.412	13.475	13.538	13.601	13.664	13.727	13.790	13.854

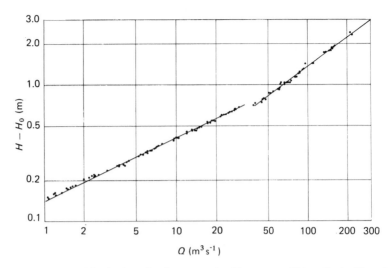

Fig. 6.15   Logarithmic plot of rating curve for Thorverton (River Exe). (Based on data from the South West Water Authority.)

Values for a and b obtained by least-squares fit of the data have been rounded off; the fitted H_0 values are in metres.

When distinctive parts of the stage – discharge relationship can be related to observed different permanent physical controls in the river channel, then separate straight-line logarithmic equations are justified and can be evaluated by least-squares fitting to apply to corresponding ranges in the stage heights.

6.5.1   Irregularities and Corrections

In the middle and lower reaches of rivers where the beds consist of sands and gravels and there is no stable channel control, the stage – discharge relationship may be unreliable owing to the alternate scouring and depositing of the loose bed material. If the general long-term rating curve remains reasonably constant, Stout's technique allows individual shift corrections to be made to evaluations of Q in accordance with the most recent gaugings by adjusting the stage readings (Fig. 6.16).

Discharges depend both on the stage and on the slope of the water surface; the latter is not the same for rising and falling stages as a non-steady flow passes a gauging station. From gaugings at a particular stage value there are thus two different values of discharge. From gaugings made on a rising stage, the corresponding discharges will be greater than those measured at the same stage levels on the falling stage. Thus, there can be produced a looped rating curve. When the river sustains a steady flow at a particular stage, an average of the two discharges may be taken, but otherwise the values of the relevant rising or falling side of the looped rating curve should be used.

Other irregularities in the stage – discharge relationship may be caused by non-uniform flow generated by interference in the channel downstream of

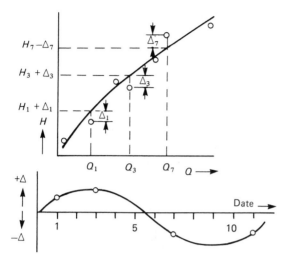

Shift correction

Date	Recorded stage	Measured discharge	Stage correction	Corrected stage	Estimated discharge
May 1	H_1	Q_1	Δ_1	$H_1 + \Delta_1$	Q_1
2	H_2		Δ'_2	$H_2 + \Delta'_2$	Q'_2
3	H_3	Q_3	Δ_3	$H_3 + \Delta_3$	Q_3
4	H_4		Δ'_4	$H_4 + \Delta'_4$	Q'_4
5	H_5		Δ'_5	$H_5 + \Delta'_5$	Q'_5
6	H_6		$-\Delta'_6$	$H_6 - \Delta_6$	Q'_6
7	H_7	Q_7	$-\Delta_7$	$H_7 - \Delta_7$	Q_7

Fig. 6.16 Stout's method.

the gauging section, thus overriding the flow control. Such interference can result from the backing up of flow in a main channel owing to flow coming in from a downstream tributary, or from the operation of sluice gates on the main channel. Vegetation growth in the gauging reach will also interfere with the Q versus H relation. In certain rivers it may be advisable to have rating curves for different seasons of the year. More permanent interference may be provided by changes in the cross-section owing to the scouring or the deposition of an exceptional flood and if discharges have been obtained for stages which include inundation of the flood plain, any further developments on the flood plain would affect these values.

6.5.2   Extension of Rating Curves

It is always extremely difficult to obtain velocity measurements and hence estimated discharges at high stages. The range of the stage – discharge relationship derived from measurements is nearly always exceeded by flood flows. Hydrologists responsible for river gauging should make determined attempts to measure flood peaks, particularly at stations where the rating at high flows is in doubt. However there are several techniques that can be adopted to assess the discharge at stages beyond the measured limit of the rating curve, but all extensions are strictly only valid for the same shape of cross-section and same boundary roughness.

Logarithmic extrapolation.   If the rating curve plots satisfactorily as a straight line on log – log paper, it may be extrapolated easily to the higher stages. However in using this method, especially if the extrapolation exceeds 20% of the largest gauged discharge, other methods should be applied to check the result. Alternatively, the straight-line equation fitted to the logarithmic rating curve could be used to calculate higher discharges, with similar reservations to check by another method.

Velocity – area method.   In addition to the rating curve, plots of A versus H and $\overline{V}$ versus H can be drawn. The overall mean velocity across the channel $\overline{V}$ is calculated from $Q = A\overline{V}$ within the range of the measured stages. The cross-sectional area curve can be extrapolated reliably from the survey data (and a change of slope will indicate any change in cross-section shape). The mean velocity curve can normally be extended with little error. Then the higher values of discharge can be calculated for the required stage from the product of the corresponding $\overline{V}$ and A.

Stevens method.   Extrapolation of the rating curve can also be made using the empirical Chezy formula (or other friction formula) for calculating open channel flow (Francis, p. 277). Under uniform flow conditions, the Chezy equation is:

$$Q = A\overline{V} = A\,C\sqrt{(RS_o)}$$

where $\overline{V}$ is the overall mean velocity across the channel, R is the hydraulic mean depth (see Fig. 6.4), S_o is the bed slope and C is a coefficient.

R is obtained for all required stages from $R = A/P$ with A and P measured in the cross-sectional survey. If $C\sqrt{S_o}$ is taken to be constant (k), then Q versus $kA\sqrt{R}$ plots as a straight line. For gauged stage values of H, corresponding values of $A\sqrt{R}$ and Q are obtained and plotted. The extended straight line can then be used to give discharges for higher stages (Fig. 6.17). The method relies on the doubtful assumption of C remaining constant for all stage values.

The Manning formula.   This can be used instead of the Chezy formula for extending rating curves, but it is also applied more widely in engineering practice for calculating flows. The formula, where quantities are in SI units, is

$$Q = A\,\overline{V} = \frac{A\,R^{\frac{2}{3}}S_o^{\frac{1}{2}}}{n}$$

It is applied in a similar way, with $S_o^{\frac{1}{2}}/n$ assumed constant, and Q being plotted against $AR^{\frac{2}{3}}$.

In both the Stevens and Manning formula methods, when a flood discharge exceeds bankful stage, the roughness factor, C or n, can be changed to

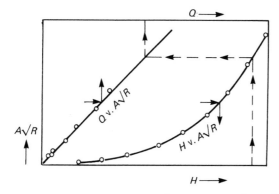

Fig. 6.17   The Stevens method. $Q = A.C\sqrt{(R.S_0)}$; $Q = C\sqrt{S_0}A\sqrt{R}$; for large H, $C\sqrt{S_0} \rightarrow$ constant.

model the different flow conditions and the separate parts of the extended flow over the flood plain are calculated with the modified formula. It is generally accepted that the Manning equation is superior to the Chezy equation, since n changes less than C as R varies.

Estimates of flood discharges at strategic locations along a river are usually made by this Manning-based method. After notable flood events, surveyors can measure the required cross-sectional area, the wetted perimeter and the bed slope of the affected reach; the peak surface water level is assessed from debris or wrack marks. Selecting an appropriate value of n (Table 6.3 or, for greater detail, see Chow, 1959) for the channel roughness, an estimated peak discharge is calculated. Considerable experience is needed in using this method since the validity of the formula depends on the nature of the flow and the appropriate corresponding line of slope.

Table 6.3   Sample Values of Manning's n

Concrete lined channel	0.013
Unlined earth channel	0.020
Straight, stable deep natural channel	0.030
Winding natural streams	0.035
Variable rivers, vegetated banks	0.040
Mountainous streams, rocky beds	0.050

6.6   Flumes and Weirs

The reliability of the stage – discharge relationship can be greatly improved if the river flow can be controlled by a rigid, indestructible cross-channel structure of standardized shape and characteristics. Of course, this adds to the cost of a river gauging station, but where continuous accurate values of discharge are required, particularly for compensation water and other low flows, a special measuring structure may be justified. The type of structure depends on the size of the stream or river and the range of flows it is expected to measure. The sediment load of the stream also has to be considered.

The basic hydraulic mechanism applied in all measuring flumes and weirs is the setting up of critical flow conditions for which there is a unique and stable relationship between depth of flow and discharge. Flow in the channel upstream is sub-critical, passes through critical conditions in a constricted region of the flume or weir and enters the downstream channel as super-critical flow. It is better to measure the water level (stage) a short distance upstream of the critical-flow section, this stage having a unique relationship to the discharge.

6.6.1   Flumes

Flumes are particularly suitable for small streams carrying a considerable fine sediment load. The upstream sub-critical flow is constricted by narrowing the channel, thereby causing increased velocity and a decrease in the depth. With a sufficient contraction of the channel width, the flow becomes critical in the throat of the flume and a standing wave is formed further downstream. The water level upstream of the flume can then be related directly to the discharge. A typical design is shown in Fig. 6.18. Such *critical depth* flumes can have a variety of cross-sectional shapes. The illustration shows a plain rectangular section with a horizontal invert (bed profile), but trapezoidal sections are used to contain a wider range of discharges and U-shaped sections are favoured in urban areas for more confined flows and sewage effluents. Where there are only small quantities of sediment, the length of the flume can be shortened by introducing a hump in the invert to reduce the depth of flow and thus induce critical flow more quickly in the contraction, but the flume must be inspected regularly and any sediment deposits cleared. Relating the discharge for a rectangular cross-section to the measured head, H, the general form of the equation is:

$$Q = K \, b \, H^{\frac{3}{2}}$$

where b is the throat width and K is a coefficient based on analysis and experiment. For the derivation of K, the reader is referred to BS 3680. With flumes built to that Standard, Q may be assessed to within 2% without the need for any field calibration. There are many different flume designs that have been built to serve various purposes in measuring a range of flow conditions. These are described fully in Ackers *et al.*, (1978).

6.6.2   Weirs

Weirs constitute a more versatile group of structures providing restriction to the depth rather than the width of the flow in a river or stream channel. A distinct sharp break in the bed profile is constructed and this creates a raised upstream sub-critical flow, a critical flow over the weir and super-critical flow downstream. The wide variety of weir types can provide for the measurement of discharges ranging from a few litres per second to many hundreds of cubic metres per second. In each type, the upstream head is

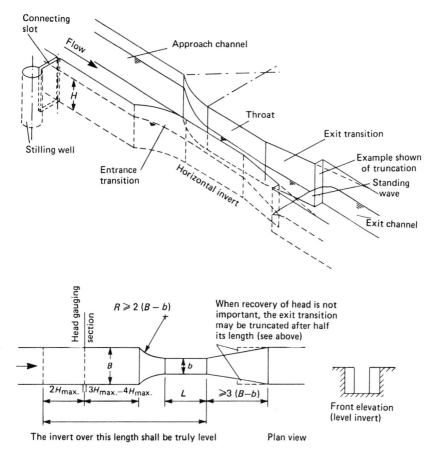

Fig. 6.18   Rectangular throated flume. (Reproduced from BS 3680 (Part 4C) 1981, by permission of BSI, 2 Park Street, London W1A 2BS, from whom complete copies can be obtained.)

again uniquely related to the discharge over the crest of the structure where the flow passes through critical conditions.

For gauging clear water in small streams or narrow man-made channels, *sharp-crested* or *thin-plate* weirs are used. These give highly accurate discharge measurements but to ensure the accuracy of the stage – discharge relationship, there must be atmospheric pressure underneath the nappe of the flow over the weir (Fig. 6.19). Thin plate weirs can be *full width* weirs extending across the total width of a rectangular approach channel (Fig. 6.19a) or contracted weirs as in Fig. 6.19b and c. The shape of the weir may be *rectangular* or *trapezoidal* or have a triangular cross-section, a *V-notch*. The angle of the V-notch, θ, may have various values, the most common being 90° and 45°.

The basic discharge equation for a rectangular sharp crested weir again takes the form:

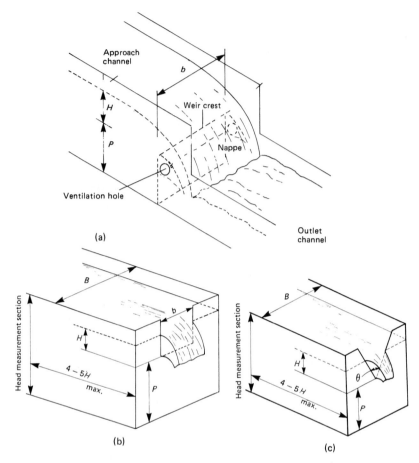

Fig. 6.19   Thin plate weirs. (Reproduced (a) from P. Ackers, *et al.* (1978) *Weirs and Flumes for Flow Measurement*, by permission of John Wiley & Sons, Inc.; (b) and (c) from BS 3680 (Part 4A) 1981, by permission of BSI, 2 Park Street, London W1A 2BS, from whom complete copies can be obtained.)

$$Q = K \, b \, H^{\frac{3}{2}}$$

but in finding K, allowances must be made to account for the channel geometry and the nature of the contraction. Such hydraulic details may be obtained from the standard texts. For the V-notch weirs, the discharge formula becomes:

$$Q = K \tan \left(\theta/2 \right) H^{\frac{5}{2}}$$

Tables of coefficients for thin-plate weirs are normally to be found in the specialist reference books (e.g. Brater & King, 1976).

For larger channels and natural rivers, there are several designs recommended for gauging stations and these are usually constructed in concrete. One of the simplest to build is the *broad-crested* (square-edged) or the *rectangular-profile* weir (Fig. 6.20). The discharge in terms of gauged head H is given by:

$$Q = K\,b\,H^{\frac{3}{2}}$$

The length L of the weir, related to H and to P, the weir height, is very important since critical flow should be well established over the weir. However, separation of flow may occur at the upstream edge, and with increase in H, the pattern of flow and the coefficient, K, change. Considerable research has been done on the calibration of these weirs. The broad-crested weir with a curved upstream edge, also called the *round-nosed horizontal-crested* weir, gives an improved flow pattern over the weir with no flow separation at the upstream edge, and it is also less vulnerable to damage. The discharge formula is similarly dependent on the establishment of weir coefficients.

A new form of weir with a *triangular profile* was designed by E.S. Crump in 1952. This ensured that the pattern of flow remained similar throughout the range of discharges and thus weir coefficients remained constant. In addition, by making additional head measurements just below the crest, as well as upstream, the *Crump* weir allows flow measurements to be estimated above the modular limit when the weir has drowned out at high flows. The geometry of the Crump weir is shown in Fig. 6.21a. The upstream slope of 1:2 and downstream slope of 1:5 produce a well controlled hydraulic jump on the downstream slope in the modular range. Improvement in the accuracy of very low flow measurement has been brought about by the compounding of

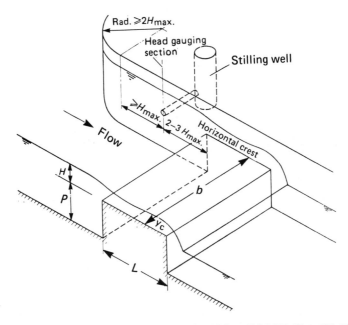

Fig. 6.20   Rectangular profile weir. (Reproduced from BS 3680 (Part 4B) 1969 by permission of BSI, 2 Park Street, London W1A 2BS, from whom complete copies can be obtained.)

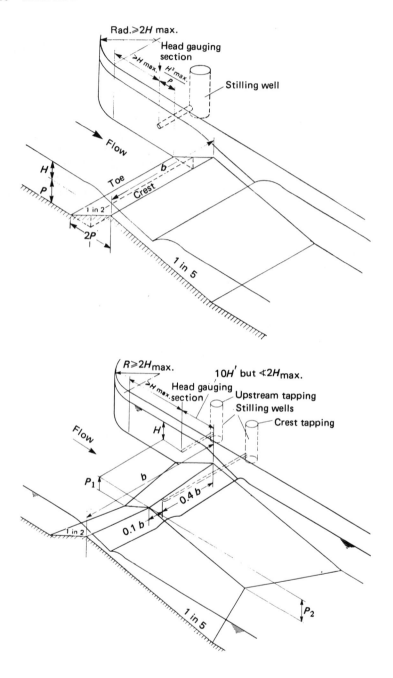

Fig. 6.21   (a) Triangular-profile weir (Crump weir). (b) Triangular-profile flat-V weir. (Reproduced from (a) BS 3680 (Part 4B) 1969, (b) BS 3680 (Part 4G) 1981, by permission of BSI, 2 Park Street, London W1A 2BS, from whom complete copies can be obtained.)

the Crump weir across the width of the channel. Two or more separate crest sections at different levels may be built with sub-dividing piers to separate the flow. Such structures are designed individually to match the channel and flow conditions. A great deal of research effort has been put into the development of the Crump weir and many have been built in the UK; readers are referred to the relevant publications of the Hydraulics Research Station and the Water Data Unit (e.g. Herschy et al., 1977).

The flat-V weir (Fig. 6.21b) is developed from and an improvement on the Crump weir. By making the shape of the crest across the channel into a shallow V shape, low flows are measured more accurately in the confined central portion without the need for compounding. The triangular profile may be the same as the Crump weir, 1:2 upstream face and 1:5 downstream, but a profile with both slopes 1:2 is also used. This weir can also operate in the high non-modular flow range and several crest cross-sectional slopes have been calibrated. An extra advantage of the flat-V weir is that it passes sediment more readily than the Crump.

All these structures have a clear upper limit in their ability to measure the stream flow. Usually as the flow rate increases, downstream channel control causes such an increased downstream water level that a flume or weir is drowned out; the unique relationship hitherto existing between the stage or upstream level and the discharge in the so-called 'modular' range is thereafter lost. It is not practicable to set crest levels in flumes and weirs sufficiently high to avoid the drowning out process at high flows since upstream riparian interests would object to raised water levels and out-of-bank flows at discharges previously within banks. These structures are generally used for measuring low and medium flows; flood flows are not usually measurable with flumes and weirs.

6.7   Dilution Gauging

This method of measuring the discharge in a stream or pipe is made by adding a chemical solution or tracer of known concentration to the flow and then measuring the dilution of the solution downstream where the chemical is completely mixed with the stream water.

In Fig. 6.22, c_o, c_1 and c_2 are chemical concentrations (e.g. g litre^{-1}); c_o is the 'background' concentration already present in the water (and may be negligible), c_1 is the known concentration of tracer added to the stream at a *constant rate q*, and c_2 is a sustained final concentration of the chemical in the well mixed flow. Thus $Qc_o + qc_1 = (Q + q)c_2$, whence:

$$Q = \frac{(c_1 - c_2)}{(c_2 - c_o)} \, q$$

An alternative to this constant rate injection method is the '*gulp*' injection, or *integration*, method. A known volume of the tracer V of concentration c_1 is added in bulk to the stream and, at the sampling point, the varying concentration, c_2, is measured regularly during the passage of the tracer cloud. Then:

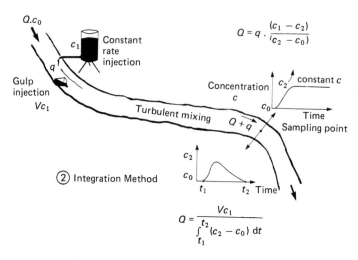

Fig. 6.22   Dilution gauging: two basic methods.

$$Vc_1 = Q \int_{t_1}^{t_2} (c_2 - c_0) \, dt$$

with the unknown Q, easily calculated.

The chemical used should have a high solubility, be stable in water and be capable of accurate quantitative analysis in dilute concentrations. It should also be non-toxic to fish and other forms of river life, and be unaffected itself by sediment and other natural chemicals in the water. The most favoured chemicals are sodium dichromate, especially in clear mountain streams (care needed not to exceed acceptable concentration limits), and lithium chloride which has simpler laboratory analysis but may also be detrimental to fish life. Chemical dyes such as Rhodamine B have also been developed as tracers with the advantage of being easily detected at very low concentrations. Applications of radioactive tracers in chemical gauging has been shown to be efficient and specially useful in measuring flows through sewage works since they are unaffected by sediment and other pollutants. The best known radio isotopes for river gauging are tritium (as tritiated water it already exists in background concentrations in surface waters) and bromine-82 which has a short half-life of 35 h and therefore does not pollute natural streams.

Dilution gauging is an ideal method for measuring discharges in turbulent mountain streams with steep gradients where current metering is impracticable. Careful preparations are needed and the required mixing length, dependent on the state of the stream, must be assessed first, usually by visual testing with fluorescein. However, the equipment is easily made portable for one or two operators, and thus the method is recommended for survey work in remote areas. The method becomes costly for large rivers and high discharges, but can be usefully applied to the calibration of gauging structures.

For more detailed considerations, the reader is referred to Herschy (1978, Chapter 4).

6.8   Modern Gauging Techniques

The Electromagnetic Method has been devised for river reaches where there is no stable stage – discharge relationship or where the flow is impeded by weed growth. An electomotive force (e.m.f.) is induced in the water by an electric current passed through a large coil buried beneath and across the river bed. The e.m.f., which is directly proportional to the average velocity through the cross-section, is recorded from electron probes at each side of the river. Interference or background noise induced by other electrical devices in the vicinity may have to be measured by noise-cancellation probes and allowed for in the calculations. This method is costly to instal and needs electricity on site at the gauging station, but the results from one or two installations in southern England show great promise (Herschy, 1978).

The electromagnetic flow measurement method and more so the following ultrasonic method are now well-established hydrometric techniques in the UK.

The ultrasonic method uses sound pulses to measure the mean velocity at a pre-scribed depth across the river. Thus it is a sophisticated form of the velocity – area method; the cross-section must be surveyed and water levels must be recorded. Acoustic pulses are beamed through the water from transmitters on one side of the river, received by sensors at the other side, and the times taken are recorded. Transmitting and recording transducers are installed on each bank with the line joining them making an angle across the river. The pulses sent in the two directions have slightly different travel times due to the movement of the water downstream. The time measurements and necessary computations based on the differences in the times of travel are made electronically on instruments housed on site. Clear water is essential for this method; the line for the acoustic pulses must not be hampered by weeds or other detritus. Where there is a wide range in stage, the transducers can be made adjustable vertically to send the acoustic beams at different depths or a multi-path system from a set of transducers may be installed. The ultrasonic method is now well proven and it gives a high accuracy; although it does not require a stage – discharge relationship, a stable bed is recommended.

The integrating-float technique developed recently (Sargent, 1981) uses the principle of moving floats (see Fig. 6.10) by releasing compressed-air bubbles at regular intervals from special nozzles in a pipe laid across the bed of the stream (Fig. 6.23a). The bubbles rising to the surface with a constant terminal speed V_r are displaced downstream a distance L at the surface by the effects of the velocity of the flow as the bubbles rise. In a unit width at a point across the channel, the discharge q over the total depth is given by:

$$q = V_r \overline{L}$$

where $\overline{L}$ is the mean surface displacement of the bubbles at that point. The total discharge of the river is obtained from:

(a) Section

(b) Plan

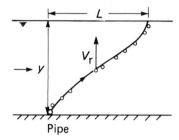

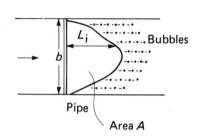

Fig. 6.23   Integrating float technique.

$$Q = \sum_{i=1}^{n} q_i \, \Delta b_i$$

where n is the number of points across the river and Δb_i is the width of a segment.

From Fig. 6.23b it can therefore be seen that:

$$Q = \sum_{i=1}^{n} V_r \overline{L}_i \Delta b_i = V_r A$$

The area A is obtained in the field by taking photographs of the bubble pattern and then using a microcomputer technique to calculate area A from the photographic prints allowing for the orientation of the camera which is sited on one bank. From experiment it has been found that a mean value for V_r of 0.218 m s^{-1}, for the particular nozzles used, in depths up to 5 m gives an error of less than 2%. When calibrated with the discharges over a weir, the integrating float technique gave a range of errors from -6 to $+10\%$ in Q. Difficulties occur when the line of bubbles is indistinct in the turbulence of high flows. The effectiveness of the method diminishes with increase in river width, but for over 50 m the water surface can be photographed in sections. The method is a very economical alternative to standard gauging procedures. Its applicability has been tested in several flow conditions but in the future its main contribution is likely to be in the measurement of controlled flows, e.g. in canals.

References

Ackers, P., White, W.R., Perkins, J.A. and Harrison, A.J.M. (1978). *Weirs and Flumes for Flow Measurement.* John Wiley. 327 pp.

Brater, E.F. and King, H.W. (1976). *Handbook of Hydraulics*, 6th ed. McGraw-Hill.

British Standards Institution (various dates). *Methods of Measurement of Liquid Flow in Open Channels*. BS 3680.

Corbett, D.M. *et al*. (1962). *Stream-Gauging Procedure*. US Geol. Surv. Water Supply Paper 888, 245 pp.

Chow, Ven Te (1959). *Open Channel Hydraulics*. McGraw-Hill, 680 pp.

Francis, J.R.D. (1975). *Fluid Mechanics for Engineering Students*. 4th ed. Arnold, 370 pp.

Grover, N.C. and Harrington, A.W. (1966). *Stream Flow, Measurements, Records and Their Uses*. Dover Publications, 363 pp.

Herschy, R.W. (ed.) (1978). *Hydrometry, Principles and Practices*. John Wiley, 511 pp.

Herschy, R.W., White, W.R. and Whitehead, E. (1977). *The Design of Crump Weirs*. Tech. Memo. No. 8. DOE Water Data Unit.

Sargent, D.M. (1981). 'The development of a viable method of streamflow measurement using the integrating float technique.' *Proc. Inst. Civ. Eng.*, **71**(2), 1–15.

7

Groundwater

In Chapter 5 on Soil Moisture, the introduction to the complexities of water storage below the ground surface confined attention to the water held in the soil matrix. Water was considered in the static state. In evaluating groundwater on a larger scale, it is necessary to deal with groundwater in motion. The degree of difficulty that arises in calculating groundwater movement is affected by whether the ground is saturated or not, since in the aeration zone the two-phase mixture of vapour and liquid in the pores or voids causes hysteresis in the pressure water content to permeability relationships. When the voids are filled completely with water, those complex relationships are no longer relevant and the more familiar concepts of single-phase liquid flow can be applied, but generally not without some simplifying assumptions.

Before proceeding to describe groundwater flow, it is pertinent to consider the sources of groundwater. Most of the water stored in the ground comes from residual precipitation at the surface infiltrating into the top soil and percolating downwards through the porous layers. Certain quantities pass into the ground along river banks at times of high flows and these generally sustain the flow by returning water to the rivers as the flow recedes. However, the longer term renewal of groundwater is brought about by infiltration of rainfall over a catchment area.

7.1 Infiltration

When a soil is below field capacity and surplus rainfall collects on the surface, the water crosses the interface into the ground at an initial rate (f_0) dependent on the existing soil moisture content. As the rainfall supply continues, the rate of infiltration decreases as the soil becomes wetter and less able to take up water. The typical curve of infiltration rate with time shown in Fig. 7.1 reduces to a constant value f_c, the *infiltration capacity*, which is mainly dependent on soil type. From Chapter 5, it will be appreciated that sandy soils have higher infiltration capacities than fine clay soils.

The rate at which water infiltrates into the soil can be measured by an infiltrometer. The simplest method adds water to the ground surface contained within a 200 mm diameter tube set vertically into the soil. The water is supplied from a graduated burette and the water depth is restored to a constant level by measured additions at regular time intervals. The rate of infiltration is then easily calculated. To prevent horizontal dispersion below the tube the infiltrometer may consist of two concentric rings. The area within each ring is flooded as before, but is is the inner ring that gives the infiltration measurement with the water draining in the outer ring preventing lateral seepage from the central core. Another method of assessing infiltration uses a watertight sample plot of ground on to which simulated rainfall of known uniform intensity is sprayed from special nozzles. The surface runoff from the plot is measured, and by sequential operations of the 'infiltrometer', assessments of the volume of water retained in the surface depressions and detained in the soil can be made. An infiltration rate curve can again be estimated up to the point when, with the soil at field capacity, the constant infiltration capacity is obtained. Full details are given by Musgrave and Holtan (1964). From either method, infiltration curves can be compiled for a range of soil types and for different vegetational covers.

When the hydrologist is required to model the infiltration process, two formulae are often used:

(a) Horton (1940) suggested that the form of the curve given in Fig. 7.1 is exponential and that this might be expected from infiltration being a decay process as the soil voids become exhausted. He proposed that the infiltration rate at any time t from the start of an adequate supply of rainfall is:

$$f_t = f_c + (f_0 - f_c)\exp(-kt) \tag{7.1}$$

The values of f_c and of k, the exponential decay constant, are dependent on soil type and vegetation.

(b) Philip has studied infiltration and soil water movement extensively for many years. (Philip, (1960) gives a review). He developed a simple formula for infiltration rate related to time:

$$f_t = \tfrac{1}{2}A\,t^{-\frac{1}{2}} + B \tag{7.2}$$

This is derived from an analytical expression for total infiltration volume in the form of a polynomial for which experiment showed that only the first two terms needed to be considered and hence only two terms appear in Equation 7.2. The values of A and B are related to the

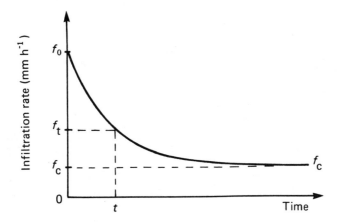

Fig. 7.1   Generalized infiltration rate.

soil properties and to the physical considerations of soil water movement.

Once recharge water has infiltrated into the soil, its passage downwards to join the groundwater storage depends on the geological structure as well as on the rock composition. Fig. 7.2 shows a section through a series of sedimentary rocks in which it is most usual to find productive *aquifers*, beds of rock with high porosity that are capable of holding large quantities of water. In general, the older the rock formation, the more consolidated is the rock material and the less likely it is to contain water. Igneous and metamorphic rocks are not good sources of groundwater unless weathered and/or fractured. The sedimentary rock strata have different compositions and porosities. In the much simplified diagram, layers of porous sands or limestones are subdivided by less porous material such as silt or clay which inhibit water movement. Semi-porous beds which allow some seepage of water through them are known as *aquitards*; they slow up percolation to the porous layers below, which are called leaky aquifers since they can lose as well as gain water through an aquitard. The clay beds which are mainly impermeable are called *aquicludes* and the porous layers between them are *confined aquifers* in which the water is under pressure. In the top sandy layer, the *water table* at atmospheric pressure marks the variable upper limit of the unconfined aquifer, although locally a lens of clay can hold up the groundwater to form a *perched water table*.

7.2   Groundwater Movement

To assess groundwater movement, the fine details of the soil or rock structures are disregarded and flow is considered on a macroscopic rather than a microscopic scale. For example, in Fig. 7.3a, a contained block of porous

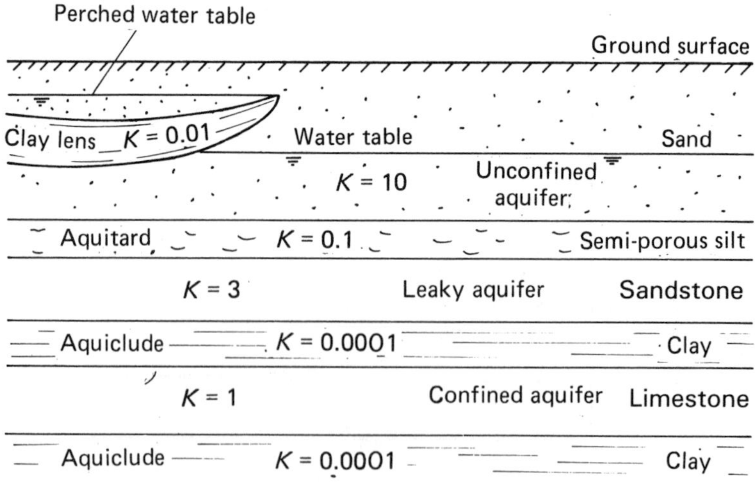

Fig. 7.2   Aquifer definitions (with sample K values md^{-1}).

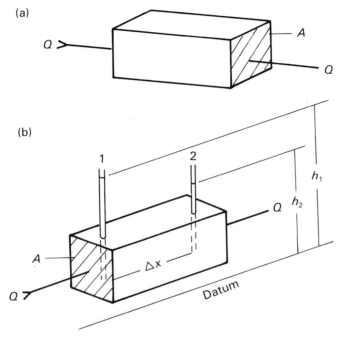

(a)

(b)

Fig. 7.3

material has a cross-sectional area A. If a flow rate Q is introduced at one end and an equal outflow Q is measured at the other, then the discharge per unit area, defined as the specific discharge, v, is given by:

$$v = Q/A \qquad \text{(in units of velocity)} \tag{7.3}$$

If manometers are fixed within the porous block (Fig. 7.3b) at a distance Δx and the difference in the water levels is $(h_1 - h_2)$, then experiment shows that v is directly proportional to $(h_1 - h_2)$ and inversely proportional to Δx. Thus $v \propto (h_1 - h_2)/\Delta x$. If h is measured positively upwards from an arbitrary datum, then $(h_1 - h_2)/\Delta x \rightarrow - dh/dx$ in the limit. This relationship of specific discharge to the *hydraulic gradient* $i = - dh/dx$ was established from the experimental findings of the French engineer Darcy, leading to the Darcy equation:

$$v = Ki = - K \, dh/dx \tag{7.4}$$

K, the proportionality 'constant', has the same units as v, i.e. those of velocity. K is known as the *hydraulic conductivity* and is essentially a property both of the porous medium and of the fluid flowing through the medium. With water as the fluid, K has high values for coarse sands and gravels (e.g. 10 to 10^3 md^{-1}) and lower values for compact clays and consolidated rocks (e.g. 10^{-5} to 1 md^{-1}). (See Fig. 7.2). Darcy's Law may be expressed in terms of the flow rate Q through a cross-sectional area A:

$$Q = KAi = - KA \, dh/dx \tag{7.5}$$

Example. If, in Fig. 7.3, $A = 1.5$ m $\times$ 1.5 m $= 2.25$ m^2, $x = 4$ m, $h_1 = 3$ m and $h_2 = 2.5$ m.
For a block of sandstone with $K = 3.1$ md^{-1}

$$Q = -KA(h_2 - h_1)/\Delta x$$
$$= -3.1 \times 2.25 (2.5 - 3)/4$$
$$= 0.87 \text{ m}^3 \text{ d}^{-1}$$

If the block were composed of silt, $K = 0.08$ md^{-1}

$$Q = -KA(h_2 - h_1)/\Delta x$$
$$= -0.08 \times 2.25 (2.5 - 3)/4$$
$$= 0.0225$$
$$= 2.25 \times 10^{-2} \text{ m}^3 \text{ d}^{-1}$$

An important extension of Darcy's Law to groundwater flow is its application in three dimensions. Equation 7.4 gives the specific discharge in the single x direction, and it is assumed that the medium does not change in character so that K remains constant in that direction. However, the ground structure may differ radically in other directions, and therefore at a point in the ground three specific discharges may be defined:

$$v_x = -K_x \, dh/dx$$
$$v_y = -K_y \, dh/dy \qquad\qquad (7.6)$$
$$v_z = -K_z \, dh/dz$$

with K having different values in the x, y and z directions.

Variations in K, the hydraulic conductivity, lead to two main classifications of porous media according to:

(a) position in a soil or geological stratum — if K is independent of location within a geological formation, the latter is said to be *homogeneous*, but if K is dependent on position the formation is *heterogeneous*; and
(b) direction of measurement — if K is independent of the direction of measurement within a geological formation, the formation is said to be *isotropic*, but if K varies with direction of measurement, the formation is *anisotropic*.

Combinations of these four major characteristics are demonstrated in Fig. 7.4, in which the relationships of the K values in the x and z directions at two sampling points are defined. Homogeneity is usually found in a single stratum of a sedimentary rock, but a sequence of different layers of rock would make for heterogeneity overall with each layer having a homogeneous K. Anisotropy (i.e. $K_x \neq K_z$) may be caused by a layering or aligning of clay lenses or minerals within an unconsolidated sediment or by faults and fractures in a solid rock providing increased specific discharge in one direction.

Darcy's Law has been shown to be applicable in saturated and unsaturated porous media (Childs, 1969), but the volume of porous medium for which it is used must be very large in comparison to the microstructure. Water usually moves slowly in the ground so that the Reynolds number, Re, is small and the flow is laminar (see Chapter 6). The linear relationship between specific discharge and hydraulic gradient holds up to Re values about 1; above this the relationship becomes non-linear and eventually the

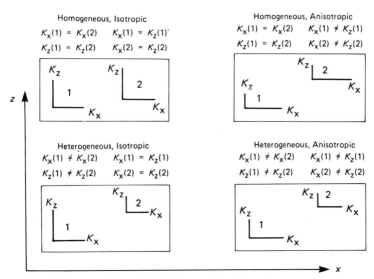

Fig. 7.4   Variations in hydraulic conductivity. (Reproduced from R. Allan Freeze & John A. Cherry (© 1979) *Groundwater*, p. 33, by permission of Prentice-Hall, Inc., Englewood Cliffs, NJ.)

flow becomes turbulent. Thus more detailed consideration of flow patterns must be made in massively fissured rocks such as limestones, through which the flow may be turbulent.

Hydraulic conductivity can be determined in many ways according to the nature of the ground material. Samples can be tested in the laboratory, or formulae involving grain size can be applied, but in practice preference is given to field measurements using timed movement of tracers, auger hole tests or well-pumping tests. Some average values for K are given in Table 7.1.

A further term much used in analysing the hydraulics of groundwater is *transmissivity* (T), which is given by

$$T = Kb$$

with b being the thickness of the saturated aquifer. It represents the rate of flow per unit width of the aquifer under unit hydraulic gradient. In Fig. 7.5, an unconfined aquifer is formed by porous material contained in an impermeable valley. If the material is homogeneous and isotropic, with K, the hydraulic conductivity of the material, the specific discharge through the aquifer in the direction of the arrow is given by:

$$v = \frac{K(z_1 - z_2)}{x}$$

where $(z_1 - z_2)/x$ is again the hydraulic gradient i, i.e. the slope of the water table.

The total flow rate through the aquifer with width y and depth b is then:

Table 7.1   Average Values of Hydraulic Conductivity

Material	Particle diameter (mm)	K (m d^{-1})
Unconsolidated		
Gravel, medium	8 −16	270
Sand, medium	0.25 −0.5	12
Silt	0.004–0.062	0.08
Clay	<0.004	0.0002
Consolidated		
Chalk — very variable according to fissures, e.g. whole aquifer		30.0
Sandstone		3.1
Limestone		0.94
Dolomite		0.001
Granite, weathered		1.4
Schist		0.2

$$Q = y b K i \quad \text{or} \quad y T i \tag{7.7}$$

Example. In Fig. 7.5, the average width of the valley aquifer (y) is 0.7 km and the length of the section (x) 2.5 km. If the average thickness of the saturated aquifer ($K = 2$ m d^{-1}) is 200 m, the transmissivity $T = Kb = 400$ m^2 d^{-1}. With values of z_1 and z_2 of 350 m and 300 m respectively, then

$$\begin{aligned}Q &= y \, T \, i \\ &= 700 \times 400 \times 50/2500 \\ &= 5600 \text{ m}^3 \text{ d}^{-1}\end{aligned}$$

When considering water movement in the ground, it is necessary to define changes in state of the aquifer, and for this a description of the storage capacity of the medium is required. The *specific storage*, S_s, is defined as the volume of water that can be released from a unit volume of a saturated aquifer by a unit reduction in hydraulic head. The dimension of S_s is (L)$^{-1}$ and S_s is more usefully represented in an aquifer of thickness b by the equation:

$$S = S_s b$$

where S is called the *storativity* (or *storage coefficient*) and is nondimensional. Values of the storativity (S) in confined aquifers range from 5×10^{-3} to 5×10^{-5} and are related to porosity and compressibility factors (the aquifer is not dewatered). In unconfined aquifers (Fig. 7.5), the storage coefficient is known as the *specific yield* (S_y), which is the volume of water released by dewatering from storage per unit surface area of the aquifer per unit decrease in water table level. Values of S_y range from 0.01 to 0.30 and relate solely to the porosity.

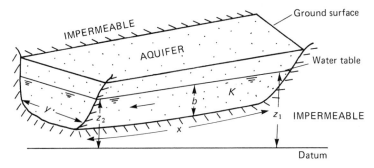

Fig. 7.5   Unconfined aquifer.

7.3   Groundwater Flow Equations

The non-steady state movement of water in the ground is governed by Darcy's Law and the continuity principle. Fig. 7.6 shows an element of saturated earth or rock with sides of length x, y and z. Considering first one-dimensional movement in the x direction, and using the principle of continuity, the following equality can be written for the difference between inflow and outflow:

Inflow = Outflow + Change in Storage

$$q_x = \left(q_x + \frac{\partial q}{\partial x} x \right) + S_s\, xyz\, \frac{\partial h}{\partial t} \tag{7.8}$$

where S_s is the specific storage and $\partial h/\partial t$ is the change in head with time. Applying Darcy's Law, Equation 7.5:

$$q_x = y\, z \left(- K_x \frac{\partial h}{\partial x} \right)$$

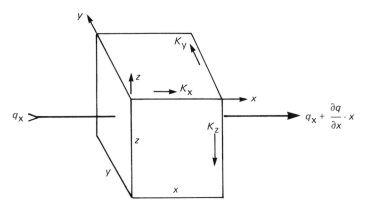

Fig. 7.6   Element of saturated rock.

Replacing the flow rates gives:

$$0 = xyz \left(-K_x \frac{\partial^2 h}{\partial x^2} \right) + S_s xyz \frac{\partial h}{\partial t}$$

Dividing through by xyz and rearranging:

$$K_x \frac{\partial^2 h}{\partial x^2} = S_s \frac{\partial h}{\partial t}$$

For horizontal, two-dimensional flow in the $x - y$ plane this equation becomes:

$$K_x \frac{\partial^2 h}{\partial x^2} + K_y \frac{\partial^2 h}{\partial y^2} = S_s \frac{\partial h}{\partial t} \tag{7.9}$$

and for describing the flow completely in the three dimensions, the equation is:

$$K_x \frac{\partial^2 h}{\partial x^2} + K_y \frac{\partial^2 h}{\partial y^2} + K_z \frac{\partial^2 h}{\partial z^2} = S_s \frac{\partial h}{\partial t} \tag{7.10}$$

with S_s the specific storage as previously defined ($S_s = S/b$). The final equation governs the changing or *transient* flow in three dimensions in an anisotropic aquifer.

For a homogeneous, isotropic material, the hydraulic conductivities (K) are equal and constant and Equation 7.10 reduces to:

$$K \left(\frac{\partial^2 h}{\partial x^2} + \frac{\partial^2 h}{\partial y^2} + \frac{\partial^2 h}{\partial z^2} \right) = S_s \frac{\partial h}{\partial t} \tag{7.11}$$

This simplifies further for *steady* flow when $\partial h/\partial t = 0$ and becomes:

$$\frac{\partial^2 h}{\partial x^2} + \frac{\partial^2 h}{\partial y^2} + \frac{\partial^2 h}{\partial z^2} = 0 \tag{7.12}$$

which is the Laplace equation. The solution of Equation 7.12 gives the hydraulic head h in terms of x, y and z. The solution of the full equation for transient flow in an anisotropic medium gives h in terms of t as well as x, y and z. Both Equation 7.10 and Equation 7.12 are difficult to solve in all but simple situations. They have tested the ingenuity of many researchers. In engineering practice, groundwater movement problems have often been treated by simplifying the boundary conditions. In addition, assumptions of an isotropic aquifer or steady-flow conditions or both are often made to facilitate the solution and yet provide an acceptable precision.

Where conditions allow a problem is reduced to two dimensions, e.g. water movements in a vertical $x - z$ plane through layered aquifers or in the horizontal $x - y$ plane at some convenient level of z. Such simplification enables satisfactory answers to be obtained to problems of well yields or in the assessments of areal groundwater resources by the construction of two-dimensional flow nets.

7.4   Flow Nets

The relative inaccessibility of groundwater compared with surface water means that only point measurements in the ground reservoir are possible. The hydrologist must rely on applying the groundwater flow equations to assess resources. The mathematical representations of groundwater flow form the basis for models of internal drainage and of changes in storage. Flow nets comprising lines of equipotential and stream or flowlines can represent groundwater movement in two dimensions. In this section, the application of nets or rigid grids is extended to analogue and digital computer models.

7.4.1   Graphical Solutions

A block of saturated land with several vertical piezometers installed is portrayed diagrammatically in Fig. 7.7a. Piezometer A from ground level 450 m penetrates to a depth of 150 m above datum and the water level rests at 375 m. Using the nomenclature introduced in Chapter 5 for soil moisture, the total potential or hydraulic head h is equal to the sum of the pressure head ψ and the elevation head z (height of pressure measuring point above datum):

$$h = \psi + z$$

Thus knowing the height of the land surface, the length of the piezometer and from a measure of the water depth d, the value of h can be obtained. For A, the piezometer length is 300 m, d is 75 m and z is 150 m:

$$h = (\psi + d) - d + z$$
$$= 300 - 75 + 150 = 375$$

In this example where a common datum and piezometer length are used, the hydraulic head h is simply the level at d m below the surface or $450 - 75 = 375$ m.

At piezometer B, where the land surface is also 450 m, d is at 150 m and therefore $h = 300$ m. If the distance Δx between A and B is 300 m, then the mean hydraulic gradient is:

$$i = \frac{h_A - h_B}{\Delta x} = \frac{375 - 300}{300} = 0.25$$

Thus at the elevation head, $z = 150$ m, there is a difference in potential from A to B and therefore there will be a component of specific discharge v_{AB} from A to B of $K i$ or $0.25\ K$ m s^{-1}, where K is the hydraulic conductivity of the medium. In the 'field' of piezometers in Fig. 7.7a, piezometers C and E from different surface levels have h values of 375 m, at piezometer D, h is 400 m, and at B and F, h values are 300 m. Within the block of land there is a three-dimensional surface defined by the hydraulic heads, and this is known as a *piezometric surface*. This passes through all the rest levels. On a plan view (Fig. 7.7b) the points A, C and E lie on an *equipotential line* (375 m). Through the points B and F on the two-dimensional plan runs the equipotential line of 300 m. Once the equipotential lines have been determined for an isotropic

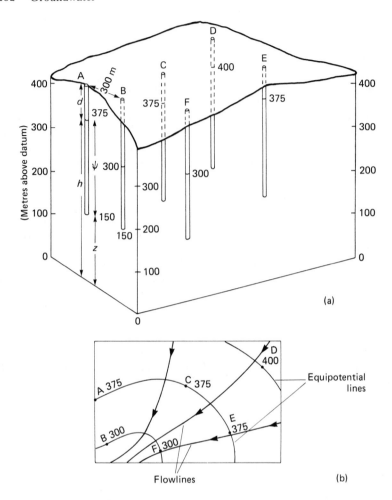

Fig. 7.7   (a) A 'field' of piezometers. (b) Plan view (*x-y* plane).

aquifer, *flowlines* may be constructed perpendicular to the equipotential lines in the direction of maximum potential gradient downwards. In the example, it is obvious that the ground water is draining to the corner of the block between *B* and *F*, and hence, three flow lines with direction arrows have been drawn on the plan. The pattern of equipotential lines and flow lines constitutes a *flow net*, of which Fig. 7.7b is a very simple example.

Flow nets drawn under certain rules allow flow rates to be calculated very simply. Fig. 7.8 shows a flow net of equipotential lines and flowlines drawn for a two-dimensional groundwater flow. The equipotentials have equal drops of head, Δh, between any adjacent pair. Taking a typical cell in which the distance between the equipotenial lines is Δx, then the velocity of flow through the cell is $V = K \Delta h/\Delta x$. For unit thickness of aquifer (perpendicular to the flow net), the flow rate through the cell bounded by flow lines Δy apart, is:

$$\Delta q = V.\Delta y.1 = K \frac{\Delta h}{\Delta x} .\Delta y = K \, \Delta h \frac{\Delta y}{\Delta x}$$

Since Δq is constant between two adjacent flowlines (no flow can cross them), all the cells between two such flowlines having the same Δh must have the same width to length ratio, $\Delta y/\Delta x$. If the flowlines are drawn so that Δq is the same between all pairs of adjacent flowlines, then the ratio $\Delta y/\Delta x$ will be the same for *all* the cells in the flow net. In addition, the spacings can be chosen such that $\Delta y = \Delta x$ and all cells then become curvilinear squares. Following such rules, then $\Delta q = K \, \Delta h$, per unit thickness of aquifer.

If there are N drops of Δh between equipotential boundaries whose potential difference is H, then $\Delta h = H/N$. If there are M 'flowtubes' between impermeable boundaries, then the total flow rate (per unit thickness of aquifer), is:

$$Q = M \, \Delta q = M \, K \, H/N$$

Summarizing the properties and requirements of flow nets in homogeneous, isotropic media:

(a) equipotential lines and flow lines must all intersect at right angles;
(b) constant-head boundaries are equipotential lines;
(c) equipotential lines meet impermeable boundaries at right angles;
(d) if a square grid is used, it should be applied throughout the flow net (although difficulties will arise near sharp corners and towards remote or infinite boundaries).

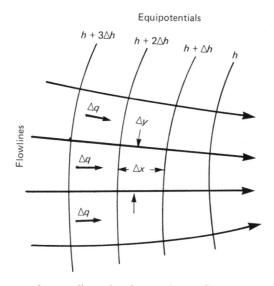

Fig. 7.8   Flow net for two-dimensional groundwater flow.

For anisotropic aquifers and heterogeneous ground conditions, flow nets can be constructed and used successfully, but complex adjustments to equipotential lines across hydraulic conductivity boundaries must be made and for these the reader is referred to the specialist texts (e.g. Cedergren, 1977; Freeze & Cherry, 1979; Todd, 1980).

7.4.2 Electrical Analogues

The formulation of a flow net that is an indirect solution of the Laplace equation for two dimensional steady water flow in an isotropic, homogeneous material:

$$\frac{\partial^2 h}{\partial x^2} + \frac{\partial^2 h}{\partial y^2} = 0 \qquad \text{or} \qquad \frac{\partial^2 h}{\partial x^2} + \frac{\partial^2 h}{\partial z^2} = 0$$

can be made by using electrical circuitry. Ohm's Law expresses the flow of electrical current I, as:

$$I = -\sigma \frac{dV}{dx}$$

where σ is the conductivity, V is the voltage and x a length in the direction of flow. This is analogous to Darcy's Law:

$$v = -K \frac{dh}{dx}$$

For equipotential lines the voltage V, corresponding to head h, is constant; insulated boundaries are equivalent to impermeable boundaries. The electrical conductivity σ is analogous to the hydraulic conducticity K.

There are two main types of electrical analogue: the simpler uses a sheet of conductive paper (such as Teledeltos paper) or a grid drawn with conducting ink on squared paper (the graphite grid resistance analogue), whereas the second, much more elaborate, depends on the assemblage of a network of resistances. In both methods, the basic properties of the problem area are modelled, boundaries defined and sources and sinks identified. Equipotentials can be traced or constructed across the area, flow lines identified and the total current representing the required discharge measured. In the more sophisticated models, changes in hydraulic conductivity (electrical conductivity) can be accommodated. Electrical analogues are of great value in assessing regional groundwater resources and in predicting the effects of over exploitation on the stability of the water table in major aquifers (Rushton & Redshaw, 1979).

7.4.3 Numerical Methods

With the widespread availability of digital computers, the h values for a flow net can be computed from analytical solutions to some simple problems, but more often for complex conditions, approximate numerical methods need to be used.

The finite-difference method is the most straightforward and easily understood

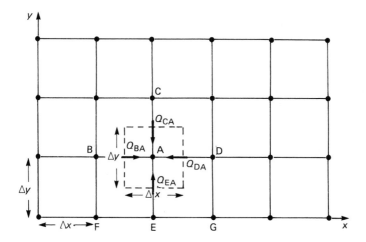

Fig. 7.9   Finite-difference method.

numerical method. For two dimensional problems, the area of the ground-water flow either in the $x - y$ plane or the $x - z$ plane is subdivided by a square grid (Fig. 7.9). The nodes of the grid are the points at which the values of head are required. At each node, the h value is assumed to be representative of a block centred on the node. If the value of h at A is unknown, it can be calculated from considerations of the flow pattern around the block. Using the lettering in Fig. 7.9, and applying Darcy's Law, flow per unit thickness from the neighbouring node B is given by:

$$Q_{BA} = K_{BA} \frac{(h_B - h_A)}{\Delta x} \Delta y$$

Similarly, values of Q from C, D and E towards A can be stated, but from the conservation of mass for steady state flow, the sum of these flows must be zero. Thus for an isotropic, homogeneous aquifer with all the Ks equal and since the grid is square ($\Delta x = \Delta y$) then:

$$h_A = \tfrac{1}{4}(h_B + h_C + h_D + h_E)$$

Such an equation can be used to determine h values at all inner nodes within a grid. The boundary nodes often have known h values for certain boundary conditions. They are the starting points of sequential h determination at all nodes across an area by the method of relaxation. In this method, after initial estimates of h have been made at all nodes, a residual error (R) at A viz:

$$R_A = h_A - \tfrac{1}{4}(h_B + h_C + h_D + h_E)$$

is removed by adjusting h_A (resulting in changes to R values at B, C, D and E). Such adjustments are made systematically until all R values have fallen to an acceptable tolerance. Treatments are available for boundaries where h is not known or where their alignments are not grid lines.

With the wide capabilities of the digital computer, complex problems incorporating anisotropic conditions and transient flows can be solved using this and similar finite difference methods. Package programs for such computations are readily available.

The finite-element method is a more recently developed numerical technique. It is being used in many branches of engineering. The computations resemble finite-differencing, but a mesh of more flexible geometry is used to model the problem area. Variable anisotropy can easily be accommodated by adjusting the local axes to the principal directions of hydraulic conductivity, which reduces computational time. The method normally gives greater accuracy than finite differences, but usually at greater expense. Considerable research is being undertaken with the finite-element method, further details are to be found in Pinder & Gray, (1977).

7.5   Groundwater Measurement

It has already been emphasised that within the hydrological cycle in temperate regions of the world, groundwater constitutes the largest storage of fresh water. In the previous sections of this chapter, the theory of groundwater movement has been described, but little attention has been given to the practical aspects of finding or measuring groundwater. The relatively slow but varying water movement through the heterogeneous mixture of unconsolidated sands and gravels and consolidated rocks ensures that a continuous slowly varying base flow is maintained in most rivers. Part of this contribution to surface flow can be measured at permanent spring sites, the methods used depending on the quantities involved. Small springs can be sampled volumetrically over a measured time interval to give discharge in $l \ s^{-1}$ or $m^3 \ s^{-1}$, but more usually small thin plate weirs are installed (see Chapter 6). The lateral contribution of groundwater through seepage to surface streams cannot be measured directly, and calculated estimates must be made using knowledge of the water table and of the flow characteristics of the geological strata.

The basic measurement for assessing groundwater is the depth of the water table in unconfined equifers or the position of the piezometric surface in confined aquifers. According to the nature of the terrain, penetration of the ground can be made by methods ranging from hand-dug wells in unconsolidated surface deposits to the high-speed drilling of deep boreholes. The latter are usually employed for the exploitation of confined aquifers at great depths. In localities where every village or separate farmstead had its own well-water supply, there are often disused wells that can be made available for water-table measurements. When the groundwater storage is to be exploited for large public water supplies, special narrow observation wells are sometimes bored in addition to the large-diameter supply wells destined to be pumped. Hydrogeologists assemble well data from all possible sources to obtain knowledge of the groundwater behaviour. In the UK, some records of well levels are available for hundreds of years, especially in the chalk lands of Southern England.

The slow movement of water in the ground results in slow natural changes in groundwater levels, and thus well measurements are in general made regularly on a monthly basis. Where more rapid movement is expected or detailed information is required, the sounding of wells may be made weekly or even over shorter periods. In the expansion of hydrometric studies in the UK for the evaluation of water resources, well-level recorders have been recommended. These are labour saving and therefore, in developed countries, provide an economical method of obtaining data and of ensuring the continuation of valuable long records at old established measuring wells. Some of the various types of level recorders used in river flow measurements (Chapter 6) are readily adapted for recording well levels.

The standing-water level in a well is dependent on the character of the well, whether it is lined or unlined, and its depth. The structure and composition of the ground are most important too. A shallow well may penetrate through only one or two layers of rock and, if it is unlined, the water level may be considered to represent the level of the local water table. A deep borehole could pass through many different geological strata and the well-water level may be affected by pressures in confined aquifers, thus giving a piezometric level rather than the level of the water table. A cased or lined well certainly gives the piezometric level representing the water pressures at the bottom of the well.

To assess natural groundwater contribution to streams and other surface channels, it is the measurements of the water table from the shallower wells that are required.

7.6   Unconfined Flow

The lateral seepage from a river bank into a river produced by unconfined flow from a porous aquifer with a well defined water table is portrayed in Fig. 7.10. The aquifer is assumed to be isotropic and homogeneous, and it is underlain by an impermeable stratum. The flow pattern in the combined unsaturated and saturated layers of such an aquifer is not so easily identified as in a completely saturated confined aquifer. The free surface of the water table has an increasingly downward slope to meet the river bank at point A, the top of a seepage surface. The water-table slope reflects a hydraulic gradient and the line of the water table describes a stream line if no percolation occurs from the ground surface. In effect, the water table is the upper boundary of a flow net in the saturated part of the aquifer. The river bed/bank interface is an equipotential line and hence all streamlines meeting it must turn to meet BC at right angles. The seepage surface AB is not an equipotential and the water-table flow line is tangential to the river bank at A (Raudkivi & Callendar, 1976).

To be able to calculate the flow to the river, several approximations are generally made. These approximations, developed by Dupuit and Forchheimer, make the major assumptions:

(a) the hydraulic gradient dh/dx is equal to the slope of the water table;
(b) the related specific discharge is constant throughout the depth of flow; and
(c) the line ED may be taken to act as a constant head boundary with water table height h_D.

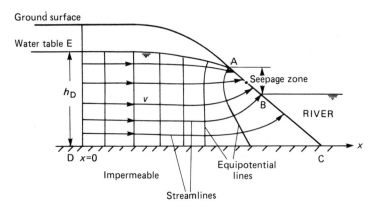

Fig. 7.10   Unconfined flow.

Thus, the streamlines are taken to be horizontal and the equipotential lines vertical.

Making these assumptions for steady-flow conditions, the discharge per unit width over the depth of the aquifer (h), hydraulic conductivity (K), is given at all sections by:

$$Q = - K h \, dh/dx \qquad (h \text{ decreasing with increasing } x)$$

In reality, there would be non-horizontal flow components near the water table and near ABC. However, making the Dupuit – Forchheimer assumption of a horizontal flow pattern, satisfactory results are obtained for the discharge when the water-table slope is small and the variation in the unconfined aquifer depth relative to that depth is small.

An extension of the seepage flow in an unconfined aquifer can include recharge from water percolating downwards through the unsaturated zone. This is a problem encountered by drainage engineers designing channels for leading off surplus water. In Fig. 7.11, a uniform rainfall rate is assumed to give a uniform infiltration rate f at the ground surface of an isotropic, homogeneous aquifer of hydraulic conductivity K. It is often required to know the depth of the water table relative to the water level in the river or channel. With steady-state conditions and adopting the Dupuit – Forchheimer assumptions, the specific discharge v at distance x, from Darcy's Law, is $v = - K \, dh/dx$. Then the flow q at x per unit width for the aquifer depth h is $q = - K h \, dh/dx$. However, from continuity in the steady state, the recharge into the aquifer from the infiltration is also given by $q = fx$. If all the flow in the aquifer is assumed to come from the infiltrated rain and none across the boundary ED where $x = 0$, then:

$$fx = - K h \, dh/dx$$
$$K h \, dh = - f x \, dx$$

Integrating between h_1 and h_2 with x going from 0 to a:

$$K (h_1^2 - h_2^2) = f a^2$$

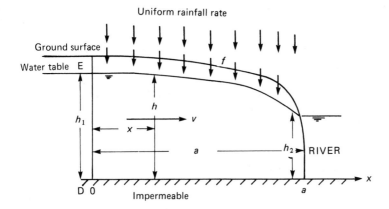

Fig. 7.11   Recharging unconfined flow.

whence:

$$h_1 = \sqrt{\left(\frac{f}{K} a^2 + h_2^2 \right)}$$

Thus, h_1, the water-table depth at distance a from the river can be found, knowing the water level in the channel, the infiltration rate and K, (ignoring the seepage surface). Further manipulation of the equations will provide the expression for Q, the baseflow to the river per unit length of channel, in terms of infiltration:

$$Q = fa$$

Example. In Fig. 7.11, it is required to know the height of the water table at 150 m from the river when irrigation water is applied at 0.8 mmd⁻¹ assuming a steady state is attained with uniform infiltration. The porous medium has K = 4.3 md⁻¹ and the drain level rests at 12 m above the impermeable stratum.

$$h_1 = \sqrt{\left(\frac{f}{K} a^2 + h_2^2 \right)}$$

$$= \sqrt{\left(\frac{0.008}{4.3} \times 150^2 + 12^2 \right)}$$

$$= 12.173 \text{ m}$$

7.7   Groundwater Exploration

The existence of large quantities of water stored in the ground gives this phase of the hydrological cycle an added importance. In addition to supplying the base flow of rivers, groundwater is a major water resource. The engineering hydrologist charged with finding reliable water supplies must therefore be well schooled in stratigraphic geology, but with the complications of some ground structures, he may find it necessary to call on the services of a specialist hydrogeologist. However, most civil engineers have had a basic training in geology and should be able to make simple surveys in order to locate worthwhile groundwater supplies.

In developed countries such as the UK, there are large-scale geological maps readily available. The stratigraphy of the whole country is well known and most of the major aquifers have been identified. For many parts of the world such information is scarce and any existing maps may be very rudimentary. Aerial photography plays a large role in surveying remote areas of developing countries. The resultant mapping from the aerial photographs provides the basic drainage patterns and may give indications of geological structure and possible groundwater sources. More recently, satellite photographs taken regularly at fixed intervals record changes in the land-surface characteristics and thereby register seasonal variations in vegetation, indicators of groundwater availability. Satellites also carry remote sensing instruments, and from ground-surface temperature differences the changing wetness of the soil can be monitored. The application of remote sensing in hydrology is still in the development stages, but the advent of such techniques will be welcomed by engineers charged with finding water in inaccessible regions.

Most exploration is however necessarily made on the ground. Unconsolidated materials composed of alluvial, glacial or aeolian deposits are the most easily exploited for groundwater. They are primarily found at or near the ground surface and their value increases with their depth. The heterogeneous composition of unconsolidated deposits, sands, gravels, silts and clays may make water quantity assessments a little complicated, but the engineering works involved in any exploitation are relatively inexpensive. The most reliable groundwater supplies from such deposits are to be found in river valleys with recharge coming from the surrounding higher land. In areas of low relief composed perhaps of glacial deposits, increased depth of unconsolidated material may be found in buried valleys excavated in previous geological ages. Consolidated rocks that make the best aquifers are sandstones and limestones. In the UK, the Bunter sandstones (Triassic) and the Chalk (Cretaceous) are the best groundwater sources, but the Oolite and Carboniferous limestones, particularly where fractured and fissured, also contribute large quantities. The harder older rocks and the crystalline igneous rocks are worth developing for water supply only where they are weathered and faulted.

Once the structural geology and rock composition of a survey area have been established, the existence of water and in particular the depth of the water table have to be proved. There are three methods currently favoured in exploring for groundwater; two are geophysical methods operated from the surface, whereas the third employs test drilling of wells or boreholes.

7.7.1   Resistivity Surveys

The resistivity of a rock formation is given by:

$$\rho = \frac{RA}{L} \text{ ohm m}$$

where R is the electrical resistance (ohm), and A (m^2) is the cross-sectional area over distance L (m). This measure varies with the composition of the strata and its water content. In porous rocks, the resistivity depends more on the water content than the material composition, decreasing with increasing water content. Fine clays tend to provide lower resistivity than coarse sands and gravels. The apparent resistivity is determined by passing an electric current into the ground through current electrodes and measuring the drop in potential across a pair of potential electrodes. The spacing of the electrodes determines the depth of penetration of the current, and relationships are obtained between electrode spacing and apparent resistivity. In surface deposits, the depth of the water table and the underlying bedrock level are readily identified. In surveying down to maximum depths of about 500 m, the resistivity plots display the limits of the various rock types. This method of subsurface surveying is widely used, and experienced operators can produce simply and cheaply the necessary information on the sites of groundwater sources.

7.7.2   Seismic Surveys

Shock or sound waves are sent through the strata and the boundaries between the different rocks are identified by the refraction of the shock waves. A small explosive charge is set off at a selected shot point at about 1 m depth and the waves are detected and timed over various distances by geophones on the ground surface. From the recorded wave arrival times, the sound velocities can be evaluated. The velocities are low (250 m s^{-1}) in unconsolidated deposits near the surface, but increase to over 5000 m s^{-1} in solid dense rocks. The sound velocities are increased by saturated strata; thus using this method too, the water table can be readily identified in homogeneous material. The method is applied easily to depths of 100 m, but is limited to the depth of the material giving the highest sound velocities, which is usually the bedrock. It can therefore be used directly for determining the thickness of unconsolidated deposits and identifying buried channels therein for reliable groundwater sources.

7.7.3   Wells and Boreholes

The most direct method of determining the subsurface geological structure is by the drilling and logging of boreholes. About 100 mm in diameter, these can penetrate to great depths with modern drilling techniques. Samples of the rocks are brought up at regular depths and where there are changes in the strata. Thus a continuous core is assembled on the surface and the rock characteristics identified, measured and documented. For the groundwater hydrologist, particulars of the significant aquifers and aquicludes provide the most valuable information.

These geological details are, however, only obtained for the borehole location; boreholes at other selected sites would give some spatial knowledge of the hydrogeology over a potential source area. In practice, exploration of possible groundwater sources generally begins with a surface survey by the resistivity or seismic methods, and the subsurface details are obtained by a few logged boreholes at selected sites. Once an extensive groundwater source in unconsolidated deposits or a potentially good yielding deeper aquifer is found, then the exploitation stage begins and pumping wells must be drilled.

7.8   Well Hydraulics

The means of procuring water from the ground depend on the physical location of the source. Shallow wells can provide small quantities when required by simply winching up a suitable container. For a continuous supply, a pump can be installed, the power for the pump and its capacity depending on the height through which the water has to be raised to the surface. Occasionally, an engineer developing a deep confined aquifer may find that the water reaches the ground surface of its own accord; the aquifer is then part of an artesian basin with the recharge area at a greater altitude providing a pressure head to the top of the well. (A noted example of an artesian basin is the Chalk syncline of the Thames Valley with wells in London penetrating the chalk aquifer which outcrops in the much higher Chiltern Hills and North Downs.)

Before establishing the attendent works for a large-scale supply of groundwater, the 'reliable yield' of the aquifer must be established by assessing the groundwater flow to the well. From the previous considerations of groundwater movement, it will be appreciated that well hydraulics depend on the confined or unconfined state of the penetrated aquifer.

7.8.1   Steady Flow in a Confined Aquifer

When the well fully penetrates a horizontal confined aquifer (Fig. 7.12), flow to the well is also horizontal from all directions (i.e. radial two-dimensional flow). To ensure a steady flow, there must be continuous recharge to the aquifer from sources distant to the well. Assuming also that the aquifer is homogeneous and isotropic and is not affected by compression in dewatering, the flow to the well at any radius r can be expressed by Darcy's Law:

$$Q = 2\pi r K b \, dh/dr \qquad\qquad (7.13)$$

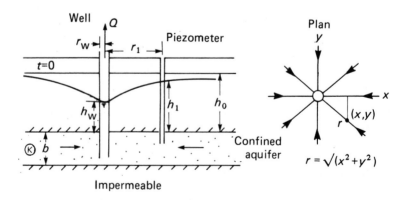

Fig. 7.12   Steady flow in a confined aquifer.

where Q is the pumping rate. Integrating over the radius distance r_w to r_1 and h_w to h_1 gives:

$$Q = \frac{2\pi Kb\,(h_1 - h_w)}{\ln\,(r_1/r_w)}$$

or:

$$Q = \frac{2\pi T\,(h_1 - h_w)}{\ln\,(r_1/r_w)} \qquad\qquad (\textit{Thiem equation})$$

with T the transmissivity of the aquifer.

By pumping the well at a steady rate and waiting until the well level h_w is constant, observation of the drawdown level h_1 at an observation well at a known distance r_1, from the pumped well, allows estimation of the transmissivity of the aquifer. In practice, the observations from two or more observation wells at different radii are more useful since head losses in the well, caused by friction in the well casing, can then be allowed for.

Example. A well in a confined aquifer was pumped at a steady rate of 0.0311 m³ s⁻¹. When the well level remained constant at 85.48 m, the observation well level at a distance of 10.4 m was 86.52 m. Calculate the transmissivity.

$$Q = \frac{2\pi T(h_1 - h_w)}{\ln\,(r_1/r_w)}$$

$$T = \frac{Q\ln\,(r_1/r_w)}{2\pi(h_1 - h_w)}$$

$$= \frac{0.311\ln\,(10.4/0.3)}{2\pi(86.52 - 85.48)}$$

$$= 0.0169 \text{ m}^2 \text{ s}^{-1}$$

or $T = 1460$ m² d⁻¹.

7.8.2 *Steady Flow in an Unconfined Aquifer*

The groundwater flow to a well in an unconfined aquifer may be complicated

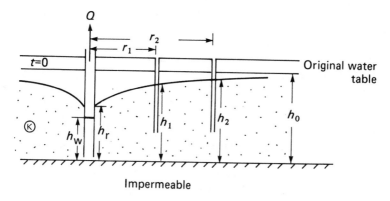

Fig. 7.13 Steady flow in an unconfined aquifer.

by the downward movement of recharge water from ground surface infiltration, but here only distant sources in the aquifer are assumed to maintain the steady-state flow. In Fig. 7.13, the water table intersects the well at h_r with the water level in the well at h_w. Between the two levels is a seepage zone. However, in estimating the flow to the well, the Dupuit – Forchheimer assumption of horizontal flow may be used, and Equation 7.13 becomes:

$$Q = 2\pi r K h \ dh/dr$$

with h the height of the water table replacing b the thickness of the confined aquifer. Integrating over the radius distance distance r_1 to r_2 with corresponding values h_1 and h_2 then:

$$Q = \frac{\pi K (h_2^2 - h_1^2)}{\ln (r_2/r_1)}$$

from which K can also be evaluated. If the squares term is expanded to give

$$2(h_2 - h_1)(h_2 + h_1)/2$$

then $K(h_2 + h_1)/2$ represents the average transmissivity T between the observation wells (piezometers), and the equation for Q becomes the same as the Thiem equation applied in confined aquifers. In applying this equation to an unconfined aquifer, the effects of the seepage from the water table into the well (h_w and r_w) are negligible if the distance of the nearest observation well r_1 is greater than 1.5 times the original water table height h_0.

7.8.3   Non-steady Flow to Wells

The steady-state groundwater flow situations analysed above, and consequent steady-state positions for the piezometric surface or water table, are only achieved after what may be lengthy periods of time after the start of pumping. During such periods, even though the pumping rate may be steady, the piezometric surface or water table will be falling with time until the steady-state position is reached. Indeed, if the aquifer were infinite in extent (with no vertical recharge), a steady state would never be achieved. The ability to analyse such non-steady groundwater flow situations for wells being pumped at a constant rate, is required.

In confined aquifers, the relief of pressure as a piezometric surface falls introduces two compressibility effects, viz, the pore water expands owing to the smaller water pressure, and simultaneously the pore space contracts owing to a greater mechanical stress from the overburden as the reduced water pressure takes less of the load. Thus water is released *from storage* over the aquifer to make up the pumped abstraction — no dewatering of the pore spaces occurs.

In unconfined aquifers, on the other hand, compressibility effects are quite negligible, but dewatering of the pore spaces does occur as the water table falls. The released water from the whole aquifer integrates to equal the constant pumping abstraction.

In both types of aquifer, the head, h, varies both with distance and time after the start of pumping. Even with a constant pumping rate, the situation is described as being 'non-steady' flow!

7.8.4   Non-steady Flow in a Confined Aquifer

It was shown earlier (Equation 7.9) that horizontal, two dimensional, non-steady flow can be given by:

$$K_x \frac{\partial^2 h}{\partial x^2} + K_y \frac{\partial^2 h}{\partial y^2} = S_s \frac{\partial h}{\partial t} \tag{7.9}$$

For a homogeneous, isotropic aquifer this can be written:

$$\frac{\partial^2 h}{\partial x^2} + \frac{\partial^2 h}{\partial y^2} = \frac{S}{T} \frac{\partial h}{\partial t}$$

($S_s = S/b$, $T = Kb$) with T and S, the transmissivity and storativity respectively. For non-steady flow to a well, this equation may be transformed to radial coordinates. Thus with $r = \sqrt{(x^2 + y^2)}$:

$$\frac{\partial^2 h}{\partial r^2} + \frac{1}{r} \frac{\partial h}{\partial r} = \frac{S}{T} \frac{\partial h}{\partial t} \tag{7.14}$$

The solution of this equation yields $h(r,t)$ (Fig. 7.14), the hydraulic head at distance r from a well at time t after the commencement of steady pumping at a rate Q. For practical purposes, what is usually required is $\{h_0 - h(r,t)\}$ which is the drawdown, $s(r,t)$, from the initial rest level head, h_0. The solution of the equation is:

$$s(r,t) = \frac{Q}{4\pi T} \int_u^\infty (\exp\text{-}u)/u \; du \qquad (Theis \text{ equation}) \tag{7.15}$$

where Q is the steady pumping rate and $u = r^2 S/4Tt$. Expansion of the integral gives:

$$s(r,t) = \frac{Q}{4\pi T} \left[-0.577216 - \ln u + u - \frac{u^2}{2.2!} + \frac{u^3}{3.3!} - \cdots \right] \tag{7.16}$$

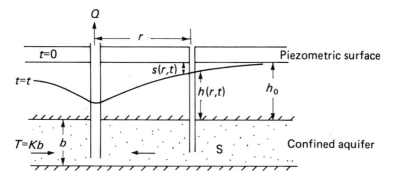

Fig. 7.14   Non-steady flow in a confined aquifer.

The expression within the brackets is usually denoted by $W(u)$ known as the *well function* so that:

$$s(r,t) = \frac{Q}{4\pi T} W(u) \quad \text{with } u = \frac{r^2 S}{4Tt} \tag{7.17}$$

Values of the well function, $W(u)$, for a range of u values are given in Table 7.2.

Table 7.2   The Well Function, $W(u)$

u	1.0	2.0	3.0	4.0	5.0	6.0	7.0	8.0	9.0
X 1	0.219	0.049	0.013	0.004	0.001				
X 10^{-1}	1.82	1.22	0.91	0.70	0.56	0.45	0.37	0.31	0.26
X 10^{-2}	4.04	3.35	2.96	2.68	2.47	2.30	2.15	2.03	1.92
X 10^{-3}	6.33	5.64	5.23	4.95	4.73	4.54	4.39	4.26	4.14
X 10^{-4}	8.63	7.94	7.53	7.25	7.02	6.84	6.69	6.55	6.44
X 10^{-5}	10.94	10.24	9.84	9.55	9.33	9.14	8.99	8.86	8.74
X 10^{-6}	13.24	12.55	12.14	11.85	11.63	11.45	11.29	11.16	11.04
X 10^{-7}	15.54	14.85	14.44	14.15	13.93	13.75	13.60	13.46	13.34
X 10^{-8}	17.84	17.15	16.74	16.46	16.23	16.05	15.90	15.76	15.65
X 10^{-9}	20.15	19.45	19.05	18.76	18.54	18.35	18.20	18.07	17.95
X 10^{-10}	22.45	21.76	21.35	21.06	20.84	20.66	20.50	20.37	20.25
X 10^{-11}	24.75	24.06	23.65	23.36	23.14	22.96	22.81	22.67	22.55
X 10^{-12}	27.05	26.36	25.96	25.67	25.44	25.26	25.11	24.97	24.86
X 10^{-13}	29.36	28.66	28.26	27.97	27.75	27.56	27.41	27.28	27.16
X 10^{-14}	31.66	30.97	30.56	30.27	30.05	29.87	29.71	29.58	29.46
X 10^{-15}	33.96	33.27	32.86	32.58	32.35	32.17	32.02	31.88	31.76

Knowing the 'formation constants' of the aquifer, S and T, and for a given Q, the drawdown $s(r,t)$ can be estimated directly for any radius r and time t. Alternatively, in pumping tests to find S and T, measurements of drawdown are taken at known times, t, at observation wells of known distance, r, and the aquifer constants S and T can be determined by more complex and indirect methods. (Freeze & Cherry, 1979).

Example. To calculate the drawdown in a confined aquifer at $r = 25$ m after 6 hours of pumping water with a constant discharge of 0.0311 m³ s⁻¹. The aquifer constants are $S = 0.005$ and $T = 0.0092$ m² s⁻¹.

$$u = r^2 S/4Tt$$

$$= \frac{25^2 \times 0.005}{4 \times 0.0092(6 \times 3600)}$$

$$= \frac{3.125}{794.88}$$

$$= 3.93 \times 10^{-3}$$

$$s(r,t) = \frac{Q}{4\pi T} W(u)$$

$$= \frac{0.0311}{4 \times \pi \times 0.0092} \times 4.97$$

$$= 1.337 \text{ m.}$$

The Theis solution of the non-steady radial flow equation has, for certain conditions, a simplified form due to *Jacob*. If u is small (<0.01) only the first two terms in the series of Equation 7.16 need be used (to within 1 % of the full series) so that:

$$s(r,t) = \frac{Q}{4\pi T} \left(-0.5772 - \ln \frac{r^2 S}{4 Tt} \right)$$

whence

$$s(r,t) = \frac{Q}{4\pi T} \ln \frac{2.25 Tt}{r^2 S}$$

A semi-log plot of $s(r,t)$ versus $\ln (t/r^2)$ allows T and S to be estimated from the slope and intercept of the resulting straight line, which is drawn with greater regard to the later points.

7.8.5 Non-steady Flow in an Unconfined Aquifer

In an unconfined aquifer, it has been seen that steady flow, based on the Dupuit – Forchheimer assumptions, becomes somewhat unrealistic close to the well. In assessing non-steady flow, dewatering of the medium results in changes in transmissivity and when the water table is lowered, the storativity is decreased. Thus the application of the non-steady flow equations becomes difficult. However, for small drawdowns over short time periods, the Theis' equation may be used for rough estimates of $s(r,t)$.

Various methods have been derived to deal with the problems of complex non-steady flow in unconfined aquifers and for these the reader is referred to the specialist texts on groundwater (e.g. Bouwer, 1978; Todd, 1980), which also deal with the added difficulties encountered with non-homogeneous and anisotropic aquifers and the common situation where leaky aquifers obtain recharge through an aquitard (Fig. 7.2). This chapter has provided only a general introduction to the complexities of groundwater movement.

References

Bouwer, H. (1978). *Groundwater Hydrology*. McGraw-Hill, 480 pp.

Cedergren, H.R. (1977). *Seepage, Drainage and Flow Nets*. John Wiley, 2nd ed, 534 pp.

Childs, E.C. (1969). *An Introduction to the Physical Basis of Soil Water Phenomena*. John Wiley, 493 pp.

Freeze, R.A. and Cherry, J.A. (1979). *Groundwater*. Prentice Hall, 604 pp.

Horton, R.E. (1940). 'An approach towards a physical interpretation of infiltration capacity.' *Proc. Soil Sci. Soc. Am.*, **5**, 399–417.

Musgrave, G.W. and Holtan, H.N. (1964). 'Infiltration'. Ch. 12 in *Handbook of Applied Hydrology*, ed. V.T. Chow. McGraw-Hill, 30 pp.

Philip, J.R. (1960). 'Theory of infiltration', in *Advances in Hydroscience*, Vol. 5, Academic Press, 216–296.

Pinder, G.F. and Gray, W.G. (1977). *Finite Element Simulation in Surface and Subsurface Hydrology*. Academic Press, 295 pp.

Raudkivi, A.J. and Callander, R.A. (1976). *Analysis of Groundwater Flow*. Arnold, 214 pp.

Rushton, K.R. and Redshaw, S.C. (1979). *Seepage and Groundwater Flow*. John Wiley, 339 pp.

Todd, D.K. (1980). *Groundwater Hydrology*. John Wiley, 2nd ed, 535 pp.

Further Reading

Bowen, R. (1980). *Groundwater*. Applied Science Publishers, 227 pp.

Heath, R.C. and Trainer, F.W. (1981.) *Introduction to Ground Water Hydrology*, Water Well Journal Pub. Co., Worthington, Ohio, 285 pp. (Reprinted Wiley, 1968.)

8

Water Quality

In recent decades, the growth of populations in the developed countries and their rise in living standards have demanded the rapid expansion of water supplies. Engineering hydrologists have been fully occupied in the assessment and development of increasing quantities of water for domestic and industrial use. While natural water resources may be plentiful, especially in the advanced countries, the need for larger storages and for greater water transference to the main centres of consumption has led to increased concern for the quality of the water. For example, the sharing of resources along large rivers serving many centres of population poses great problems. Water for consumption may be taken directly from the river and waste water from the town or city returned to the river. If this is repeated at each large centre from the upper reaches of a river to its mouth then the same water will be treated and re-used several times. Water engineers are charged with the duty of ensuring that the water supplied is non-toxic and of a sufficiently high standard for human consumption and that the waste waters are treated to remove pollutants to bring the water to an acceptable quality for return to the natural river.

In developing countries, particularly in tropical regions, the problem of water quality is usually even more acute, since the collection and treatment of waste water is not so far advanced. In small settlements, sources of drinking water and disposal of sewage are sometimes scarcely separated, and in congested cities lacking adequate drainage, the dangers from water-borne diseases are compounded.

Increasing industrialization of the world's communities has also led to greater pollution of natural sources of water. Some of the large industrial consumers, such as the electricity-generating stations, may only modify the river temperature, but many manufacturing industries, such as paper-making, may strongly pollute the natural streams. Particular attention is now focused on the waste water from factories and industrial sites, since many toxic materials cannot be abstracted by the normal processes used in treatment works. Industrial companies are obliged by law to ensure that their waste water is of a sufficiently high quality to be released to natural surface drainage or be piped into the domestic sewerage system.

In this modern, rapidly changing world, the hydrologist must therefore be concerned with water quality in addition to the traditionally recognized evaluations of water quantities.

8.1   Features of Water Quality

The chemical composition of water, H_2O, is one of the first formulae learnt in chemistry, and its existence in the gaseous, liquid and solid states according to temperature is readily understood. However, it is in its liquid form that the quality of water is of most importance both for the nature of the pollutants it may carry and for the greater use that it affords.

The principal features of water quality in the streams, rivers and lakes with which the water engineer is most concerned may be considered in three main groups — physical, chemical and biological (Tebbutt, 1977).

8.1.1   Physical Features

Solids form the most obvious extraneous matter to be carried along by a flowing river. The quantity and type of solids depend on the discharge and flow velocity. They range from tree trunks, boulders and other trash dislodged and carried away by floods to minute particles suspended in a tranquilly flowing stream. The solid pollutants of a river derive from organic and inorganic sources. When evaluating quality for the potential use of water, solids are measured in mg l^{-1}.

Colour, taste and odour are aesthetic properties of water that are judged subjectively and are caused by dissolved impurities either from natural sources, like the peaty waters from upland moors, or from the discharge of noxious substances into the water course by man.

Turbidity is the term for the cloudiness of water due to fine suspended colloidal particles of clay or silt, waste effluents or micro-organisms, and is measured in turbidity units (FTU) based on the comparison of the scattering of light by a water sample with that of a standard suspension of formazin.

Electrical conductivity (EC) is a physical property of water that is dependent on the dissolved salts. Thus its measurement in microsiemens per centimetre (μS cm^{-1}) gives a good estimate of the dissolved solids content of a river.

Temperature is a standard physical characteristic that is important in the consideration of the chemical properties of water. Its measurement, in °C, in natural rivers is also necessary for assessing effects of temperature changes on living organisms.

Radioactivity in water bodies has increasing attention as its harm to life in all forms becomes more recognized, but its measurement remains a specialized procedure adopted when dangerous doses are suspected.

8.1.2   Chemical Features

Water chemistry is a very extensive subject since water is the most common solvent and many chemical compounds can be found in solution at the temperatures of naturally occurring water bodies. Only a selection of the more significant chemical features will be mentioned here.

pH is a measure of the concentration of hydrogen ions (H^+) and indicates the degree of acidity or alkalinity of the water.

pH = $\log_{10} 1/(H^+)$ and on the scale from 0 to 14 a pH of 7 is indicative of a neutral solution. If the pH is less than 7 then the water is acidic, and if the pH is greater than 7, the water is alkaline.

Dissolved oxygen (DO) plays a large part in the assessment of water quality, since it is an essential ingredient for the sustenance of fish and all other forms of aquatic life. It also affects the taste of water, and a high concentration of dissolved oxygen in domestic supplies is encouraged by aeration. Values of dissolved oxygen are given in mg l^{-1} (O_2).

Biochemical oxygen demand (BOD) is a measure of the consumption of oxygen by micro-organisms (bacteria) in the oxidation of organic matter. Thus a high BOD (mg l^{-1} (O_2)) indicates a high concentration of organic matter usually from waste water discharges.

Nitrogen may be present in water in several forms: in organic compounds (usually from domestic wastes), as ammonia or ammonium salts, in nitrites or fully oxidized nitrates. Measures of nitrogen [mg l^{-1} (N)] give indications of the state of pollution by organic wastes with larger quantities in the nitrate form being an indication of oxidation (purification).

Chlorides, most often occurring in the NaCl common salt form, are found in brackish water bodies contaminated by sea water or in groundwater aquifers with high salt content. The presence of chlorides (mg l^{-1} Cl) in a river is also indicative of sewage pollution from other chloride compounds.

8.1.3   Biological Features

The existence of plant and animal life in rivers and other water bodies is a prime indicator of water quality and it has a different significance for the river engineer and the water supply engineer. The former welcomes the fish population, particularly the lucrative salmon and trout, and the smaller creatures and plants that provide their food. The water engineer, however, finds that growths of algae and populations of small aquatic animals can cause serious problems in pipes, reservoirs and other control works.

Of much greater importance in assessing quality for domestic supplies, is the content of micro-organisms in the water. Many harmful diseases are transmitted by water-borne organisms either within parasitic carriers, like the *Schistosoma* causing bilharzia, or as free-swimming pathogenic bacteria and viruses. These can be isolated and identified only by microscopic examination of water samples. On a routine basis, the common organism *Escherichia coli* (*E. coli*) found in all human excreta is taken as an indicator of sewage pollution. The measure of concentration in a water sample is the most probable number (MPN) per 100 ml which is derived statistically from a number of samples. All supplies of water destined for human consumption must have regular bacteriological examination.

8.2   Measurement of Water-Quality Variables

8.2.1   Sampling

In describing methods of water sampling, attention is focused here on river waters, since these are more the concern of the hydrologist. Choice of sampling site may be governed by an abstraction point or a discharge point associated with an industrial user or waste water treatment works. However it is often practicable to take water-quality samples at a river gauging station. Ideally, a single sample from the well mixed waters downstream of a weir would suffice to give a good representation of the water quality of a small river. At a current meter station, the river should be sampled at several points across the channel and in deep rivers (over 3 m) at 0.2 and 0.8 depths. For shallower streams one sample in the vertical at 0.6 depth should be adequate. (These depths correspond to the points giving the mean flow velocity in a vertical section — see Chapter 6.) Once the flow characteristics of the river are known, the sampling scheme in Fig. 8.1 can be recommended.

The timing and frequency of sampling also need consideration, particularly if there is a regular pattern of flow control or of effluent discharge from industries above the sampling point. It is often worthwhile having a concentrated sampling period when the river regime is steady to establish a regular norm. Then anomalous conditions of flood flows with their increased load of suspended solids, or of unusual influxes of pollutants from accidental spillages, when sampled, can be related in perspective to average water-quality values. Seasonal changes must also be identified in any water-quality variations. In addition to the establishment of the average water-quality characteristics of a river, it is becoming important for Water Authorities, in the interests of environmental conservation, to be vigilant at all times in maintaining satisfactory river water-quality values. Hence there has been a rapid development of continuous monitoring of water quality, which will be described later.

For taking single samples of the river water, standard instruments have been devised (Fig. 8.2a). The displacement sampler, shown in Fig. 8.2b, recommended for the determination of DO can be used for general sampling in open channels. The inlet is opened when the container is at the required depth and the water is fed into the bottom of the bottle. When the whole container is full the water flows from the exit, and the flow should continue until the bottle contents have been changed several times before the sampler is removed. Numerous sampler designs have been adopted in conjunction with studies of suspended solids, the latter having received particular attention in the United States. A selection of these is shown in Fig. 8.3. In each container there is a glass or polythene bottle.

When obtaining samples for chemical analysis, great care must be taken against contaminating the water sample; all containers, even the simple bucket dipping into a turbulent well mixed stream, must be washed out with the flowing river water before being used for a sample. The temperature of the water must be taken at the time of sampling. Sampling bottles must be carried in suitable crates and delivered to the laboratory the same day. Delay

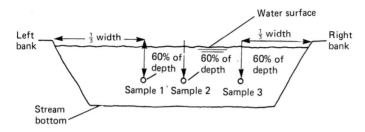

(a) Selected sampling points for a deep, wide stream with a uniform depth

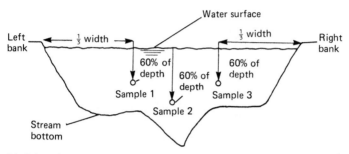

(b) Selected sampling points for a deep, wide stream with a non-uniform depth

Fig. 8.1   (Reproduced from N.L. Nemerow (1974) *Scientific Stream Pollution Analysis*, by permission of Hemisphere Publishing Corporation.)

in carrying out the analyses can result in spurious values, since some of the chemical properties of the water can be altered by the changing conditions in storage.

8.2.2   Physical Measures

Some of the physical measurements, such as water temperature and electrical conductivity, are carried out in the field at the time of sampling. For most of the variables, more sophisticated treatment in the laboratory is necessary. Rating scales, based on the impurities in the water, have been devised for the aesthetic features, viz colour, taste, odour and turbidity. The determination of total suspended solids requires the filtering of the water sample, the drying of the solids in an oven at 105 °C for an hour and subsequent weighing. If the suspended material is very fine, its separation is made in a centrifuge before drying and weighing.

8.2.3   Chemical Analyses

Although there are now instruments to measure some chemical properties directly, for example the pH meter, most measures of the chemical contents of water must be made by laboratory analyses of samples. It is far beyond the scope of this chapter to describe the analytical techniques for all the chemicals likely to be found in river water, but a brief outline of the main types of

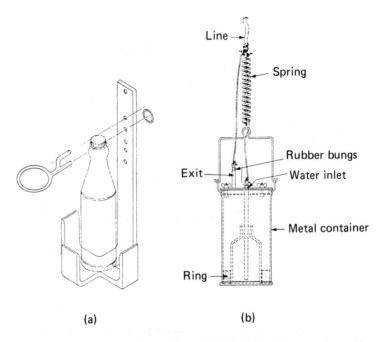

Fig. 8.2   Water quality samplers. (a) Sample bottle holder for manual sampling. (b) Displacement sampler for dissolved oxygen determination. [(a) reproduced from UNESCO-WHO (1978) *Water Quality Surveys*, by permission. (b) reproduced from Department of the Environment (1972) *Analysis of Raw, Potable and Waste Waters*, by permission of the Controller, Her Majesty's Stationery Office, © Crown copyright.]

analysis used would benefit the hydrologist. The analytical methods may be grouped under four headings: titrimetric, colorimetric, spectral and potentiometric (Barnes *et al.*, 1981). Titrimetric analysis, the well known balancing of reactions using coloured indicators can give satisfactory results down to 1 mg l⁻¹ in determining, for example, DO or alkalinity. Colorimetric analysis uses the property of a solution that its colour intensity is related to the concentration of the compound forming the solution. There are several experimental techniques used in colorimetry that are reliable for measuring such chemicals as ammonia and chlorine as long as they are not mixed with other compounds with similar coloration. Both of these analytical techniques are relatively economical in practice.

Spectral and potentiometric methods of analysis employ the more sophisticated equipment of physical chemistry. In the spectral method, the light intensity of the particular metal ions in a flame (e.g. the sodium yellow) is measured and compared with the light intensity from a known standard solution of the metal ions. In its simple form, the method is applicable for sodium, potassium, lithium, calcium and magnesium. A more advanced technique, atomic absorption spectrophotometry, can measure small concentrations well below 1 mg l⁻¹ of many other heavy metals (copper, lead,

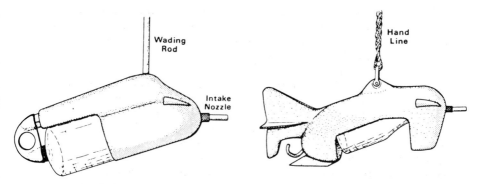

Fig. 8.3 Samplers for suspended solids. (Reproduced from K.J. Gregory & D.E. Walling (1973) *Drainage Basin Form and Process*, by permission of Edward Arnold.)

zinc, etc.), but the instrument is costly and only found in advanced well equipped laboratories. The potentiometric method relates to the simple electric cell. A metal electrode set into a solution of a metallic salt generates a difference in electrical potential. The concentration is related to, and obtained by measuring, the potential difference between the metal electrode and the unknown concentration of metallic ions in the solution. The pH meter is based on this principle.

For greater detail of analytical methods, particularly in respect of organic material, the reader is referred to the specialist texts (DOE, 1972; Barnes *et al.*, 1981).

8.3   Water Quality Records

Before the England and Wales Water Act, 1973, measurements of water quality in these countries were made by several organizations. The River Authorities responsible for water resources, land drainage, fisheries and the overall well-being of the main rivers, made regular spot samples. These were usually carried out by fisheries departments in the rural areas and pollution departments in the urban and industrial areas. The Water Boards and Water Companies employed their own chemists and bacteriologists to sample the raw water and check on the treatment processes for the domestic piped water supply. The local authorities responsible for sewage disposal maintained their own laboratories and staff for assessing the quality of the final effluent for discharge to the rivers.

The whole organization of water quality measurement in the UK is now the responsibility of the Regional Water Authorities and there is a tremendous impetus in this work. The water quality sampling stations on the main rivers have been continued and the number of stations greatly expanded throughout the country. Analytical work has been centralized, where possible, in each of the Water Authorities and laboratories have been upgraded with the latest equipment and instruments. Many of the analyses are still directed towards the monitoring of water supplies and waste water effluents, but increasing attention is being paid to the continuous measurement of water quality in the rivers.

Some of the water quality sampling stations on the main rivers have now many years of records. The number of measured variables has increased as further details are required. In the UK these records are sent to a central authority, the Water Data Unit (Department of the Environment), and are available there for wider study. Some of the data are published regularly and a sample is given in Table 8.1. The 1978 records of a selection of water quality measurements are shown from stations on six main rivers. The choice of rivers ranges from the Tees, rising in the highest parts of the Pennine moors in north-east England, through the industrially affected Trent and Don, the Bedford Ouse and Thames representative of Lowland Britain, to the Exe, rising on Exmoor in the extreme south-west of the country. The values of the water quality variables are annual averages computed from sample measurements, or continuous records where such recording instruments are installed. The variables are those of greatest importance in the assessment of river water quality and most commonly measured by the Authorities. The poor quality of the Don stands out, with high conductivity and the highest concentrations of suspended solids (SS), BOD, NH_3 and chlorides. The River Exe flows through a truly rural area and its waters are of the highest quality, needing little treatment before supplying the City of Exeter. The Bedford Ouse and Thames also drain predominantly rural countryside, but their velocities are lower and they receive the effluents of large cities and towns as well as the water from intensively fertilized arable lands.

In the interests of river quality on a nationwide scale, central government has instigated overall surveys from time to time. The latest report for England and Wales was published for the year 1980 by the National Water Council (NWC, 1981). It records steady increase in the length of unpolluted water courses since a similar survey in 1958. Of the 39 880 km of rivers and canals surveyed in 1980 in England and Wales, 69% are described as of good quality according to the NWC river classification and only 2% are of bad quality, i.e. waters grossly polluted and likely to cause nuisance. The improvements in water quality are attributed to increased investment in new and upgraded sewage treatment works and trunk sewers. Another reason for the improvements has been the increased waste water treatment by industry, and in some areas, the decline in industrial production has resulted in less pollution reaching the rivers.

8.4   Water Quality Requirements

The standard of water quality varies considerably according to the intended water use. The industrial requirements for cooling water depend on the plant construction; steel and concrete must not be exposed to corrosive elements, so the pH should be greater than 7 and the chloride content can become troublesome over 200 mg l^{-1}. The quantity of DO should be negligible and

Table 8.1 Water Quality Statistics for Selected Rivers in the UK (1978): Annual Average Values

(Reproduced from Department of the Environment (1980) *Water Data 1978.*)

River	Measuring authority	Temp. (°C)	pH	EC (µS cm⁻¹)	SS (mg l⁻¹)	DO (mg l⁻¹ O)	BOD (mg l⁻¹ O)	NH₃ (mg l⁻¹ N)	Nitrate (mg l⁻¹ N)	Chloride (mg l⁻¹ Cl)
Tees	Northumbrian	8.4	7.8	355	19.8	11.19	2.4	0.198	3.73	33.3
Trent	Severn – Trent	11.3	7.8	1062	24.9	10.85	4.2	0.254	8.72	133.5
Don	Yorkshire	12.1	7.5	1433	35.4	8.28	4.6	4.771	7.15	218.8
Bedford Ouse	Anglian	11.0	8.1	845	20.8	9.77	3.1	0.151	10.04	63.5
Thames	Thames	11.2	8.0	553	22.1	9.47	—	0.376	7.37	40.3
Exe	South-West	10.9	7.3	167	8.6	11.00	1.9	0.056	2.53	18.1

the biological content should be removed by chlorination (James, 1976). Where water is consumed in an industrial process, there is usually a maximum desired limit for total dissolved solids; a low pH of 6.0 – 6.8, indicative of a 'soft' water, is a requirement for laundries. Most water used in agriculture in the UK becomes a source of pollution since it is mainly applied to the cleansing of dairies and animal housing, and its initial quality is not of great consequence. However when used for irrigation, dissolved solids and some metallic elements may be detrimental to the crops grown.

The most stringent criteria applied to water quality are those considered in the provision of domestic supplies. All piped water supplied by Water Authorities and Water Companies in the UK is wholesome and safely potable. The actual quality of the water varies over the country according to the source of supply and its treatment, but the consumer is safeguarded against toxic substances and bacterial infection. With the increase in the quantities of complex chemicals in agricultural fertilizers, herbicides and pesticides, domestic detergents and petrochemical by-products being introduced into the environment, there has been an increase in concern in recent years for the safety of water supplies.

The World Health Organization (WHO) recognizes the dangers on the international scale and has published recommended standards for drinking water (WHO, 1984). In general, the 'highest desirable' levels and the 'maximum permissible' levels of concentration of the various water quality features have been laid down. In addition to the characteristics already described, the concentration of toxic chemicals, e.g. some of the heavy metals and the complex organic compounds, are specifically quoted.

There has been considerable discussion among British water engineers and chemists on the desirability of adopting the international standards as statutory requirements in this country since much of the domestic water supply comes from relatively clean upland streams or deep groundwater sources and the origins of possible pollutants are well noted (Goodman,

Table 8.2   International Water Quality Guidelines
(Reproduced from WHO, 1984.)

Contaminant	Unit	Guideline value	Comment
Microbiological			
Faecal coliforms	no./100 ml	0	For all drinking water.
Coliform organisms	no./100 ml	3	In occasional samples of distribution system.
Coliform organisms	no./100 ml	10	In unpiped supplies, not to occur repeatedly.

Inorganic

Arsenic	mg/l	0.05
Cadmium	mg/l	0.005
Chromium	mg/l	0.05
Cyanide	mg/l	0.1
Fluoride	mg/l	1.5 Natural or additive.
Lead	mg/l	0.05
Mercury	mg/l	0.001
Nitrate	mg/l (N)	10
Selenium	mg/l	0.01

Organic

Aldrin & dieldrin	μg/l	0.03
Benzene	μg/l	(10)
Benzo(a) pyrene	μg/l	(0.01)
Carbon tetrachloride	μg/l	(3)
Chlordane	μg/l	0.3
Chloroform	μg/l	(30)
2,4-D	μg/l	100
DDT	μg/l	1
1,2-dichloroethane	μg/l	(10)
1,1-dichloroethene	μg/l	(0.3)
Heptachlor	μg/l	0.1
Hexachlorobenzene	μg/l	(0.01)
Gamma-HCH (lindane)	μg/l	3
Methoxychlor	μg/l	30
Pentachlorophenol	μg/l	10
Tetrachloroethene	μg/l	(10)
Trichloroethene	μg/l	(30)
2,4,6-trichlorophenol	μg/l	(10)

( ) = Estimated values, could differ by two orders of magnitude.

Aesthetic quality

Constitute or characteristic	Unit	Guideline value
Aluminium	mg/l	0.2
Chloride	mg/l	250
Copper	mg/l	1.0
Hardness	mg/l (CaCO₃)	500
Iron	mg/l	. 0.3
Manganese	mg/l	0.1
pH		6.5–8.5
Sodium	mg/l	200
Solids-total dissolved	mg/l	1000
Sulphate	mg/l	400
Zinc	mg/l	5.0

1975). However certain proposed quality levels have been set, related to the category of the water sources and the degree of treatment they require. There are four categories of raw water defined as follows:

A$_1$ Quality requiring only simple rapid filtration and disinfection.

A$_2$ Water requiring normal treatment (prechlorination, coagulation, flocculation, filtration and final chlorination).

A$_3$ Water requiring intensive physical and chemical treatment (additional processes of settlement, slow filtration, activated carbon treatment and more thorough disinfection).

A$_4$ Worst category requiring biological treatment to remove ammonia or need for desalination – advanced treatment to make it potable.

A selection of the UK proposed values is given for the first three categories of raw water treatment in Table 8.3 which illustrates how the more stringent treatments can accept poorer quality water to bring it to an acceptable potable standard. In providing safe drinking water, the strict standards for the biological features must be observed. For all bacteriological coliforms, in 95 % of the samples there should be 0/100 ml and no more than 10/100 ml in the remaining 5 % of the samples. For *E. coli*, the comparable numbers allowed are 0/100 ml and no more than 2/100 ml.

Considerable research is in progress with regard to many other elements and compounds, both organic and inorganic, that may now be found in rivers from which domestic supplies are extracted. Investigations are also being carried out on the more sophisticated treatment processes required for waste waters to remove the toxic materials before discharge of effluents into the surface streams. Standards for waste water effluents are made in the UK by considering each potentially polluting discharge in relation to the local surface water conditions. For treated domestic sewage effluents, the Royal Commission on Sewage Disposal recommended that the suspended solids content should not exceed 30 mg l^{-1} and the BOD of the sewage effluent should not exceed 20 mg l^{-1}. It is assumed that the dilution of the effluent in the stream would give a resulting BOD concentration of less than 2 mg l^{-1}. The permitted quality of effluent can be varied according to the amount of dilution in the stream but the resulting BOD should not exceed 4 mg l^{-1}. These recommendations are flexible and can be applied to meet the needs of local conditions (Water Research Centre, 1979).

8.5   Monitoring

The advances made in scientific instrumentation have encouraged the development of automatic water quality systems, which are gradually replacing laborious manual methods and providing continuous observations. With appropriate sensors taking measurements once an hour or even once every 15 min and the information stored on a data logger, a serious deterioration in water quality is recorded and downstream users can be forewarned. Depending on weather conditions, and season, rivers have a certain self-purification capacity in the assimilation of organic matter, and by having a continuous

Table 8.3 UK Proposed Raw Water Quality Levels

Figures in brackets are concentrations of elements that would be harmful to humans and therefore should not be exceeded. (Reproduced from A.H. Goodman (1975) in *Symposium on Maintenance of Water Quality*, IWES.)

	Raw water treatment category					
	A_1		A_2		A_3	
	Recommended	Normal	Recommended	Normal	Recommended	Normal
pH	6.5 – 8.5	4.0 – 9.2	4.0 – 9.2		4.0 – 9.2	
Odour, taste			Not objectionable			
Temperature (°C)	22.5	25	22.5	25	22.5	25
Ammonia (mg l⁻¹ NH₄⁺)	0.5	1.0	0.5	1.0	0.5	1.0
Chlorides (mg l⁻¹)	200	(600)	200	(600)	200	(600)
Phosphates (mg l⁻¹)	0.30	0.75	0.50			
Cyanide (mg l⁻¹)		0.05		0.05		0.05
Copper (mg l⁻¹)	0.05		0.25		0.25	
Zinc (mg l⁻¹)	5		10		15	
Cadmium (mg l⁻¹)		(0.005)		(0.005)		(0.005)
Chromium (mg l⁻¹)		(0.025)		(0.025)		(0.05)
Lead (mg l⁻¹)		(0.05)		0.05		0.05
Arsenic (mg l⁻¹)		(0.05)		(0.05)		(0.05)

measurement of the water quality, the ability of a stream to take further effluents can be assessed. Thus a full knowledge of the ever-changing state of the rivers is essential for water resources management.

Sensors for the water quality variables have developed gradually over the years. pH meters were used in the 1930s in both water and waste water treatment works; then in the 1960s, several authorities began to install recording thermometers and sensors for DO and suspended solids (SS). An automatic analyser for ammonia was also introduced. Gradually a water quality monitoring station was established with the river water being pumped to the sensors from a suitable sampling point and with all the instruments housed in a building on the river bank. One of the first of these monitoring stations was on the River Trent (Lester and Woodward, 1972).

Encouraged by the Water Resources Board, the former central authority, several commercial firms built water quality monitors to recommended specifications. The different makes with the variables they measure are given in Table 8.4. Certain companies will include sensors for a wide range of variables to suit clients' requirements, but all include the five standard measures, DO, temperature, SS, EC and pH in the basic package. To emphasize the advantages of water quality monitoring over manual sampling, comparable results are given for a month from the River Trent (Table 8.5). It will be noted that the monitor records the extremes, i.e. the maximum and minimum values, which have been missed by the spot sampling, but that the average values are not too dissimilar.

In recent years, there have been extensive studies of the benefits and shortcomings of water quality monitoring. A Harmonized Monitoring Scheme has been initiated by the Department of the Environment to encourage a national standard in the measurements, the chemical analyses and the statistical analyses of the data to ensure comparability of results over the country. A great diversity in methods of sampling and in chemical analyses exists from one authority to another and there remain considerable differences of opinion on the methods and techniques to use. Further research on the problems involved and trials of new instruments continue.

Table 8.4   Water Quality Monitors

Maker	Variables measured
Electronic Instruments Ltd. (EIL)	1) DO, temp., SS, EC, pH 2) DO, temp., SS, EC, pH, NO_3, Cl, etc.
Protech Advisory Services Ltd.	1) DO, temp., SS, EC, pH 2) DO, temp., SS, EC, pH, Cl, NO_3, cyanide, etc.
Plessey Co. Ltd.	DO, temp., SS, EC, pH
Quality Monitoring Instruments Ltd.	DO, temp., SS, EC, pH
Technicon Instruments Ltd.	DO, temp., SS, EC, chemical oxygen demand, metals, phenols, phosphate, sulphate, etc.

Table 8.5   Statistics from Manual Sampling and Automatic Monitoring, R. Trent, April 1969

(Reproduced from W.F. Lester and G.M. Woodward (1972) *Water Pollution Control*, **71**, 289–298.)

	Maximum	Minimum	Average
Temperature (°C)			
Manual	16.5	12.0	14.9
Monitor	18.0	8.0	13.6
SS (mg l^{-1})			
Manual	97	12	29
Monitor	210	9	34
EC (μS cm^{-1})			
Manual	1100	710	890
Monitor	1100	580	870
DO (mg l^{-1} (O))			
Manual	7.8	7.6	7.7
Monitor	10.4	5.0	7.5
Ammonia (mg l^{-1}N)			
Manual	2.2	0.7	1.3
Monitor	2.9	0.4	1.4

8.6   Modelling

The need to forecast from inadequate data, a problem that is always present for the hydrologist, also pertains in the realms of water quality. The contemporary solutions to this problem are found in mathematical modelling aided by the capabilities of the digital computer. Water resources engineers having chosen for their rivers a satisfactory hydrological model from the large number of available mathematical representations of catchment runoff, are now turning their attention to modelling the complex physical chemical and biological processes at work in river waters. This new challenge is stretching the ingenuity of numerous researchers and some well founded models are appearing for forecasting water temperatures, DO and BOD (Rinaldi *et al.*, 1979). The natural development of successful modelling is the application of the model to aid in the control of water quality which then becomes an integral part of the whole of water management within a designated area (Young & Beck, 1974).

References

Barnes, D., Bliss, P.J., Gould, B.W. and Vallentine, H.R. (1981). *Water and Waste Water Engineering Systems*. Pitman, 513 pp.

Department of the Environment (DOE) (1980). *Water Data 1978*. Water Data Unit, Reading.

Department of the Environment (DOE). (1972). *Analysis of Raw, Potable and Waste Waters*. HMSO, 305 pp.

Goodman, A.H. (1975). 'Raw water quality criteria', in *Symposium on Maintenance of Water Quality*. IWES, 142 pp.

Gregory, K.J. and Walling, D.E. (1973). *Drainage Basin Form and Process*. Arnold, 456 pp.

James, A. (1976). 'Water quality', Chapter 7 in *Facets of Hydrology*, ed. J.C. Rodda, John Wiley, 368 pp.

Lester, W.F. and Woodward, G.M. (1972). 'Water quality monitoring in the United Kingdom, *Water Pollution Control*, **71**, 289–298.

National Water Council (NWC). (1981). *River Quality: The 1980 Survey and Future Outlook*. National Water Council, December 1981, 39 pp. and maps.

Nemerow, N.L. (1974). *Scientific Stream Pollution Analysis*. Scripta Book Company (Hemisphere Publishing Corporation), 358 pp.

Rinaldi, S., Soncini-Sessa, R., Stehfest, H. and Tamura, H. (1979). *Modelling and Control of River Quality*. McGraw-Hill, 380 pp.

Tebbutt, T.H.Y. (1977). *Principles of Water Quality Control*. Pergamon Press, 201 pp.

UNESCO–WHO (1978). *Water Quality Surveys. Studies and Reports in Hydrology 23*, UNESCO–WHO, 350 pp.

Water Research Centre (1979). *Emission Standards in Relation to Water Quality Objectives*. WRC, Technical Report 17, 72 pp.

World Health Organization (1984). *Guidelines for Drinking-water Quality*, Vol. 1 Recommendations, WHO, Geneva.

Young, P. and Beck, M.B. (1974). 'The modelling and control of water quality in a river system, *Automatica*, **10**, 455–468.

9

Data Processing

9.1 General Considerations

Advances in scientific hydrology and in the practice of engineering hydrology are dependent on good, reliable and continuous measurements of the hydrological variables. The measurements are recorded by a wide range of methods, from the simple writing down of a number by a single observer to the invisible marking of electronic impulses on a magnetic tape. The young engineer may be faced with the problem of using information assembled by any of the many different methods, and indeed the newly trained hydrologist often finds that his first job entails the organization of a data receiving and processing system. Although the most advanced techniques are used in the developed countries, many emerging nations of the third world employ only direct manual methods.

This chapter sets out to cover the wide field of data processing and quality control of the observations, and does so by concentrating on the two main hydrological variables, rainfall and river flow. Handling basic records of other variables is also the concern of the hydrologist, especially when out on site in remote areas, but many of the principles and techniques outlined for rainfall and river flow are applicable to such other variables, e.g. evaporation and groundwater measurements. However, on a national scale, the routine meteorological measurements used in evaporation calculations are generally processed by the experts in the meteorological service and the groundwater measurements are handled by headquarters geologists. Similarly, water-quality measurements tend to the centralized in analytical laboratories where there is concern to establish uniform analytical procedures to ensure homogeneous and comparable data.

9.2 Rainfall Data

Rainfall data are assembled from the measurements made with the range of instruments described in Chapter 3.

Storage gauges. Observations are made at 0900 h GMT or LMT each day (Daily gauges) and are allocated to the previous day, or at 0900 h on the first day of each month (Monthly gauges) and regarded as the previous month's total. The rainfall measurement is noted immediately in a pocket register or on a special form (Fig. 9.1). In the UK if the observer contributes the records directly to the Meteorological Office or to a regional Water Authority, the necessary forms will be provided; such documentation should always be

METEOROLOGICAL OFFICE DATA SHEET

Metform 1091
[Revised 1982]

REGISTER OF RAINFALL, 198.......

Station name .... Met.O.Stn No.

County .... Complete site details should be recorded overleaf. WA Stn No. ....

Regular observation time in winter .... [clock time] and during British Summer Time at .... [clock time]

THE MORNING MEASUREMENT MUST BE ENTERED TO THE PREVIOUS DAY – See instructions overleaf

RAINFALL DATE	Jan	Feb	Mar	Apr	May	Jun	Jly	Aug	Sep	Oct	Nov	Dec	LEAVE BLANK	RAINFALL DATE
	mm	mm	mm	mm	mm	mm	mm	mm	mm	mm	mm	mm		
1														1
2														2
3														3
4														4
5														5
6														6
7														7
8														8
9														9
10														10
11														11
12														12
13														13
14														14
15														15
16														16
17														17
18														18
19														19
20														20
21														21
22														22
23														23
24														24
25														25
26														26
27														27
28														28
For *Leap Years 29		*												29
30														30
31														31
TOTALS														TOTALS

This summary is for local reference only – it should NOT be forwarded to the Meteorological Office unless specifically requested.

Dd 8229223 2/82 WS883

ANNUAL TOTAL

Fig. 9.1 Recommended form for daily rainfall measurements. (Reproduced by permission of the Controller, Her Majesty's Stationery Office, © Crown copyright.)

supplied to observers by an authority requiring the information. At the end of each month, the values are entered on to a data sheet and sent to the relevant collecting centre: a station reporting rainfall only often uses a postcard of the type shown in Fig. 9.2. If observations for one or more days have been missed, these occasions are carefully noted on the form and card and the accumulated total measured at the next observation is indicated with the missing days bracketed together. Observers are encouraged to measure and note, in 'Remarks on the Days', heavy falls of rain over short periods. It is sometimes difficult to make the measurements exactly at 0900 h at remote monthly gauges, and a helpful observer will note the actual time of observation. This is important on very wet days when some of the day's rain after 0900 h could be allocated in error to the previous month.

Autographic gauges. The Dines Tilting Syphon Rain Recorder draws a pen trace on a daily chart or, if fitted with a strip chart, on a larger time scale chart which may run for a week or a month (Fig. 9.3). A daily storage gauge kept near the Dines recorder provides the 'true' observation and its measure-

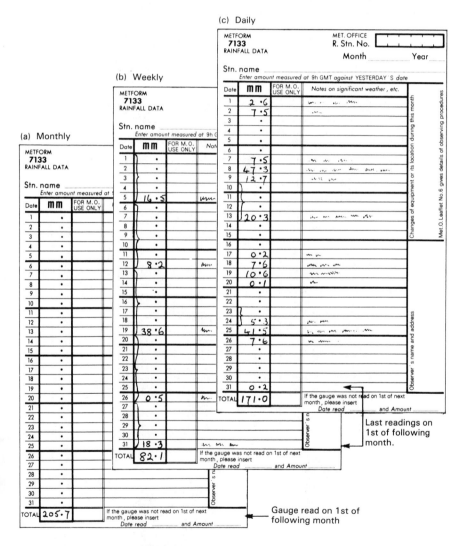

Fig. 9.2 Rainfall postcards. (Reproduced by permission of the Controller, Her Majesty's Stationery Office, © Crown copyright.)

Fig. 9.3 Dines tilting syphon rain recorder charts. (Apparatus reproduced from K. J. Gregory & D. E. Walling (1973) *Drainage Basin Form and Process*, by permission of Edward Arnold. Drum chart by permission of Casella London Ltd.)

ments are submitted as for storage gauges. The autographic charts (hyetograms) require careful processing by experienced analysts. Basic data, i.e. rainfall amounts occurring in each hour and their actual duration, can be abstracted and recorded on hourly tabulation forms for a month at a time. Information more useful to the hydrologist is the maximum depths of rain falling in fixed time periods (5, 15, 30, 60 and 120 min) or the minimum duration of falls of 5, 10, 15, 20 and 25 mm, which can be abstracted and entered on a special form (Fig. 9.4). From these data, peak rainfall amounts over selected time periods can be identified and, given sufficient years of records, return periods of critical falls can be evaluated (see Chapter 10). Similar analyses of strip charts provide information on the high rainfall intensities over periods of minutes required in urban drainage design.

Tipping-bucket gauges.   The mechanism of these gauges (Chapter 3) provides counts equivalent to 0.2 mm, 0.5 mm or 1.0 mm of rain over fixed time intervals such as 1 min., 5 mins. or 15 mins., which are recorded on magnetic tape or in solid state memory systems. To guard against data loss by instrument failure, the monthly storage total is measured separately and if required is submitted as with storage gauges. Otherwise, the data are read from a counter which may be interrogated by telephone for immediate use, or the magnetic tape is processed on a digital computer to give amounts in required time intervals.

9.3   Rainfall Data Quality Control

The worth of rainfall data depends primarily on the instrument, its installation, its site characteristics and its operation by a responsible observer. It is essential for a hydrologist using the data to have direct knowledge of a rain gauge station, and authorities are recommended to keep a 'history' of each station. Such information may be compiled and recorded on a form such as shown in Fig. 9.5 on which will be noted a space for a photograph of the rain gauge site from a known orientation. A well documented up-to-date history of a rain gauge station is invaluable in assessing the reliability of the rainfall measurements and is an important first step in quality control.

Before using or storing rainfall records, their validity should be checked immediately on receipt at the central authority office. In the UK, Water Authorities assemble the records in their areas and a certain amount of careful scrutiny takes place in each office although in general the UK Meteorological Office prefers to receive the observer original values. Some authorities are adopting rigorous checks on the data, but the most comprehensive quality control is carried out by the Meteorological Office. This central government authority for meteorological affairs has a very high international reputation and it has particularly made its mark by leading the world in the development of a computer system for handling the records from over 6000 rain gauge stations (Bleasdale & Farrar, 1965). The careful manual methods of station comparison and regional mapping to identify errors and anomalies have been gradually automated and adapted to computer methods (Shearman, 1975). The general procedure is outlined here,

Metform 7159 (Revised 1983)

RAINFALL STN NO	MONTH	YEAR
		198

Stn name..

INDIVIDUAL STN THRESHOLD COPIED FROM I.S.G.D.

5 min	15 min	30 min	60 min	120 min
mm	mm	mm	mm	mm

Col.	1 Minimum duration 5·0mm	2 Minimum duration 10·0mm	3 Minimum duration 15·00mm	4 Minimum duration 20·0mm	5 Minimum duration 25·0mm	6 Greatest amount in 5 min		7 Greatest amount in 15 min		8 Greatest amount in 30 min		9 Greatest amount in 60 min		10 Greatest amount in 120 min		
DATE	hr : min	hr : min	hr : min	hr : min	hr : min	mm and tenths	Commenced Time GMT HHmm	mm and tenths	Commenced Time GMT HHmm	mm and tenths	Commenced Time GMT HHmm	mm and tenths	Commenced Time GMT HHmm	mm and tenths	Commenced Time GMT HHmm	DATE
1	:	:	:	:	:											1
2	:	:	:	:	:											2
3	:	:	:	:	:											3
4	:	:	:	:	:											4
5	:	:	:	:	:											5
6	:	:	:	:	:											6
7	:	:	:	:	:											7
8	:	:	:	:	:											8
9	:	:	:	:	:											9
10	:	:	:	:	:											10
11	:	:	:	:	:											11
12	:	:	:	:	:											12
13	:	:	:	:	:											13
14	:	:	:	:	:											14
15	:	:	:	:	:											15
16	:	:	:	:	:											16
17	:	:	:	:	:											17
18	:	:	:	:	:											18
19	:	:	:	:	:											19
20	:	:	:	:	:											20
21	:	:	:	:	:											21
22	:	:	:	:	:											22
23	:	:	:	:	:											23
24	:	:	:	:	:											24
25	:	:	:	:	:											25
26	:	:	:	:	:											26
27	:	:	:	:	:											27
28	:	:	:	:	:											28
29	:	:	:	:	:											29
30	:	:	:	:	:											30
31	:	:	:	:	:											31

Tabulated by............& checked by...........at.........; Rec'd at Met.O................ Pre-vet by...........; PCK Data Entry by ....date.........; Verified by ....date.......

Fig. 9.4 Forms for data from autographic gauges (Metform 7159 (1983) reproduced by permission of the Controller, Her Majesty's Stationery Office, © Crown copyright.)

PARTICULARS OF RAIN GAUGE STATIONS

STATION .... STN. NO. ....

CATCHMENT .... N. G. REF. ....

AVERAGE 1941-70 .... HEIGHT ....
(based on ....)

RECORD ....

OBSERVER ....

....

....

GAUGE   Type .... Diam of Funnel ....

Height of top of gauge above ground ....

HISTORY ....

....

....

....

PHOTOGRAPH   SITE COMMENTS

....

....

....

....

....

....

....

....

....

....

....

Mean	J	F	M	A	M	J	J	A	S	O	N	D	YEAR
1941-70														

Fig. 9.5   Example of a rainfall station history.

but the program details are constantly modified and improved to take advantage of the increased facilities of new computer technology as it comes into service.

9.3.1   The Meteorological Office Quality Control Program

Rainfall depths in the UK are now measured in millimetres and tenths of a millimetre (e.g. 16.4 mm), but some records are still received in inches and hundredths (e.g. 0.43 in); such daily values are converted to mm, and the monthly total is the sum of the daily mm values.

There are several types of errors that can occur on an observer's postcards. On a first inspection, some of these may be identified and corrected at once, some are noted and marked and others may remain undetected. They are:

(a) Misreadings, misplaced decimal points, copying errors and arithmetic mistakes.

(b) Accumulated readings over several days entered as if a one-day total.

(c) Correct readings entered on wrong days, persistently or only occasionally.

(d) Inadvertant omissions of observations made but not noted.

(e) Occasional errors due to temporary disturbance of the gauge or its exposure.

(f) Systematic errors caused by gradual alteration of exposure over a long period (years) or a leaking gauge with increasing losses.

The powerful computing system of the UK Meteorological Office, COSMOS, comprises three linked computers, a CDC Cyber 205 and two IBM 3081Ds. There are a large number of peripherals – disk drives, magnetic tape drives, printers and plotters and terminals (Meteorological Office, 1987a). One of the IBM 3081D machines is regularly available for data processing services and the rainfall quality control is carried out on a time sharing option by the Archive Data Management System (ADMS) (Meteorological Office, 1987b). This is a suite of over 30 separate programs designed to carry out complex operations concerned with data handling, making appropriate comparisons with data from neighbouring stations or producing maps or tabulated values.

Data are received by the rainfall section of the Meteorological Office in four ways:

(a) Monthly returns from observers,e.g. postcards.

(b) Computer tapes from Water Authorities.

(c) Cachettes from solid state event recorders.

(d) Computer copying from other data sets, e.g. climatic network returns.

The assemblage of the rainfall data into appropriate dataset files is carried out by nine of the ADMS programs. Before and during the keying-in of the postcard observations, minor difficulties such as poor legibility or missing zeros are identified and made good. Errors in monthly totals cause rejection of records and these are corrected and resubmitted. Monthly returns for hourly rainfall, Water Authority data on tapes, the automatically logged data on solid state event recorder cachettes and the transfer of rainfall values from other data sets are all handled by the data entry set of programs in ADMS.

Twelve of the ADMS programs are concerned with quality control.

Towards the end of a month, when about 70% of the station records for the previous month have been counted on the daily rainfall file, the stations are separated into five designated areas and a first quality control program is run. The program processes all daily gauges, identifies spurious data, makes appropriate alterations and 'flags or marks' corrected values. Printouts and daily rainfall maps are produced for the five areas to aid subsequent manual quality control.

The basis for checking the daily rainfall at a station is interpolation between rainfall values at surrounding stations (Shearman, 1975). A maximum of eight neighbouring stations evenly distributed within an area of radius 25 km around the station being tested are selected first; suspect values are then discarded leaving a maximum of six consistent daily totals. Each of the daily totals is converted to a percentage of the relevant station's annual average rainfall and the interpolated percentage rainfall (R) for the testing station is given by:

$$R = \frac{\Sigma(R_i/D_i^2)}{\Sigma(1/D_i^2)}$$

where R_i = rainfall percentage of annual average at station i
and D_i = distance between station i and testing station (km)

The estimated value from the interpolated percentage and the average annual rainfall for the station is compared with the recorded value and the difference (DIFF) is considered insignificant and the record acceptable if DIFF $\leqslant 2.5$ mm and DIFF $\leqslant 2 \times$ error in estimate.

This checking procedure is applied to all the daily values at each rain gauge station and it identifies and automatically corrects:

(a) wrong days, with an additional day-shift mechanism to allocate the rainfall to the correct day;
(b) indicated accumulations with a simple proportioning routine to allocate the measured total among the rain days;
(c) unindicated accumulations which requires a searching procedure before applying the apportioning routine;
(d) transposed values; and
(e) erroneous daily totals due to incorrect time of observation, particularly when there has been heavy rainfall.

The 'flagged' amended daily values then undergo manual quality control with the aid of the plotted maps, a chronicle of the past month's weather and the Daily Weather Reports, particularly when the weather has been showery or when a wet spell has extended over the end of a month. The daily quality control program is run again as required on receipt of late batches of post-cards or other returns.

A comparable quality control program is run for monthly data from all gauges, daily and monthly, and again maps are plotted, for each month, on a regional basis giving actual amounts or percentages of average annual rainfall as further aids to manual quality control.

An annual quality control program is run each year as a check on the sequential homogeneity of the data and to make sure that daily or monthly amendments have resulted in acceptable annual totals.

A number of miscellaneous programs in ADMS take care of subsidiary operations such as the updating of station information on the Rainmaster file, the provision of backup data tapes, the feedback of data to Water Authorities and the interchange of information with other Meteorological Offices.

9.4   Determination of Rainfall Averages

The average annual rainfall calculated from the rainfall (and other forms of precipitation) measured at a rain gauge station is a most important hydrological statistic. Sometimes called the rainfall *normal*, it is a measure of the general rainfall characteristics of the site. Formerly, the absolute value of the normal in inches or millimetres was used alone by engineers in assessing rainfall quantities, often for lack of specific data over durations other than a year, and many empirical formulae were based on this annual value. With increased data handling facilities, engineering hydrologists are encouraged to use original basic data for solving problems, but the average annual rainfall remains a valuable statistical tool as demonstrated in the previous section.

The value of the average annual rainfall depends on the number of years available for its computation and also the period of the years used. The first standard normals evaluated by the UK Meteorological Office, after careful consideration, took the 35 years of the period 1881 – 1915 (Carruthers, 1945). The next standard period covered the following 35 years 1916 – 1950.

More recently, the length of the period for calculating standard averages has been reduced to 30 years to conform to international practice, and the current period for which the statistics are published is 1941–70.

9.4.1   Standard Period Averages for Short-Period Stations

Naturally, it is not possible to calculate the standard average annual rainfall for a prescribed period for all active rainfall stations. Many good rainfall stations do not have such a long life, or they may also span several years unrelated to the standard period. In order to determine a standard normal for each rainfall station, an objective method has been developed by the Meteorological Office (Meteorological Office, 1963). The method is presented for the 1916 – 50 period in its basic manual form, but it is readily adapted for operation on a digital computer with peripheral plotting facilities. A convenient sized area is selected and here the method is demonstrated for the Wear and Tees river basins, part of the Northumbrian Water Authority's area.

The procedure is carried out in several distinct stages:

(a) An *index map* of the area is prepared showing the location of all stations with satisfactory 1916 – 50 average annual rainfalls (Fig. 9.6).

(b) A *second index map* is prepared on foil or tracing paper showing the location of all short-period and unsatisfactory rainfall stations.

(c) For each of the years from 1916, the annual values of the satisfactory

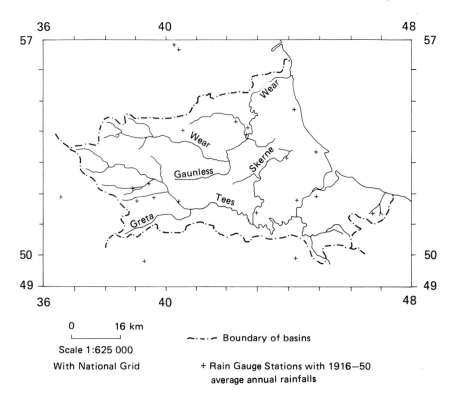

Fig. 9.6   Wear and Tees river basins index map. (Reproduced from Meteorological Office (1963) *Estimation of Standard Period Averages for Stations with Incomplete Data*, Hydrological Memoranda 5, by permission of the Controller, Her Majesty's Stationery Office. © Crown copyright.)

stations on the first index map, expressed as percentages of the 1916 – 50 station averages, are plotted on a map and *isopercental* lines drawn (thus there is a series of annual isopercental maps, the example in Fig. 9.7 being for 1949).

(d) For each short-period station, using the second index map overlay, the estimated percentages are read from the maps for each of the years of observations at the station. If for any one year the estimated percentage = p and the measured rainfall = r (mm), then *one* estimate of the standard period average

$$R = \frac{100\,r}{p}\ (\text{mm})$$

(e) The series of estimates for each short-period station is examined for steady upward or downward trend and for abrupt changes in values. (The station 'history' is consulted for possible reasons.) Unsatisfactory estimates are rejected (see Table 9.2).

(f) The remaining estimates for any one station fluctuate randomly about a mean value. Those estimates falling outside ± 2 standard deviations

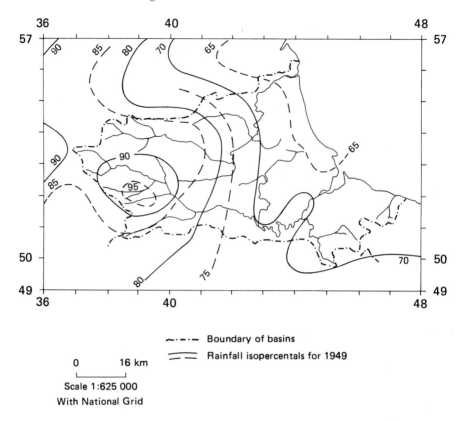

Fig. 9.7   Wear and Tees river basins. An annual isopercental map. (Reproduced from Meteorological Office (1963) *Estimation of Standard Period Averages for Stations with Incomplete Data*, Hydrological Memoranda 5, by permission of the Controller, Her Majesty's Stationery Office. © Crown copyright.)

(SD) are rejected and the mean value recalculated.

(g) The final mean value of the estimates is taken to give the final estimate of the 1916 – 50 average annual rainfall for the given station. The standard error of this mean is also calculated as another criterion of the accuracy.

Three examples of the application of the method are given in Table 9.1. For Station 1 with a record from 1941 – 56 all the estimates of R fall within the limits of 2 SD from the mean. The highest and lowest values providing the range of values are marked. Thus the arithmetic mean of the 16 estimates, 765.9 mm, is an acceptable estimate of the 1916 – 50 average annual rainfall. At Station 2 running from 1936 – 1956, the estimate R from the 1945 annual measurement is well beyond the control limits and so is discarded. The mean 1916 – 50 estimate, 784.3 mm, is then computed from the other 20 years. Station 3 has a long record from 1930 – 1956, but after 1949 the estimates of R drop in value, the reliability of the station has fallen off and the observations cease to be acceptable. The final mean estimate, 964.8 mm, is taken from the 18 years 1932 – 49 omitting the first two years, which are outside the control limits.

9.5   River Flow Data

A series of river discharge values requires the processing of two sets of measurements. The stage height, or river level, is generally the foundation of the record and this needs transforming into discharge by a stage – discharge relationship (Chapter 6). The discharges used to establish that relationship for a river gauging station may be obtained by velocity – area measurements with the computations usually done manually after each gauging (but a number of gaugings may be processed by computer). Alternatively, they may be obtained with a calibrated gauging structure for which the stage – discharge relationship should remain constant. Once a rating curve or rating table has been determined for a river gauging station, its checking involving further measurements and calculations is an occasional necessity depending on the stability of river conditions. It is the regular and continuous measurements of stage that require the setting-up of a consistent data processing procedure to produce discharge values from which the basic river flow statistic, the *daily mean discharge*, is obtained.

9.5.1   Stage Records

The river level measurements are made in various ways and all are related to the stage gauge. The older methods lead to manual processing, still of use in the developing countries, whereas modern recording methods are geared to processing on a digital computer.

Spot readings.   This is the simplest way of compiling a river flow record. An observer reads the level on the stage gauge at fixed hours of the day and writes it down on a suitable form. The frequency of observation depends on the size of the river and the catchment area. Large rivers of continental proportions with a slow response to changing conditions may require only one reading a day. In countries resembling England and Wales where the drainage basins are under 10 000 km^2, spot readings would be recommended four times a day, but when there are rapid changes in river levels, observations are made more frequently. For storm events, especially over small catchments, a continual watch on the river is necessary to ensure the reading of the maximum level on the stage gauge to give the peak flow rate.

Autographic charts from level recorders provide a continuous graphical record of the river stage. Most of the clockwork instruments produce weekly charts, and these records constitute nearly all the information on river flow in the UK before about 1960. Before the introduction of automated computer pro-

Table 9.1   1916–50 Average Annual Rainfall Estimates for Short-Period Stations

Year	Station 1			Station 2			Station 3		
	r	p	R	r	p	R	r	p	R
1930							1053.5	113.0	932.3/X
31							926.5	99.5	931.2/X
32							914.1	93.2	980.8
33							796.5	82.2	969.0
34							941.8	97.0	970.9
35							1055.1	112.0	942.1
36				867.4	111.0	781.4	1044.3	105.0	956.5
37				882.7	107.5	821.1	801.1	84.0	953.7
38				799.6	100.8	793.3	1119.1	115.7	967.2
39				818.6	104.5	783.3	1062.5	111.4	953.8
1940				792.5	107.5	737.2	820.2	85.0	964.9
41	741.2	95.5	776.1	764.0	96.7	790.1	761.0	79.0	963.3
42	545.8	76.5	713.5/	643.4	82.5	779.9	755.9	78.5	962.9
43	680.0	86.5	786.1	566.7	77.0	736.0	1022.9	105.0	974.2
44	730.3	102.0	716.0	858.0	107.5	798.1	1031.5	106.5	968.5
45	643.1	88.5	726.7	560.1	91.8	610.1/X	921.8	96.5	955.2
46	689.9	94.3	731.6	812.3	105.5	770.0	965.5	100.0	965.5
47	825.8	100.8	819.2	683.3	90.5	755.0	1087.4	112.0	970.9
48	859.8	107.0	803.6	693.7	94.4	734.9	1044.7	106.5	980.9/
49	594.1	72.0	825.1/	605.8	69.5	871.7/	907.8	94.0	965.7
1950	811.5	113.0	718.1	896.9	114.0	786.8	1042.4	111.0	939.1
51	892.0	111.2	802.2	1032.5	126.5	816.2	1022.9	112.5	909.2
52	677.7	84.0	806.8	759.5	100.0	759.5	704.3	75.1	937.8
53	582.9	75.8	769.0	603.8	79.2	762.4	721.9	77.0	937.5
54	859.5	117.2	733.2	916.2	115.0	796.7	1174.0	127.0	924.4
55	538.5	70.9	759.5	642.9	78.3	821.1	686.3	76.0	903.0
56	808.5	105.4	767.1	759.5	96.0	791.1	932.2	103.3	902.4

	Station 1		Station 2		Station 3	
	$n = 16$		$n = 21$		$n = 20$	
	Total	12254.0	Total	16295.9	Total	19229.5
	Mean	765.9	Mean	776.0	Mean	961.5
	Range	111.6	Range	261.6	Range	49.7
	SD	38.8	SD	50.0	SD	13.8
Mean ± 2 SD		688.3	Mean ± 2 SD	676.0	Mean ± 2 SD	933.9
		843.5		876.0		989.1
			$n = 20$		$n = 18$	
			Total	15685.8	Total	17366.0
			Mean	784.3	Mean	964.8

cessing in the 1960s, all the hydrometric data from the thousands of gauging stations in the USA were abstracted from charts and calculated by teams of hydrological assistants.

9.5.2   Daily Mean Discharge — Manual Processing

If there is only one or several spot readings in the day and there has been little change in river levels, a *daily mean stage* is obtained by taking the single reading with a datum correction if necessary or the arithmetic mean of the several readings with datum corrections. The *daily mean discharge* is then read from the rating table for the daily mean stage.

If there are considerable changes in river levels during the day identified by several spot readings, these should be plotted on a graph and the points joined by smooth curves. The averaging then proceeds as for an autographic chart (see below).

If an autographic chart is available, the stage hydrograph is subdivided into convenient discrete time periods and a mean stage height is drawn for each period. For example, in Fig. 9.8, the continuous stage record for a large

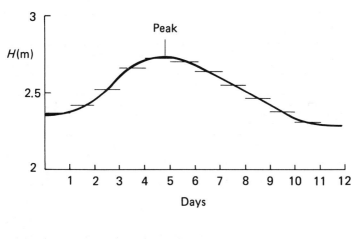

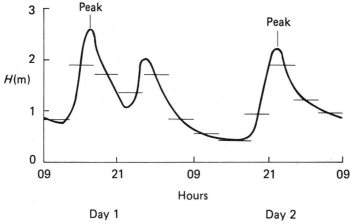

Fig. 9.8   Autographic charts and mean levels for large and small catchments.

River Authority ....

River ....   Station ....

Month .... 19......

1 Date	2 Mean Daily Flow	3 Peak flow during day	4 Minimum flow during day	5 Variation from Natural Flow +or −	6 Gross Mean Daily Flow
1					
2					
3					
4					
5					
6					
7					
8					
9					
10					
11					
12					
13					
14					
15					
16					
17					
18					
19					
20					
21					
22					
23					
24					
25					
26					
27					
28					
29					
30					
31					
TOTAL					

Fig. 9.9   River flow data form. (Reproduced by permission of the Institute of Hydrology.)

catchment area extends over many days; the daily mean stages can be easily assessed and drawn for each day, and the corresponding daily mean discharges obtained from the rating table. For a small catchment, the greater and more frequent level fluctuations are considered over, say, 4-h periods (even shorter periods may be advisable). After any necessary datum corrections, the 4-h mean stages are converted to equivalent discharges and the daily mean discharge is obtained by averaging the six 4-h discharges. (This averaging of the short-period discharges is more accurate than averaging the corresponding stages and then converting the latter value to discharge since the stage – discharge relations are not straight lines.) The peak level, and hence peak discharge, is also an important feature to be abstracted from charts and noted. A recommended form for tabulating the daily mean discharges and peak flows in Fig. 9.9 shows how river authorities can be requested to submit manually processed data to a central authority.

The engineers of the US Geological Survey developed an instrument to convert mechanically a continuous stage graph into a record of daily mean discharges. This discharge integrator simplified and expedited the manual processing of the vast number of records in the USA but its use has been superseded by the introduction of digital recorders (Grover and Harrington, 1966). Modern chart readers are described in a later section.

9.5.3   Digital Recorders

Digital recorders produce a series of stage measurements taken automatically at fixed time intervals. Most of the current river stage measurements in developed countries are now recorded on punched paper tape of which there are two main types.

16-channel paper tape used with the Fischer and Porter recorder. Across the 54 mm wide tape there are aligned 4 sets of 4 bit binary coded digits so that one line of punched holes records a 4 digit value of the stage height from 0.001 m to 9.999 m (Fig. 9.10). The recording is usually made every 15 min, but the instrument can be geared to punch at any chosen interval between 5 min and 60 min. For a normal setting, a tape that has run for a month contains nearly 12 000 decimal digits from 96 readings per day. Each month the tape is passed through a 'translator' which can be interfaced with a microcomputer for processing the data.

5-channel paper tape, from an Ott recorder for example (Fig. 9.10), is compatible with most large computers, but after exposure to changing atmospheric conditions in the field, it is sometimes found necessary to copy the records on to fresh paper tape or magnetic tape before processing. A set of identifying and explanatory information needs to be prepared on paper tape as an introduction to each set of records. Such an input can take the form

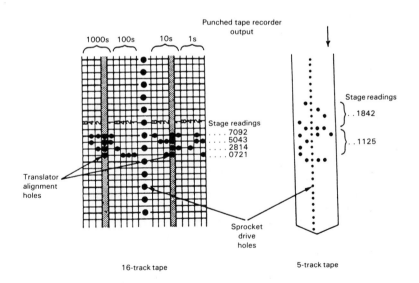

Fig. 9.10   Punched paper tapes from digital stage recorders. (Reproduced from Herschy, R.W., ed. *New Technology in Hydrometry*, Adam Hilger Ltd, Bristol, 1986.)

Table 9.2   Ott Paper Tape Input Format (5-Track Ferranti Code)

Note: tapes should normally start during the first few days of a month and finish during the first few days of the following month. All readings must be at 15-min intervals. (Reproduced from T.J. Marsh, M.L. Lees & I. Littlewood (1980) *Surface Water Data Processing*, DOE Water Data Unit.)

Description	No. of digits	Range of values
Identifier	2 (alphabetic)	OT
Station number	6	001001–399999
No. of stage – discharge formula or rating tables	1	1–9
Time on	4	0000–2345 hours
Day	2	01–31 (date)
Month	2	01–12 (month)
Year	2	00–99 (19--)
Time off	4	0000–2345 hours
Day	2	01–31 (date)
Month	2	01–12 (month)
No. of readings	4	0000–9999
Up (Time of change of formula or table)	4	0000–2345
to (Day of change of formula or table)	2	01–31
9 (Month of change of formula or table)	2	01–12
times (New formula or table no.)	1	1–9
End of opening parameters	1	+
Stage data		
First reading	4	0000–9999 (millimetres)
:		
:		
Last reading	4	0000–9999
End of tape		

shown in Table 9.2. The stage records usually overrun the end of a month and great care must be taken to ensure the correct noting of the tape changing time and date.

9.5.4   Daily Mean Discharge — Computer Processing

Most authorities in the UK operating river flow gauging stations have their own computers for handling the records, but the Institute of Hydrology's Surface Water Archive is a central body that now receives all river flow measurements. Following a great deal of study and after several years of experience, a comprehensive computer processing package has been established to deal with the various data types received each month (Marsh *et al.*, 1980). The package contains a series of computer programs to calculate a whole range of statistical information required by the water industry. Program one produces the daily mean discharges. Each month it processes the 'translated' data from the Fischer and Porter tapes, the Ott 5-channel paper tapes and 8-channel paper tapes that have been punched from manual tabulations. The procedure is outlined in the flow chart in Fig. 9.11. The station details that contain the stage – discharge relationships as a formula or a rating table and the catchment area of the gauging station are stored on magnetic tape and fed into disk storage before the program is run. The daily mean discharges, highest and lowest values during each day and the corresponding stage values are printed out for the month in the format shown in Table 9.4. The other programs in the processing package assemble monthly and annual data for storing on a Long Period File. Areal rainfalls from the Meteorological Office are added to the Long Period File to form a complete set of rainfall and river flow information for each gauged catchment.

With the rapid development of increased capacity microcomputers and the widening use of solid state recording devices, the UK river flow data system is undergoing many modifications. A hydrological data processing and analysis system, HYDATA, has been developed by the Institute of Hydrology for use on modern IBM-compatible personal computers. It is capable of storing data for up to 1000 stations received from loggers, digitizers or floppy disks with analysis programs written in FORTRAN. The system has been well tested and applied in many countries worldwide.

For smaller users with a BBC microcomputer, a very helpful package for hydrologists has been produced in the University of Southampton (Clarke, 1987).

9.6   Quality Control of River Flow Data

As with the systematic checking of the daily rainfall measurements by the Meteorological Office, the Surface Water Archive pays particular attention to the quality of the river flow records. In the complex path from the initial water level observations to the production of daily mean discharges, four major phases are identified during which data quality can be checked.

9.6.1   The Field Data Collection

Of first importance are the field observations whether they are made manually or automatically. Observers must be very careful in reading the water level at the stage gauge and in recording immediately the correct level and time. All instrumentation should be checked regularly with all parts of the recorders, clockwork, mechanical or electronic, receiving routine maintenance and testing for correct settings. The production of reliable river flow records is dependent on the quality of the basic measurements.

9.6.2   The River Stage Data

The processing package developed by the Water Data Unit includes checking routines within its first program — the validation of the data (Fig. 9.11). For each river flow station, the expected response of the catchment to rainfall inputs dependent on the catchment characteristics can be assessed and thus a

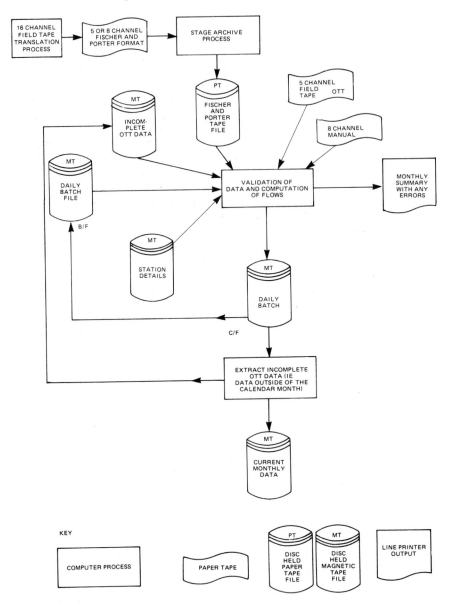

Fig. 9.11 Flowchart for computing daily mean discharges. (Modified from T.J. Marsh, M.L. Lees & I. Littlewood (1980) *Surface Water Data Processing*, DOE Water Data Unit, by permission of the Institute of Hydrology.)

limit to the difference between consecutive stage readings can be fixed. This is included on the Station Details File. The computer routine inspects the sequence of figures recorded on the punched tapes and identifies the following types of error:

Table 9.3 Standard Monthly Printout

(Reproduced from T.J. Marsh, M.L. Lees & I. Littlewood (1980) *Surface Water Data Processing*, DOE Water Data Unit.)

WATER DATA UNIT

PUNCHED TAPE PROCESSING FOR RIVER FLOWS

STATION PEEBLES   TABLE F   JANUARY 1980   01/80

STATION NUMBER 021003   CATCHMENT AREA 0694.0 SQ.KM.

RIVER TWEED

| | DISCHARGE | | | | | STAGE | | | | |
DAY	MEAN DAILY FLOW TCMD	MEAN DAILY FLOW CUMECS	HIGHEST CUMECS	LOWEST CUMECS	NO OF READS	HIGHEST	TIME/READ NO.	LOWEST	TIME/READ NO.	MEAN
1	1433	16.59	17.92	15.04	96	534	09 00	468	08 30	502
2	1296	15.00	16.98	14.43	96	512	08 45	455	14 00	467
3	2827	32.96	59.22	17.15	96	1235	06 00	516	09 00	807
4	2651	30.34	48.21	24.74	96	1079	09 00	680	08 30	782
5	2031	23.52	24.94	21.40	96	684	11 30	611	06 30	654
6	1825	21.13	21.16	20.93	96	614	09 00	601	15 00	605
7	1722	19.94	20.93	18.91	96	606	09 00	552	08 15	579
8	1483	17.17	18.71	15.97	96	522	09 00	488	08 30	516
9	1289	14.93	15.97	14.07	96	448	09 00	447	08 30	465
10	1164	13.48	14.07	12.92	96	447	09 00	421	08 00	433
11	1102	12.76	12.92	12.57	96	421	09 00	423	08 05	417
12	1053	12.19	12.57	11.86	96	412	09 00	396	21 15	403
13	1018	11.79	13.43	11.49	96	444	08 45	387	08 45	394
14	1254	14.52	16.10	12.53	96	491	13 45	412	08 45	456
15	992	11.50	12.53	10.65	96	412	09 00	366	08 45	387
16	886	10.26	10.65	9.879	96	366	09 15	346	07 30	356
17	824	9.545	9.841	9.322	96	345	09 00	331	08 30	337
18	775	8.980	9.322	8.258	96	331	09 00	301	14 15	321
19	729	8.441	8.817	8.087	94	317	19 45	296	05 45	306
20	718	8.312	8.431	8.190	96	306	09 00	299	09 45	302
21	1054	12.20	16.93	8.293	96	511	06 30	302	08 45	399
22	1359	15.73	16.31	14.76	96	508	09 00	462	08 45	483
23	1151	13.32	14.76	11.90	96	462	09 00	397	08 45	430
24	947	10.97	11.90	10.26	96	397	09 00	356	08 30	374
25	890	10.57	10.57	9.954	96	364	21 30	348	08 30	357
26	804	9.312	9.954	8.293	96	348	09 00	302	08 45	330
27	761	8.838	9.285	7.986	96	334	21 45	293	14 15	316
28	956	10.69	11.49	9.285	96	387	01 15	330	11 15	361
29	1747	20.22	26.22	10.37	96	830	08 45	359	11 30	574
30	2261	26.18	32.75	21.02	96	830	09 00	603	08 45	705
31	1563	18.09	21.07	14.90	96	604	09 00	465	08 45	537

SUMMARY FOR MONTH

MONTHLY TOTAL	468.998 CUMEC DAYS
MONTHLY TOTAL	40506.200 TCM
MONTHLY MEAN	15.129 CUMECS
MONTHLY MEAN	1306.445 TCPD
MEAN/SQUARE KM.	21.800 L/S/SQKM
MEAN RUN OFF	53 MM
MAX DAILY MEAN	2847.000 TCPD
MIN DAILY MEAN	778.000 TCPD
HIGHEST FLOW	59.220 CUMECS
LOWEST FLOW	7.986 CUMECS
PART TOTAL	CUMEC DAYS

MONTHLY MEAN STAGE	463
MAX DAILY MEAN STAGE	807
MIN DAILY MEAN STAGE	302
HIGHEST STAGE	1235
LOWEST STAGE	293

(a) More than four readings missing (sometimes at the change of Ott tapes), a non-numeric character in a reading, and a reading without four digits. Single errors in a reading will be corrected by interpolation from adjacent readings but with two or more consecutive errors, the day's readings will be rejected for manual inspection.

(b) The difference between consecutive readings greater than the allowed limit (from the Station Details File). For a sharp rise followed by a sharp fall (or the reverse) an interpolated value is inserted from the adjacent readings. Two or more successive sharp falls or a single sharp fall which then steadies or an isolated sharp rise which steadies cause the whole day's readings to be rejected. If there are four successive sharp rises the record is accepted, but a fifth sharp rise or fall will invalidate the day's readings.

(c) The reading higher than the upper limit given for the station. The reading is accepted but not counted as valid.

(d) The reading less than or equal to datum. Single low readings are replaced by interpolation between adjacent readings. Two or more readings less than datum are accepted and the corresponding discharges will be zero.

The checking of river stage records is best carried out by hydrologists with a knowledge of the river's behaviour; gauging authorities submitting their data for national archives are asked to indicate errors and anomalies. Thus unusual river conditions resulting in extra high or low levels when reported can allow the central quality controllers to update the Station Details File. A more detailed account of the stage checking technique is given in the Institute of Hydrology Report No. 15 (Plinston & Hill, 1974).

9.6.3   The Stage – Discharge Relation

The quality of the stage – discharge relationship plays a large part in the final status of the daily mean discharge record. Rating formulae for gauging structures are generally considered reliably constant but the $Q - H$ relation may be modified considerably by the growth of weeds on the weir or flume. Regular maintenance is essential, especially in the growing seasons. Velocity – area stage – discharge relationships rely on further instrumental measurements so errors in the velocity sampling must be avoided. Once a stage – discharge curve has been established, any further gaugings deviating markedly from the curve must have obvious physical causes or errors in the computations must be suspected. Details of expected and desirable accuracies are to be found in Herschy (1978).

9.6.4   The Daily Mean Discharges

The daily mean discharges (m^3 s^{-1}), which represent the flow volumes in a day (m^3) averaged over the number of seconds are the final product from the data processing program. They should be checked after every monthly print-out (Table 9.3) and also on an annual basis. Discrepancies may be found by the following checks:

(a) Comparison of the daily mean discharges with the stage hydrograph traces from an autographic recorder chart. A computer plot of the daily mean discharges would help in this check.

(b) That the correct stage – discharge relationship has been used for the data. Different relationships may be required for different periods of the year.

(c) The range of flows produced is within the calibration range for the gauging station.

(d) The discharges converted into runoff (mm) from the catchment are commensurate with the catchment rainfall and with the runoff from neighbouring catchments.

A quality-control computer routine is specially designed to deal with known conditions and events. It would require a different structure for other countries and climates. The Water Data Unit suite of processing programs is essentially compiled to handle the river records from stations in the British Isles. The application of computer techniques in processing hydrometric data in the USA has to take account of a much wider range of natural conditions and many more gauging stations, some located in remote areas (Isherwood, 1970).

9.7   Automatic Chart Reading

Large quantities of historic hydrometric data recorded on autographic charts have remained in storage without ever being used, and hydrologists wishing to extend more recent digital records backwards in time for special analysis have encouraged the development of automatic processing of such charts. Several manufacturers stimulated by the demand have produced digitizers; the most successful instrument, owing to its adaptability and robustness, is the D-mac 'pencil' follower (Fig. 9.12). The chart is fixed to a flat reading table. A viewing cursor with cross wires is encircled by a coil of wire connected to a power source. When the cursor is positioned over a selected point on a chart, a following servo-mechanism in the table connected to a drive mechanism records the x and y coordinates of the point by the operation of a foot switch. The scales of the x and y axes have to be stipulated before each chart is read. The required points on the curves are selected by the operator. Additional information can be entered on an auxiliary keyboard and all the data are recorded on either punched cards or 5-channel paper tape for computer processing.

Both rainfall hyetograms and river level hydrographs have been digitized by such semi-automatic chart readers, but the operation is still labour intensive and thus has not been applied to routine processing on a large scale. Special studies have arranged the digitizing of selected records when the value of the abstracted data has been assured.

More recently, the Image Analysis Group of Oxford University's Nuclear Physics Laboratory has developed the Precision Encoding and Pattern Recognition (PEPR) system and this has been used to analyse autographic rainfall records. The system requires the record to be on 16 mm microfilm;

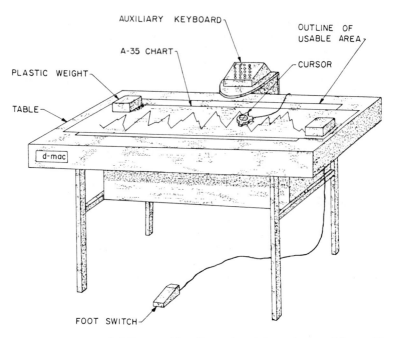

Fig. 9.12   D-mac pencil follower table. (Reproduced from *Automated Stream Flow Computations* (1974) by permission of Environment Canada.)

thus all rain-recorder charts have to be photographed, and this has involved a considerable amount of pre-scrutiny and manual inking-in of faint traces etc. In collaboration with the UK Meteorological Office some long series of charts have been digitized into time/amount computer datasets, which omit dry periods, but do identify periods of defective record to ensure that they are not regarded as dry. The chart data are adjusted so that the 0900 h to 0900 h totals agree with the standard storage rain gauge over that period. The system was capable of being adapted to analyse data from a number of different types of daily chart but there were complications in microfilming the long rolls of open time-scale strip charts.

The need for chart readers for hydrometric data will decrease as the number of digital recorders is increased. Practical hydrologists and river engineers do, however, like to see a chart record on site if only for an immediate check on the changing state of the river and on the proper functioning of the digital recorder.

9.8   Data Transmission

The processing of hydrometric data which has been described so far in this chapter has depended on the assembling of records by conventional means. The records may be sent to the processor by post (the monthly rainfall postcard or the river level logger tape or solid state event recorder cachette).

Some authorities may arrange an independent delivery service for bulk quantities of records. The data processing is geared to the reliable continuous production of information for planning and designing schemes. For operational use in times of emergency, hydrometric measurements require more direct transmission to central headquarters. For flood-warning purposes, special rain gauges and river level sensors can be interrogated by telephone or accessed on a frequent basis by a mini-computer controller. These are usually additional facilities to the regular recordings. Alarm gauges registering a measurement over a specified danger limit can instigate an alarm call. Short-wave radio is also used for real-time transmission of observations.

With the recent development of more advanced transmission systems via satellites, the possibility of reducing the labour-intensive regular assemblage of recorded data has been suggested (Herschy, 1980). A pilot study by the former Water Data Unit investigated initially the sending of critical measurements at two rainfall stations and river levels from three river gauging stations in Great Britain. In addition, the discharge measured by ultrasonic gauging was transmitted from one of the river gauging stations. The data were fed into a *Data Collection Platform* (DCP) from interfaced digital recorders, then, encoded and transmitted to a satellite which relayed the signals to a ground control station. There the data were decoded for distribution to the user. The DCP consists of electronic equipment including a microprocessor, a crystal clock and radio transmitter. An aerial and source of power, battery or mains, are also required. The microprocessor can accommodate data from six sensors, i.e. six different series of measurements, and the whole DCP can be housed on the bench alongside an orthodox data-logging system in a gauging hut or provided with a weather-proof container at a remote rain gauge station.

The potential application of this type of equipment for the regular assemblage of hydometric data will be assured in the future as the developments in microtechnology continue.

9.9   Data Archives and Publication

The significance of the term 'archive' has changed dramatically with the conversion of manual data processing to computer data processing. For example, all the older meteorological measurements tabulated on the specially designed forms are stored in box files and occupy considerable shelf space at headquarters. These are the historic data, much valued by students of climatic change or by hydrologists seeking evidence of past extreme rainfalls. Data from a network of long period rainfall stations has now been transferred to magnetic tape archives held by the UK Meteorological Office. (Shearman, 1980). However, instead of being able to research the tabulated data, making assessments and judgements as he abstracts the figures manually, the potential user must be able to define his requirements precisely from a system embodied in computer software and dependent on a machine operator.

The Surface Water Archive has also a growing archive of a great range of data required by the water industry. In addition to the river discharges compiled in some detail, water quality variables, water consumption and demand statistics are also stored. A comprehensive suite of computer programs has been written to abstract information to fulfil enquiries.

As a result of UK central government reorganization, the country's data archives for surface water, ground water, water quality and other records pertaining to the water industry are now accommodated by the Institute of Hydrology at Wallingford. The main collection of rainfall and evaporation records remain in the Meteorological Office national archives retained under Public Records Acts.

The storage of data for the 8300 river flow stations in the USA has also stimulated a study of the problem by the Geological Survey, the Federal organizing agency, (Benson & Carter, 1973). Considerable attention is now being paid to the actual need for the data, and techniques for solving problems in ungauged catchments by using information from neighbouring catchments continue to be investigated.

Most hydrometric data gathering agencies are national or local government authorities and it is their duty to make the data available to the public. The USA has an excellent record of data publication; their series of Water Supply Papers is internationally renowned. In the UK, the long series of volumes of *British Rainfall* containing annual rainfall totals for all acceptable stations and a selection of shorter duration information began in 1861 and achieved its centenary with little change in form. Publication of *British Rainfall* ceased with the 1968 volume; it has been replaced by an annual paperback, *Monthly and Annual Totals of Rainfall for the United Kingdom*, containing a smaller variety of tabular data. A new series of yearbooks entitled *Hydrological Data: UK* commenced publication by the Natural Environment Research Council in 1981, and these contain surface water and groundwater data.

Although data publication is highly desirable, and this principle is upheld internationally, the cost of maintaining the previous high standards of production is becoming prohibitive. The time taken for a yearbook to appear also means that the data are then only of historic value, useful for assessing past conditions. Operationally, hydrometric data are now disseminated to users either on magnetic tape for the customer's own analysis or as standard computer printouts of required information to order. In the future, it is expected that the customer will be able to interrogate a central archive from a telecommunications terminal in his own office.

References

Benson, M.A. and Carter, R.W. (1973). *A National Study of the Streamflow Data-Collection Program*. US Geol. Surv. Water Supply Paper 2028, 44 pp.

Bleasdale, A. (1969). *The Processing of Hydrometeorological Data in the Meteorological Office*. Water Resources Board, Conference on Data Retrieval and Processing, Paper No. 17.

Bleasdale, A. and Farrar, A.B. (1965). 'The processing of rainfall data by computer.' *Meteorol. Mag.*, **94**, 98–109.

Carruthers, N. (1945). 'The optimum period for a British Rainfall Normal,' *Quart. Jl. Roy. Met. Soc.*, **307** & **308**, 144–150.

Clarke, D. (1987). *Scigraf and Regress*. John Wiley.

Grover, N.C. and Harrington, A.W. (1966). *Stream Flow, Measurement, Records and Their Uses*. Dover, 363 pp.

Herschy, R.W. (1978). *Hydrometry, Principles and Practices*. John Wiley, 511 pp.

Herschy, R.W. (1980). 'Hydrological data collection by satellite.' *Proc. Inst. Civ. Eng.*, Part 1, **68**, 759-771.

Herschy, R.W. (ed.) (1986). *New Technology in Hydrometry*. Adam Hilger Ltd, Bristol, 240 pp.

Isherwood, W.L. (1970). 'Use of digital computers in processing, publishing and analysing hydrologic data.' *Nordic Hydrology*, **1**, 38-55.

Marsh, T.J., Lees, M.L. and Littlewood, I. (1980). *Surface Water Data Processing*. DOE Water Data Unit, 51 pp.

Meteorological Office (1963). *Estimation of Standard Period Averages for Stations with Incomplete Data*. Hydrological Memoranda No. 5, 24 pp.

Meteorological Office (1987a). *Annual Report 1986*. HMSO, 56 pp.

Meteorological Office (1987b). 'Rainfall Quality Control in Met. O. 3a (1)'. Internal Memorandum.

Plinston, D.T. and Hill, A. (1974). *A System for the Quality Control and Processing of Streamflow, Rainfall and Evaporation Data*. NERC Institute of Hydrology Report No. 15.

Shearman, R.J. (1975). 'Computer quality control of daily and monthly rainfall data.' *Meteorol. Mag.*, 104, 102-108.

Shearman, R.J. (1980). 'The Meteorological Office archive of machinable data.' *Meteorol. Mag.*, **109**, 344-354.

Part II
Hydrological Analysis

10

Precipitation Analysis

The previous chapters have been concerned with the measurement of the very many hydrological variables. The description of recommended instruments and their installation for precipitation observations occupied Chapter 3. An initial appraisal of the measurements has been made in Chapter 9, where techniques of data handling and quality control have been described.

This chapter is concerned with the next stage in the preparation of information on rainfall and snowfall for use by the engineer. It is assumed that the data for computation have been produced by recognized observations and that they have passed the standard quality tests.

10.1 Determination of Areal Rainfall

In describing the measurement of precipitation, emphasis has been made that at present it is essentially a point-sampling procedure; rainfall over an area has to be *estimated* from these point measurements (in the future direct areal rainfall *measurements* by radar may be more widely available). The depths of rainfall and water equivalent of snowfall correspond to the volume of precipitation falling on the area of a gauge orifice. Several such measurements are made over a catchment area or drainage basin and the total quantity of water falling on the catchment is evaluated. Sometimes this is expressed as a volume (m^3) for a specified time period, but more usually as a mean depth (mm) over the catchment area. This value is sometimes called the general or average rainfall over an area, but it is now increasingly referred to as the *areal* rainfall or *areal* precipitation and the term 'average rainfall' is restricted to long-term average values.

The areal rainfall is required for many hydrological studies, and it is most important to have the limits of the catchment carefully defined. For the drainage area down to the river gauging station the 'water parting' (watershed) or boundary of the catchment must be known and plotted on a topographical map as accurately as possible. This may require investigations in the field when the maps are inadequate, and due regard must be paid to man-made water-courses, such as drainage ditches and leats, that may cross a natural watershed boundary. Problems with catchment boundary definition arise in marshy areas with indeterminate drainage and there may be seasonal differences for such areas in some climates. Knowledge of the geology of the catchment is also necessary since the topographical divide may not be the true water parting. It must always be remembered that the determination of the catchment area may be a possible source of error in assessing water resources.

There are many ways of deriving the areal precipitation over a catchment from rain gauge measurements. The standard methods and their simpler modifications will be outlined first and a selection of the more sophisticated techniques will follow.

10.1.1   The Arithmetic Mean

This is the simplest objective method of calculating the average rainfall over an area. The simultaneous measurements for a selected duration at all gauges are summed and the total divided by the number of gauges. The rainfall stations used in the calculation are usually those inside the catchment area, but neighbouring gauges outside the boundary may be included if it is considered that the measurements are representative of the nearby parts of the catchment.

The arithmetic mean gives a very satisfactory measure of the areal rainfall under the following conditions:

(a) The catchment area is sampled by many uniformly spaced rain gauges.
(b) The area has no marked diversity in topography, so that the range in altitude is small and hence variation in rainfall amounts is minimal. The arithmetic mean is readily used when short-duration rainfall events spread over the whole area under study and for monthly and annual rainfall totals.

If a long-term average for the catchment area is available, then the method can be improved by using the arithmetic mean of the station values expressed as percentages of their annual average for the same long period, and applying the resultant mean percentage to the areal average rainfall (mm).

If accurate values of the areal rainfall are obtained first from a large number of rainfall gauge stations, then it may be found that measurements from a smaller number of selected stations may give equal satisfaction. In the Thames Basin of 9981 km^2, it was found that the annual areal rainfall could be determined by taking the arithmetic mean of 24 well distributed and representative gauges, to within $\pm 2\%$ of the value determined by a more elaborate method using 225 stations (IWE, 1937).

10.1.2   The Thiessen Polygon

Devised by an American engineer (Thiessen, 1911), this is also an objective method. The rainfall measurements at individual gauges are first weighted by the fractions of the catchment area represented by the gauges, and then summed. On a map of the catchment with the rain gauge stations plotted, the catchment area is divided into polygons by lines that are equidistant between pairs of adjacent stations. A typical configuration for well distributed gauges is shown in Fig. 10.1. The polygon areas, a_i, corresponding to the rain gauge stations are measured by planimeter and the areal rainfall $\overline{R}$ is given by:

$$\sum_{i=1}^{n} \frac{R_i a_i}{A}$$

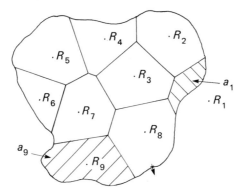

Fig. 10.1   Thiessen method.

where R_i are the rainfall measurements at n rain gauges and A is the total area of the catchment. In the illustrated example, there are nine measurements contributing to the calculation even though gauge 1 is outside the catchment boundary. The area a_1 within the catchment is however nearer to gauge 1 than to the neighbouring gauges 2, 3 and 8, and is therefore better represented by measurements at gauge 1.

The area fractions a_i/A are called the Thiessen coefficients and once they have been determined for a stable rain gauge network, the areal rainfall is very quickly computed for any set of rainfall measurements. Thus the Thiessen method lends itself readily to computer processing. However, if there are data missing for one rain gauge station, it is simpler to estimate the missing values and retain the original coefficients rather than to redraw the polygons and evaluate fresh Thiessen coefficients. If however a rain gauge network is altered radically, then the Thiessen polygons have to be redrawn, the new areas measured and a new set of coefficients evaluated.

The Thiessen method for determining areal rainfall is sound and objective, but it is dependent on a good network of representative rain gauges. It is not particularly good for mountainous areas, since altitudinal effects are not allowed for by the areal coefficients, nor is it useful for deriving areal rainfall from intense local storms. To overcome some of the shortcomings, investigations into the use of height-weighted polygons combining altitudinal and areal effects on the rainfall measurements have been made, but the supposed small improvements in results do not generally justify the extra work and complexities involved.

10.1.3   The Isohyetal Method

This is considered one of the most accurate methods, but it is subjective and dependent on skilled, experienced analysts having a good knowledge of the rainfall characteristics of the region containing the catchment area. The method is demonstrated in Fig. 10.2. At the nine rain gauge stations, measurements of a rainfall event ranging from 26 to 57 units are given. Four *isohyets* (lines of equal rainfall), are drawn at 10-unit intervals across the

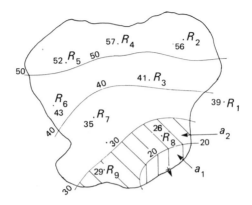

Fig. 10.2   Isohyetal method.

catchment interpolated between the gauge measurements. Areas between the isohyets and the watershed (catchment boundary) are measured with a planimeter. The areal rainfall is calculated from the product of the inter-isohyetal areas (a_i) and the corresponding mean rainfall between the isohyets (r_i) divided by the total catchment area (A).

In the illustration, there are five subareas and the areal rainfall $\bar{R}$ is given by:

$$\sum_{i=1}^{5} \frac{a_i r_i}{A}$$

In drawing the isohyets for monthly or annual rainfall over a catchment, topographical effects on the rainfall distribution are incorporated. The isohyets are drawn between the gauges over a contour base map taking into account exposure and orientation of both gauges and the catchment surface. It is in this subjective drawing of the isohyets that experience and knowledge of the area are essential for good results. The isohyetal method is generally used for analysing storm rainfalls, since these are usually localized over small areas with a large range of rainfall amounts being recorded over short distances.

The UK Meteorological Office has used the isohyetal method for many years and recently has developed and encouraged the application of a modified version. Isopercental lines are drawn instead of isohyets. For a required areal rainfall for a given event, the measurements at each station are plotted as a percentage of the station standard long-term average *annual* rainfall. Then the lines of equal percentages (isopercentals) divide the catchment into areas that are measured, applied to the mean percentages between the isopercentals and an overall areal percentage is obtained. This is applied to the standard long-term *average annual areal rainfall* for the catchment, previously derived, to give the required areal rainfall for the *event*. This improved technique is more reliable and objective and can be carried out by a computer with plotting facilities. However, although it is readily applicable in countries with many long homogeneous rainfall records from which reliable long-term average annual areal rainfall values for all catchments can be evaluated, it is not so useful in developing countries where records are limited.

10.1.4   The Hypsometric Method

This composite method for evaluating areal rainfall takes account of catchment topography in addition to the measured rainfalls (WMO, 1965). It is to be recommended for small or medium sized representative catchments in hilly regions where detailed experimental studies are required. The technique is shown in the coaxial diagram in Fig. 10.3. The rainfall observations are plotted against station height in the top left-hand quadrant and the hypsometric curve, the accumulated area of the catchment with increasing height, is plotted in the opposite quadrant. A straight line is drawn through the rainfall values and a curve of precipitation is drawn from points on that

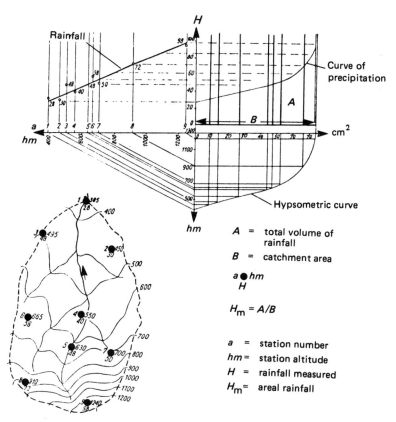

Fig. 10.3   Hypsometric method. A = total volume of rainfall. B = catchment area. (Reproduced from *Guide to Hydrometeorological Practices, 2nd edn.* (1970) by permission of the World Meteorological Organisation.)

line related to the hypsometric curve via the axes of altitude. The area under the curve of precipitation gives the total volume of rainfall over the catchment and when this is divided by the catchment area, the areal rainfall is obtained. Changes in rain gauge stations are easily incorporated into the method. The main difficulties are the initial measurements for the construction of the hypsometric curve. The application of the technique assuming a linear relationship between altitude and rainfall smooths the rainfall observations but may induce errors for some rainfall events.

10.1.5   The Multiquadric Method

The principles inherent in the hypsometric method of areal rainfall determination can be carried a step further by the three dimensional mathematical description of the rainfall surface. In Fig. 10.4, the diagram represents the volume of precipitation falling on part of a catchment area. The element $OABC$ in the $x - y$ plane represents part of the plan area of the catchment and the z values are the depths of rainfall for a given time period. Thus the coordinates (x, y, z) define the position on the rainfall surface of the rainfall measurement z by rain gauge (x, y). Given a network of rainfall observations, the (x, y, z) surface can be described mathematically in several ways. Two of the most common techniques are by fitting polynomials or harmonic (Fourier) series to the given (x, y, z) points. A third method using multiquadrics has also been applied successfully to rainfall surfaces and the areal rainfalls subsequently obtained for quantities ranging from 2-min falls to annual totals (Shaw & Lynn, 1972).

Using the latter method, the equation for the rainfall surface is given by summing the contributions from right circular cones placed at each rain gauge station. Thus:

$$z = \sum_{i=1}^{n} c_i \left[(x - x_i)^2 + (y - y_i)^2 \right]^{\frac{1}{2}} = \sum_{i=1}^{n} c_i a_{(x_i, y_i)}$$

where n is the number of data points (x_i, y_i). There are n coefficients c_i that require estimation. The coefficients are obtained from the n equations for z_1 to z_n, which written in matrix form are:

$$\underline{Z} = \underline{C} \ \underline{A}$$

where $\underline{Z}$ and $\underline{C}$ are row vectors, and $\underline{A}$ is a square matrix with elements $a_{(x_i, y_i)}$. Post-multiplying by the inverse of $\underline{A}$ gives the coefficients:

$$\underline{C} = \underline{Z} \ \underline{A}^{-1}$$

The equation for z conditioned by these c_i fits all the data points exactly. When the equation for the rainfall surface has been so defined, the volume of rainfall is obtained by integration over the area of the catchment and the areal rainfall results from dividing this value by the catchment area. In practice, the catchment area is subdivided into several rectangles over which the integrations are made and the volumes summed to give the total catchment volume of rainfall.

The method can be used efficiently for all time periods and the computer

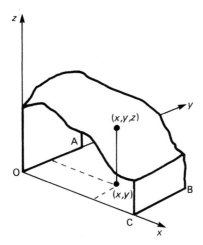

Fig. 10.4   Multiquadric method.

program can be linked to a contour package to produce machine plotted isohyetal maps of the rainfall event. The multiquadric program can also be used for raingauge network design by trials, varying the number of gauges and their distribution over a catchment.

10.2   Depth – Area Analysis

In designing hydraulic structures for controlling river flow, an engineer needs to know the areal rainfall of the area draining to the control point. Sometimes it is only the average river flow being considered, but more often the works are intended to control flood flows and then a knowledge of heavy rainfalls is required.

The technique of relating areal rainfall depths to area by analysing several storms gives depth – area relationships for different specific durations. Hence in a region where particular types of storms are experienced, the areal rainfall expected from a given catchment area for a duration to suit the catchment response can be taken from those depth – area relationships for that region.

In the UK, storm patterns are very variable and design rainfalls for different durations and different regions over the whole country have been computed by the Meteorological Office for the *Flood Studies Report* (to be detailed in a later chapter). However regional differences in storm rainfalls can be identified and, of course, are recognized in other climatic regimes. Hence it is always recommended that engineering hydrologists be aware of the prevalent type of storm which is likely to affect their study area.

To analyse a significantly heavy storm rainfall, the procedure outlined in Fig. 10.5 is followed. It is advisable to choose single-cell storm patterns for the analysis. From the measurements made at all the rain gauges in the area, the pattern of the storm is plotted by drawing the isohyets. The areas enclosed by the isohyets are measured by planimeter (col. 2) and the calculations for each isohyet (row) carried out as indicated in the following columns. Although the average rain between isohyets is taken as the arithmetic mean, the average rain enclosed by the top isohyet has to be estimated. The areal rain for each enclosing isohyet (col. 7) is then plotted against the logarithm of each area (to avoid an unduly long area axis). The depth – area relationship for the duration of the selected storm is thus obtained. It is necessary to analyse several similar duration storms experienced over the area to provide many more data points on the graph and maximum depth values for the range of areas can be read from an enveloping curve to give design data.

The analysis is time consuming and is dependent on there being a sufficient number of storm records over the area of interest. To give a quick answer in the design situation, many engineers have devised formulae to fit the rainfall depth – area relationship. These have been well documented by Court (1961) and generally give the areal rainfall enclosed by isohyet i, $\overline{R}_i$ in terms of the maximum rainfall measured, R_{max}, and the area, A_i. Assuming circular isohyets, the formulae take the form:

$$\overline{R}_i = R_{max} \exp\left(-kA_i^n\right)$$

or

$$\overline{R}_i = R_{max} - cA_i^m$$

where the coefficients k, n, c and m take values according to size of area and duration of storms. It has been shown in the USA (Hershfield & Wilson, 1960) that by comparing extreme rainfalls from tropical and non-tropical storms, the shape of the depth – area curve is not a function of storm type. However, short-duration storms tend to have steep rainfall gradients and hence cover smaller areas than storms of longer duration. This is well

Table 10.1   Area Enclosed by 150 mm Isohyet

($A = 10$ km²: areal rain = 155 mm)

	Station	Area weighting coefficient	6 h	12 h	18 h	24 h
1	1	0.7 Acc.rain	30	125	128	160
2		0.7 × Acc.rain	21	87.5	89.6	112
3	2	0.3 Acc.rain	27	120	124	153
4		0.3 × Acc.rain	8.1	36	37.2	45.9
5	Sum of weighted rain		29.1	123.5	126.8	157.9
6	Acc. rain adjusted to areal mean		28.6	121.2	124.4	155
7	Increment values		28.6	92.6	3.3	30.5
8	Max. possible depth for each duration		92.6	121.2	126.4	155

**Sample storm pattern for
a given duration**

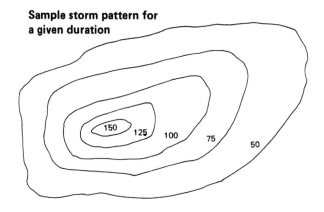

(1)	(2)	(3)	(4)	(5) = (4) x (3)	(6)	(7) = (6) ÷ (2)
Isohyet (mm)	Total area enclosed (km²)	Area bet. isohyets	Av. rain between isohyets	Vol. of rain between isohyets	Cum. vols.	Areal rain (mm)
150	10	10	155	1550	1550	155.0
125	25	15	137.5	2062.5	3612.5	144.5
100	45	20	112.5	2250	5867.5	130.4
75	75	30	87.5	2625	8487.5	113.2
50	110	35	62.5	2187.5	1067.5	97.0

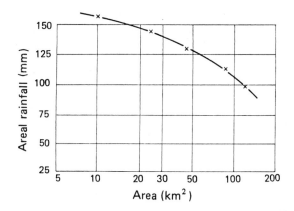

Fig. 10.5   Rainfall depth – area analysis.

recognized in the UK, where intense thunderstorms affect limited areas, whereas prolonged heavy rainfall from an occluded front, for example, usually covers a wide area. Thus no general formula can be valid for all areas and all durations.

10.3   Depth – Area – Duration (DAD)

In order to combine considerations of areal rainfall depths over a range of areas and varying durations of heavy falls, a more complex analysis of storm patterns is required. The technique of depth – area – duration analysis (DAD) determines primarily the *maximum* falls for different durations over a range of areas.

The data required for a DAD analysis are shown in Fig. 10.6. To demonstrate the method, a storm lasting 24 h is chosen and the isohyets of the total storm are drawn related to the measurements from 12 recording rain gauge stations. The accumulated rainfalls at each station for four 6-h periods are given in the table. It is not usual to find so many autographic records over a storm area, and total falls from storage gauges must be apportioned according to the traces of recording gauges to provide the breakdown into values for shorter durations. To provide area weightings to the gauge values, Thiessen polygons are drawn round the rainfall stations over the isohyetal pattern.

Table 10.2   Area Enclosed by 125 mm Isohyet

($A = 25$ km^2: areal rain = 144.5 mm)

Station	Area weighting coefficient		6 h	12 h	18 h	24 h
1	0.4		30	125	128	160
		0.4 × Acc.rain	12	50	51.2	64
2	0.3	Acc.rain	27	120	124	153
		0.3 × Acc.rain	8.1	36	37.2	45.9
3	0.1	Acc.rain	21	106	108	130
		0.1 × Acc.rain	2.1	10.6	10.8	13.0
4	0.2	Acc.rain	19	87	90	114
		0.2 × Acc.rain	3.8	17.4	18	22.8
Sum of weighted rain			26	114	117.2	145.7
Acc.rain adjusted to areal mean			25.8	113.1	116.2	144.5
Incremented values			25.8	87.3	3.1	28.3
Max. possible depth for each duration			87.3	113.1	118.7	144.5

Method repeated for areas enclosed by all the isohyets

First of all, the areal rainfall depths over the enclosing isohyetal areas are determined for the total storm, as in the previous section. The duration computations then proceed as in Table 10.1, where the area enclosed (10 km²) by

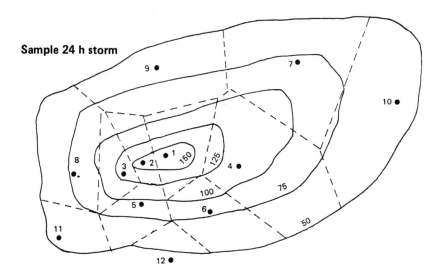

_____ Total storm isohyet

– – – – Thiessen polygons

Station	Accumulated rain from beginning of storm (mm)			
	6 h	12 h	18 h	24 h
1	30	125	128	160
2	27	120	124	153
3	21	106	108	130
4	19	87	90	114
5	17	70	72	90
6	14	55	60	80
7	14	53	58	78
8	15	60	62	76
9	17	42	50	65
10	10	30	40	60
11	12	40	45	55
12	12	35	40	45

Fig. 10.6   Depth – area – duration analysis.

the 150 mm isohyet is considered first. The areal rainfall over the 10 km² for the whole storm is 155 mm. Stations 1 and 2 are contained by the 150 mm isohyet and, by measurement, the proportions of the area represented by the two stations are 0.7 and 0.3, respectively. The values of rainfall in the four durations at the two stations are then multiplied by the corresponding areal weighting coefficients (rows 2 and 4 in the Table). The sum of the weighted values for each duration is then set in row 5, and in row 6, the sums are adjusted to the areal mean for the storm (i.e. scaled by 155 to 157.9). Incremental values for each of the four 6-h periods of the 24-h storm are calculated for row 7, and it will be noted that the highest fall occurs in the second 6-h period. The final row 8 then gives the maximum areal rainfalls for each duration, 6, 12, 18 and 24 h, by combining the appropriate sequences of incremental values. The maximum 6-h fall has already been noted, the maximum 12-h fall comes from the first two increments, but the maximum 18-h fall is the sum of the last three increments.

The second stage in the analysis is given in Table 10.2. The area enclosed by the next isohyet is treated similarly. Here, there are four gauges, stations 1, 2, 3 and 4, whose Thiessen polygons contribute to the area contained by the 125 mm isohyet. Their weighting areas are measured as 0.4, 0.3, 0.1 and 0.2, respectively. The area enclosed by the isohyet is 25 km², and the corresponding areal rain for the total storm is 144.5 mm. The maximum possible areal rainfall depth over the area of 25 km² for each duration is given in the final row.

The computations are continued by repeating the method for the areas

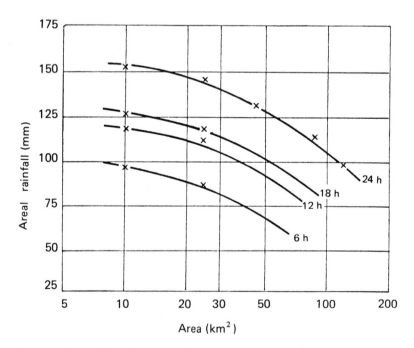

Fig. 10.7   Rainfall depth – area – duration curves.

Table 10.3   Maximum DAD Data (mm)

Area (km²)	Duration (h)					
	3	6	9	12	18	24
10	95	110	116	130	150	180
25	92	100	108	124	130	150
50	86	94	100	115	etc.	etc.
100	78	88	etc.			
150	68	etc.				
etc.						

enclosed by all the isohyets and the results plotted as shown in Fig. 10.7. As many storms as possible in the region under investigation should be analysed in this way. Naturally, the storms will have different total durations and then the range of durations analysed can be varied. The areal extent of each storm is taken to be limited by the lowest valued isohyet that bounds the storm area. Its value will be determined by the peak fall and the shape of the area covered. If the storm is widespread, several storm centres may be identified and then each centre is analysed separately.

Eventually a bank of data is compiled, the highest maximum DAD values can be plotted on a graph and the enveloping curves drawn for each duration. Additionally, a table of maximum areal rainfalls according to area and duration can be compiled (Table 10.3).

A major assumption is made in this DAD analysis, namely, that the pattern of the storm does not change with time. This is certainly not true for most storms and ideally, isohyets should be drawn for each duration required and new Thiessen part areas found each time. However, this adds considerably to the amount of work involved. DAD analysis can now be done on a computer and the technique based on the work of the former US Weather Bureau is fully described in the WMO Manual (WMO, 1969). Some choice as to how much of the analysis needs to be done by hand and how much can be computerized is left to the hydrologist and is dependent on the rainfall data available. Further studies to automate this tedious analysis have been made in South Africa (Pullen *et al.*, 1966), and a computer program has been prepared to provide DAD data using the muliquadric method of areal rainfall determination (Jones, 1973).

The *Flood Studies Report* (NERC, 1975) produced from a nationwide study of UK rainfall records a composite diagram (Fig. 10.8) giving relationships between point rainfall measurements and areal rainfall for a range of areas and durations. By applying an Areal Reduction Factor (ARF) to a rain gauge observation over a given duration, the areal rainfall over a selected area is obtained. Table 10.4 shows the ARFs for rainfall durations of 1 min to 25 days and areas 1 to 30 000 km². This is only applicable in the UK, but it demonstrates the results of widespread rainfall analyses extending the DAD method for storms and incorporating considerations of rainfall over wider areas and longer durations.

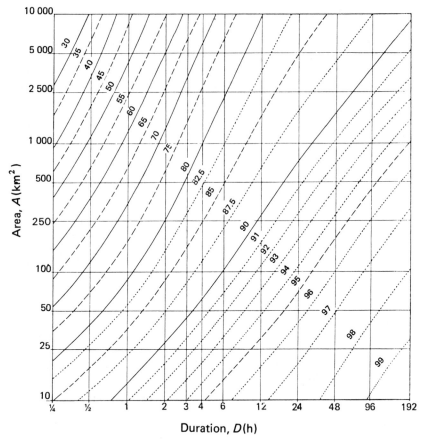

Fig. 10.8   Areal reduction factor (ARF), percentage, related to area A and duration D (for the UK). (Reproduced from National Environment Research Council (1975) *Flood Studies Report*, Vol. II, by permission of the Institute of Hydrology.)

10.4   Global Distribution of Precipitation

Engineering hydrologists are specialists and in the practice of their profession, they may not embark upon the wide range of activities that engage the attention of the general civil engineer. However, the hydrologist's skills are usually employed in the feasibility study stage of most engineering schemes, and international consultants may need hydrological knowledge of any part of the world. Thus an appreciation of the hydrological conditions pertaining in the major climatic regions of the world is essential.

The global distribution of precipitation amounts can be found in all good atlases and the causes of the seasonal fluctuations are well explained in elementary climatological or geographical textbooks. However, the definitions of the major global precipitation regimes merit inclusion here in Table 10.5. The regimes are illustrated by bar graphs in Fig. 10.9. A single rainfall

Table 10.4 Relation of ARF with Duration (D) and Area (A) for the UK

Duration D	Area A (km²)									
	1	5	10	30	100	300	1000	3000	10 000	30 000
1 min	0.76	0.61	0.52	0.40	0.27	—	—	—	—	—
2 min	0.84	0.72	0.65	0.53	0.39	—	—	—	—	—
5 min	0.90	0.82	0.76	0.65	0.51	0.38	—	—	—	—
10 min	0.93	0.87	0.83	0.73	0.59	0.47	0.32	—	—	—
15 min	0.94	0.89	0.85	0.77	0.64	0.53	0.39	0.29	—	—
30 min	0.95	0.91	0.89	0.82	0.72	0.62	0.51	0.41	0.31	—
60 min	0.96	0.93	0.91	0.86	0.79	0.71	0.62	0.53	0.44	0.35
2 h	0.97	0.95	0.93	0.90	0.84	0.79	0.73	0.65	0.55	0.47
3 h	0.97	0.96	0.94	0.91	0.87	0.83	0.78	0.71	0.62	0.54
6 h	0.98	0.97	0.96	0.93	0.90	0.87	0.83	0.79	0.73	0.67
24 h	0.99	0.98	0.97	0.96	0.94	0.92	0.89	0.86	0.83	0.80
48 h	—	0.99	0.98	0.97	0.96	0.94	0.91	0.88	0.86	0.82
96 h	—	—	0.99	0.98	0.97	0.96	0.93	0.91	0.88	0.85
192 h	—	—	—	0.99	0.98	0.97	0.95	0.92	0.90	0.87
25 days	—	—	—	—	0.99	0.98	0.97	0.95	0.93	0.91

station represents conditions in each region and the long-term monthly mean rainfalls are shown for January through to December. Under the station name is given the average annual rainfall in inches and millimetres. All the examples are in the Northern Hemisphere. Illustrations to show the seasonal patterns for the Southern Hemisphere would have the peaks for regions B1, B2 and C, falling in the southern summer months, and for regime E in the southern winter months. Variations in the patterns of rainfall within the regimes can be caused by local climatic effects and the amounts can be enhanced by orographic rains. In the regions experiencing rainfall all the year, certain seasonal effects can often be distinguished. For example, in the equatorial regions certain stations may have regular maximum months due to onshore monsoonal winds and in mid-latitudes, stations in the centre of the continents tend to have summer maxima with the smaller quantities in the winter being in the form of snow. In the desert and semi-arid areas, it is worthwhile noting that the rainfall usually occurs in heavy local storms. Indeed in any analysis of precipitation records, due regard should be paid to the causes of the rainfall and snowfall. A proper appreciation of the associated meteorological conditions should prevent erroneous conclusions being drawn from analyses and from any attempted extrapolations for design criteria.

10.5 Rainfall Frequency

The analysis of rainfall occurrence depends fundamentally on the length of

Table 10.5   Global Precipitation Regimes

(Reproduced from S. Pettersen (1958). *Introduction to Meteorology*, 2nd edn., by permission of McGraw-Hill Book Company.)

A. *Equatorial* 　　Within a few degrees N and S of Equator	Rain all year 2 max. seasons
B. *Tropical* 　　Between Equator and Tropics of Cancer and 　　Capricorn 　　(1) Inner zone to 10° N and S 　　(2) Outer zone 10° N and S to the Tropics	Summer rain Winter dry Two summer max. Single summer max. longer dry season
C. *Monsoon* 　　Within and outside the Tropics in the 　　Indian subcontinent on E side of continents	High summer max. Long dry season
D. *Subtropical* 　　Tropics to 30° N and S. W side of 　　continents 　　　At the margins — Poleward 　　　　　　　　　　Equatorward	Dry — desert areas very little rain Winter rain Summer rain
E. *Mediterranean* 　　Within 30° to 40° N and S. W side of 　　continents	Winter rain Summer dry
F. *Mid-latitudes*	Rain at all seasons
G. *Polar*	Precipitation at all seasons Very small amounts

the rainfall duration for which the information is required. In describing the measurement of precipitation, it has been emphasized that most data are provided by daily gauges, but it is the autographic gauges or rain recorders that identify the incidence of rain and give measures of rainfall quantities related to time. The daily storage gauges give rainfall totals without information as to its time of occurrence. Thus it is logical to consider rainfall frequencies for periods of a day and longer separately from shorter term duration falls derived from autographic data.

Considerable differences in rainfall quantities and the pattern of their occurrence are experienced in different climatic regimes. In general, the greater the annual rainfall amounts the less variable they are from one year to the next. In the semi-arid regions of the world with very limited amounts, the rainfall is irregular and unreliable. The seasonal pattern of rainfall has also significant effects on the analysis of its frequency. For the highly seasonal rains of the Tropics or Mediterranean regions it may be advisable to omit considerations of the dry months. When precipitation occurs all the year round, frequency analysis is more straightforward, and attention may be

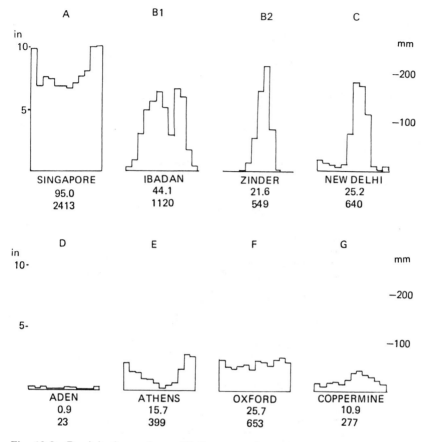

Fig. 10.9 Precipitation regimes. All diagrams refer to calendar years.

focused more readily on the frequency differences between the rainfall measured over different time periods. The pattern of occurrence of rainfall quantities for the day, the month and the year are shown in Fig. 10.10 for an imaginary rainfall station in the UK.

As many years of measurements as possible are used to determine rainfall frequencies. With a record of only 20 years for example, the determination of a representative frequency pattern for annual rainfall is unsatisfactory. In constructing the frequency bar graphs, the rainfall class limits must be chosen to give about 15 – 20 classes. (Detailed methods of class number selection are given in standard statistical texts, e.g. Brooks & Carruthers, 1953). In Fig. 10.10a, the daily bar graph shows the number of days with rainfalls grouped in 2 mm classes and it will be noted at once that the maximum frequency of occurrence is in the first interval from 0 to 2 mm. This reflects the large number of days without rain, which gives the characteristic shape of the J-distribution drawn on the diagram. If N is the total number of days of record and F is the number of days with falls less than x mm then

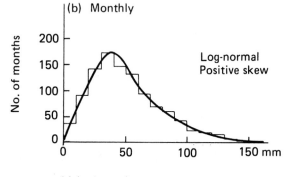

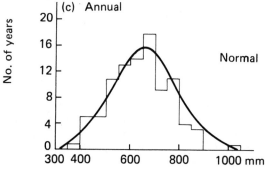

Fig. 10.10   Rainfall frequencies (UK).

$N - F$ = No. of days with falls $\geqslant x$. The equation of the J-distribution is assumed to be exponential and is given by:

$$N - F = k \exp(-jx)$$

or

$$\log_e(N - F) = a - jx$$

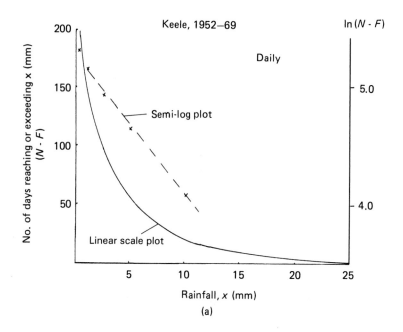

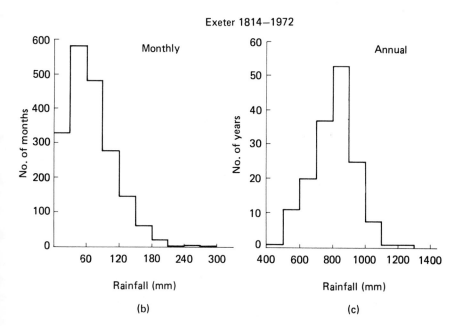

Fig. 10.11 (a) Average annual frequency of daily rainfall at Keele (1952–69). 171 days with nil rainfall. $(N\text{-}F) = 171 \exp(-0.22x)$. $\ln(N\text{-}F) = 5.14 - 0.22x$. (b) Rainfall frequencies at Exeter (1814–1972).

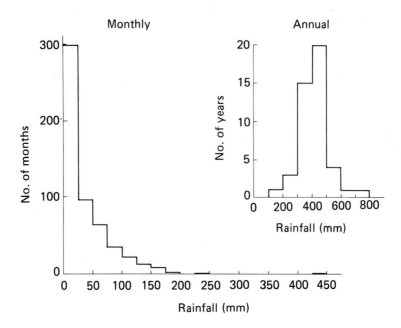

(a) Northern Transvaal 1926–1970 (South Africa, District 50) Regime B2 (21–24°S 27–32°E) Annual Average 420 mm

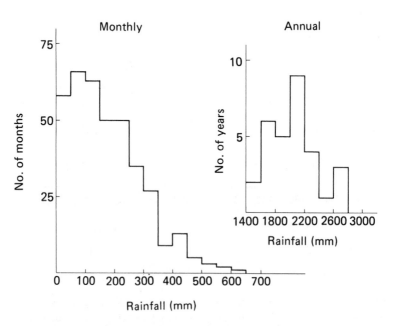

(b) Caucagua 1942–1974 (Venezuela) Regime B1 10° 17′ N 66° 23′ W 62 m O.D. Annual Average 2154 mm.

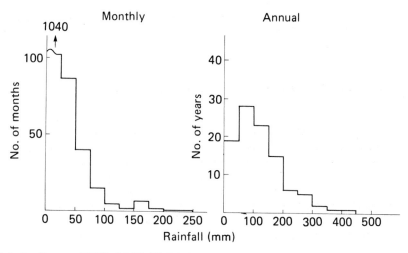

(c)   La Serena 1869–1968 (Chile) Regime E. 29° 54′ S 71° 15′ W. 132 m
OD. Annual average 126 mm

Fig. 10.12   Rainfall frequencies.

(An example is given in Fig. 10.11a, which shows plots on both linear and
semi-log scales.)

The monthly rainfall values (mainly comprising totals of daily rainfalls)
reach a maximum frequency within a few classes from the origin
(Fig. 10.11b), but the lowest class from 0 to 10 mm is still represented since it
is possible to have NIL rainfall in a month even in the UK. The continuous
frequency distribution of monthly rainfall taking all the months together
may be represented by the log-normal distribution (see Appendix). For some
hydrological analyses, the frequencies of January, February, etc rainfalls are
required. The log-normal distribution will also fit individual monthly rainfall totals abstracted from long UK records, each month having its individual
distribution parameter values.

Annual rainfall totals in the UK tend to reach maximum frequencies near
the middle of the range of values and approach the normal distribution. The
continuous curve of Fig. 10.10c does not correspond too well to the bar
graph; in the adoption of the normal distribution for annual rainfall, the
goodness-of-fit of the data should be tested. Examples of monthly and annual
frequencies for the long-term rain gauge station at Exeter are given in
Fig. 10.11b and c.

The application of the normal frequency distribution and its statistical
properties to rainfall values can be most helpful in assessing probabilities of
occurrence for water resource evaluation. Transformation of the original
non-normal rainfall data is often made to convert the distribution to
normality. Stidd (1953) has shown that the cube-root is a valid transformation for from daily up to annual rainfall totals.

In other climates, frequency distributions can be quite different from those found for rainfall values in the UK. Examples from a selection of regimes are shown in Fig. 10.12.

10.6 Intensity – Duration – Frequency Analysis

The analysis of continuous rainfalls, usually lasting for periods of less than a day, entails the abstraction of rainfall depths over specified durations. Assessment of occurrences is made by *intensity – duration – frequency* analysis. The general form of the relationship of rainfall quantity, R, with duration, t, of a continuous rainfall is shown in Fig. 10.13a. For a specific period of rain, a storm for example, the average intensity I is given by R/t. The units of this computed average rate-of-rainfall are usually millimetres per hour (mm h^{-1}). A general plot of I against t is shown in Fig. 10.13b, and a standard form of the relationship often used is $I = a/(t + b)$. In taking values

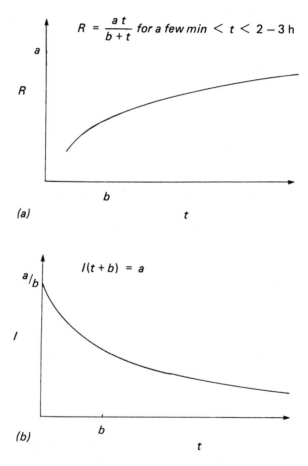

$$R = \frac{a\,t}{b + t} \text{ for a few } min < t < 2 - 3\,h$$

(a)

$$I(t + b) = a$$

(b)

Fig. 10.13   Rainfall intensity.

of duration from autographic charts, it is not always easy to determine from the pen trace when the rainfall can be deemed to have started particularly for light rainfall. For the precise definition of rainfall, it is usual to take the trace of the chart showing a rate greater than 0.1 mm h^{-1}. For engineering purposes, the commencement and termination of storm rainfalls, falls of high intensity, are more easily defined. The abstraction of short-period intense rainfalls from a set of daily charts over a number of years is a tedious business, and, as with depth – area analysis, many empirical formulae have evolved, mainly for application in the design of storm sewers for urban areas.

In evaluating the frequency of intense rainfalls, mention must be made of two classic studies. The later one of these, by Dillon (1954), took 35 years of

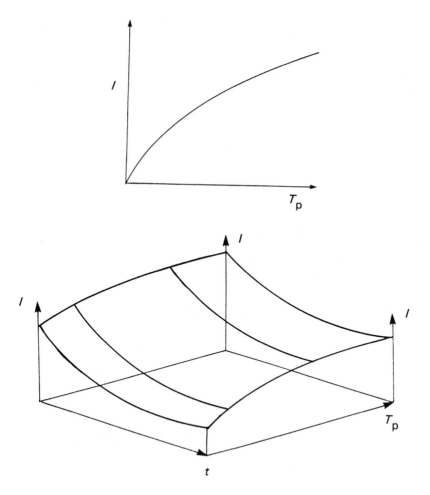

Fig. 10.14   (a) Rainfall intensity frequency. For a given duration, the average intensity, $I = a/F^b = aT_p^b$, where F = frequency, T_p = return period. (b) Intensity – duration – frequency. $I = ct^{-d}T_p^f$.

autographic recordings at Cork and identified the relationship of intensity and frequency of occurrence. In general terms, the higher the frequency of occurrence of a storm of given duration, the smaller the average intensity. This is shown in Fig. 10.14a where the inverse of the frequency, the return period, is plotted against intensity. Thus, the greater the return period (or recurrence interval) of a storm of given duration, the higher the average intensity. Combining the intensity, duration and frequency (return period) and making all three variable, gives the three-dimensional relationship of Fig. 10.14b, where $I = c\,t^{-d}\,T_p^{f}$ in which t is duration, T_p is return period and c, d and f are constants for a particular data set. Dillon obtained the simplified relationship for Cork:

$$I = 6\,t^{-3/5}\,T_p^{1/5}$$

with T_p related to frequency in 35 years. Fig. 10.15 is an example of a set of intensity – duration – frequency relationships for selected return periods.

The best known study of the frequency of short-period continuous rainfall in the UK is that by Bilham published in 1936 and reissued by the Meteorological Office in 1962. Bilham, who was a civil engineer, assembled 10 years of autographic data from 12 stations representative of lowland England and Wales with average annual rainfall under 35 inches (889 mm) and from their analysis developed the formula:

$$n = 1.25\,t\,(R + 0.1)^{-3.55}$$

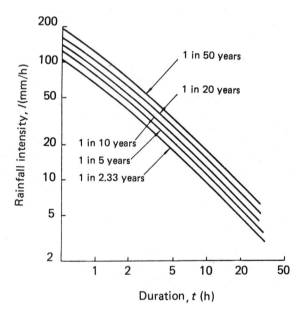

Fig. 10.15   Rainfall intensity – duration – frequency curves for Medan, Sumatra. (Reproduced from J. Wild & J.K. Hall (1982) *Proc. Inst. Civil Eng.*, **73**, 85–108, by permission.)

where n is the number of events in 10 years, R is the rainfall depth in inches and t is the duration in hours. The equation was valid for durations from 5 min to 2 h. With the accumulation of longer series of records for a wider area, the Bilham formula was updated and simplified by Holland (1967) to give $n = tR^{-3.14}$, using the same units and with the duration extended to 25 h.

Following the publication of the UK *Flood Studies Report* (NERC, 1975), the design engineer in the UK is provided with the frequencies or return periods of rainfalls for any time period from 1 min to 48 h. Data both from actual rate-of-rainfall measurements and from autographic records from stations over the whole of the UK have been analysed in great detail. From estimated point values of rainfall depth – duration – frequency, comparable areal values can be obtained by applying the ARF (see Fig. 10.8). In this major study, which will receive more detailed consideration in Chapter 17, the frequencies of short-duration continuous rainfalls were combined with the data from daily storage-gauge observations to increase the length of records available, but due regard was taken of the varying characteristics of the different measurements.

There are few countries in the world with such comprehensive sampling of rainfall as the UK, and the use of such data in the detailed Flood Studies Report (NERC, 1975) has been widely acclaimed. The techniques for rainfall analysis demonstrated in that report are proving to be models for many practising hydrologists. Though the techniques by themselves are not all new, their judicious and ordered application to the British rainfall data provides a modern alternative to the earlier exemplary studies of the former US Weather Bureau and the US Department of Agriculture that are to be found in most textbooks.

For countries without adequate rainfall observations, Bell (1969) has derived helpful generalized rainfall – duration – frequency relationships from the detailed measurements made in the USA and using data from USSR, Australia and South Africa. His analyses are confined to rainfalls of up to 2 h duration, since he assumes that intense rainfalls are most often caused by convective cells of such short duration that occur in many parts of the world. The ratios of the 1-h rainfalls for selected return periods to the 10-year, 1-h rainfall are consistent for each return period. Similarly, the ratios of the 1-h rainfalls for selected return periods to the 2-year, 1-h rainfall can also give a good constant relationship though with greater variability. If the first set of ratios is plotted on Gumbel (extreme value) paper, it can be defined in terms of return period as follows:

$$\frac{R_T^t}{R_{10}^t} = 0.21 \ln T + 0.52 \text{ for } 2 \leqslant T \leqslant 100 \text{ years}$$

where the ratio is of the T year, t-min depth (R_T^t) to the 10-year, t-min depth (R_{10}^t). Depth – duration ratios as a function of duration are given by:

$$\frac{R_T^t}{R_T^{60}} = 0.54 \, t^{0.25} - 0.50 \text{ for } 5 \leqslant t \leqslant 120 \text{ min}$$

where R_T^t is again the T-year, t-min rainfall and R_T^{60} is the T year, 1-h fall.

The two equations may be combined to give a generalized relationship for rainfall depth – duration – frequency:

$$R_T^t = (0.21 \ln T + 0.52)(0.54\, t^{0.25} - 0.50)R_{10}^{60}$$

for $2 \leqslant T \leqslant 100$ years and $5 \leqslant t \leqslant 120$ min. Thus the rainfall depth for any duration and any return period within the limits given can be estimated from a knowledge of the 10-year, 1-h rainfall. This generalized relationship provides a very useful method of estimating design storm data for countries with limited data. The comparable generalized formula based on the 2-year, 1-h rainfall is:

$$R_T^t = (0.35 \ln T + 0.76)(0.54\, t^{0.25} - 0.50)R_2^{60}$$

for $2 \leqslant T \leqslant 100$ years and $5 \leqslant t \leqslant 120$ min, but it is considered less reliable, particularly for the longer-return periods.

10.7   Extreme Values of Precipitation

The hydrological extremes of floods and droughts, with their acute effects on human affairs, inevitably provide a large proportion of the work of engineering hydrologists. The simple causes of these notable events are a great surplus of water on the one hand and a great dearth on the other. The first, giving floods, can result from extreme rainfalls, the rapid thawing of a large accumulations of snow, or a combination of both, depending on location and season. Rainfall maxima will be considered here, but the behaviour of floods will be dealt with in the analysis of river flows (Chapter 12). Droughts may also be considered by analysing low river flows, but in many areas, droughts result in dry river beds. However, the study of rainfall deficiencies can give more generally meaningful measures of drought, especially to agriculturalists, and some of these drought studies will be described in this chapter.

10.7.1   Rainfall Maxima

The world's highest recorded rainfalls for a wide range of durations were assembled by Jennings (1950), and these formed the basis for what has become a well known log – log plot of rainfall in inches against durations from 1 min to 24 months. The records in the higher durations stemmed from Cherrapunji, India, with a maximum 2-year total of 1605 inches (41 000 mm) in 1860–61. In more recent years, some of these records have been broken by measured falls from tropical cyclones in the mountains of La Reunion, an island in the Indian Ocean, and of Taiwan (Paulhus, 1965). A revised graph with the values in millimetres is given in Fig. 10.16. The points lie approximately on a straight line and thus drawing an enveloping line just through the top values gives a unique relationship between the maximum values, R, and duration, D. Comparable maximum values have also been abstracted for the UK (IOH, 1968). It will be noted that the similarly drawn straight line has nearly the same slope and only differs from the world formula in the constant of proportionality.

Table 10.6   Maximum Daily Rainfall

World		inches	mm
Cilaos, La Reunion	Mar. 15 1952	73.62	1870
Belouve, La Reunion	Feb. 27 1964	66.49	1689
Aurere, La Reunion	April 7 1958	62.33	1583
Paishih, Taiwan	Sept. 10 1963	49.13	1248
Halaho, Taiwan	Sept. 10 1963	46.98	1193
Baguio, Phillipines	July 14 1911	45.99	1168
Cherrapunji, India	June 14 1876	40.70	1034
UK			
Martinstown, Dorset	July 18 1955	11.00	279
Bruton, Somerset	June 28 1917	9.56	243
Cannington, Somerset	Aug. 18 1924	9.40	239
Longstone Barrow, Exmoor	Aug. 15 1952	9.00	229

Although high values of annual and monthly precipitation are welcomed by water resource engineers, they do not usually give rise to floods and therefore do not demand special attention. It is the high rainfalls of shorter durations that are closely watched with respect to the formation of floods. Since most rainfall data are in the form of daily totals, considerable attention has been given to maximum daily falls. Table 10.6 lists some of the world maxima, all from tropical regions, and the highest daily falls measured in the UK. A most useful study of heavy daily falls in the UK was made by Bleasdale (1963) for the years 1893 – 1960. He listed the 142 occasions on which more than 5 inches (127 mm) had been recorded, and noted that all the falls greater than 4 inches (102 mm) numbered over 450. From this

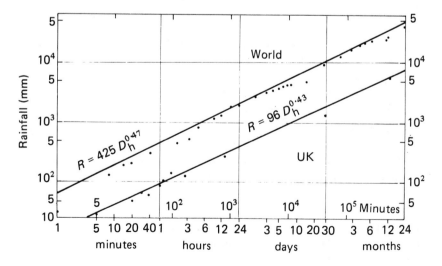

Fig. 10.16   World and UK highest recorded rainfalls.

study, Bleasdale concluded that no part of the UK could be completely immune from a daily fall of at least 102 mm. The study of heavy daily falls was extended by Rodda (1967) who took 121 stations distributed over the UK, nearly all with 50 years or more of daily data. For each record, the annual maximum daily falls were fitted to the Type 1 Gumbel extreme-value distribution. For that distribution, the probability of the annual maximum fall, x, in any year being less than or equal to a magnitude x is given by:

$$P(X \leqslant x) = \exp\left(e^{-(\alpha + x)/c}\right)$$

in which the parameters are given by: $\alpha = \gamma c + \mu$, where μ is the mean of the annual maximum falls; $\gamma = 0.57721$; and $c = \sigma \sqrt{6}/\pi$ where σ is the standard deviation of the annual maximum falls.

The daily rainfalls for selected return periods were then plotted on maps of the country and appropriate isohyets drawn (Rodda, 1973). A more useful result for the engineer from this study proved to be the relationship of the one-day maximum rainfall for given return periods to the average annual rainfall of a station. This is known in Fig. 10.17. Knowing the average annual rainfall from the map, Fig. 10.18, the maximum one-day rainfall in a year may be obtained for return periods of 2, 10, 50 or 100 years.

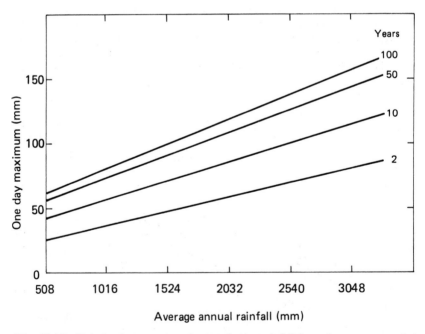

Fig. 10.17   Relation between one-day maximum rainfall for a given return period and average annual rainfall. (Reproduced from J.C. Rodda (1967) *J. Hydrol*, **5**, 58–69, by permission of Elsevier Publishing Company.)

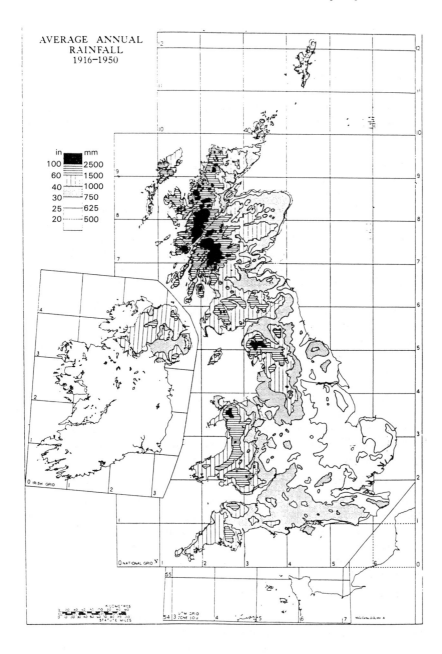

Fig. 10.18 Average annual rainfall, 1916–1950. (Reproduced by permission of the Controller, Her Majesty's Stationery Office, © Crown copyright.)

10.7.2   Probable Maximum Precipitation

As more and more rainfall measurements become available throughout the world with the increasing number of rain gauges and with the lengthening of existing series of observations, records of maximum rainfall for the various durations continue to be broken. The question arises as to whether records will always continue to be superseded or whether there is a physical upper limit to rainfall. This concept of a finite limit has been named the probable maximum precipitation (PMP), the existence of which has aroused much controversy. The PMP will of course vary over the Earth's surface according to the climatic or precipitation regime. It would be expected to be much higher in the hot humid equatorial regions than in the colder regions of the mid-latitudes where the atmosphere is not able to hold so much moisture. The concept of PMP, to which no return period or recurrence interval can be attached, is attractive to the engineer responsible for the designing of an impounding dam that must never fail nor be overtopped by flood flows. A definition of PMP has been given by Wiesner (1970) as 'the depth of precipitation which for a given area and duration can be reached but not exceeded under known meteorological conditions'. When a value of PMP is required for a specific project, it is usual to estimate it by several methods and then make an engineering judgement on the value to be used.

There are two main groups of methods for the evaluation of PMP. One uses statistical techniques applied to the measurements of extreme rainfalls. The second group studies the storm mechanisms causing heavy rainfalls. These latter studies are generally undertaken by meteorologists.

Statistical methods.   As an alternative to taking a maximum rainfall for a selected duration from Fig. 10.16, the annual maxima for the necessary duration from long autographic rainfall records are fitted to an extreme value distribution. Then for such a distribution, it can be shown that R_T (the rainfall depth in the given duration of return period T) is:

$$R_T = \bar{R} + K(T)\,\sigma$$

where $\bar{R}$ is the mean of the annual maximum rainfalls, σ is their standard deviation and $K(T)$ is known as the frequency factor and is a function of T.

For a PMP, a maximum value of $K(T)$ has to be specified to give a value of R_T that would not be exceeded. Although in principle $K(T)$ gets larger indefinitely with increasing T, it may be considered that $K(T)$ has some maximum value, K_m. Thus

$$PMP = \bar{R} + K_m \sigma$$

It has been suggested that $K_m = 15$ gives an estimate of PMP (Hershfield, 1961). This is a very useful and simple method for short-duration storms over small areas, but is dependent both on the length of records available for $\bar{R}$ and σ and on the assumption of the maximum value of the frequency factor.

A second statistical approach uses DAD analysis. For each of the storms used for the DAD determinations, a representative surface dew point is evaluated for the area. Then the maximum DAD values are further maximized by scaling them up by the ratio of the maximum precipitable water content possible at the highest dew point that can occur in the region W_{max} (usually taken to be 78 °F, 26 °C) to the precipitable water content (at the dew point) W_{actual}. Thus a PMP is obtained for a required duration from the relevant DAD values multiplied by the moisture adjustment factor W_{max}/W_{actual}.

Meteorological methods. These require considerable meteorological information; they extend the principle of maximizing rainfall values by taking the maximum precipitable water possible and substituting it into individual storm analyses. Where there are areas that have not experienced high-intensity storms, knowledge of extreme storms is transposed from neighbouring areas. Storm transposition takes into account storm movement, orographic effects and precipitable water content of the atmosphere from dew point measurements.

PMP is also estimated by computing precipitation from storm models. The adoption of a storm model identifies the significant factors in an extreme situation so that these can be maximised in an actual storm to given an estimate of PMP. For details of these techniques, the reader is referred to Wiesner (1970).

10.7.3   Droughts

The low extreme of precipitation is more easily defined and recognized, but the lack of rainfall has varying significance in the different climatic regimes in the world. In some of the arid zones, for example, there may be several years in which no measurable precipitation occurs and the flora and fauna are adapted to these (normal) conditions. The shortage of rainfall in normally humid parts of the world can, however, result in serious water deficiencies, as for example in a failure of the monsoon in parts of India. Thus drought has widely different connotations according to location and consequences.

A drought is usually considered to be a period in which the rainfall consistently falls short of the climatically expected amount, such that the natural vegetation does not flourish and agricultural crops fail. Since such extremes are rarely experienced in the UK, low rainfalls more seriously affect water supplies for industry and domestic purposes. In the UK with regular rainfall all the year round, the occasions of shortages were strictly defined in quantitative terms by the former British Rainfall Organization as follows:

(a) An *absolute* drought is a period of at least 15 consecutive days none of which has recorded as much as 0.01 inch (0.3 mm) of rain.
(b) A *partial* drought is a period of at least 29 days during which the average daily rainfall does not exceed 0.01 inch (0.3 mm); but these criteria are no longer used in UK Meteorological Office publications.

They can be applied by inspection of records at daily rainfall stations and are of some assistance to farmers and water supply engineers. However, an improvement in data presentation indicative of rainfall deficiencies is given in Fig. 10.19 showing accumulated departures from mean daily rainfall over a year. A dry year (1963) with an overall deficit of 213 mm is compared with an average year (1961). A water engineer could usefully start such a diagram at the beginning of a water year and keep a day-to-day account of the departures from normal. This type of diagram was published in *British Rain-*

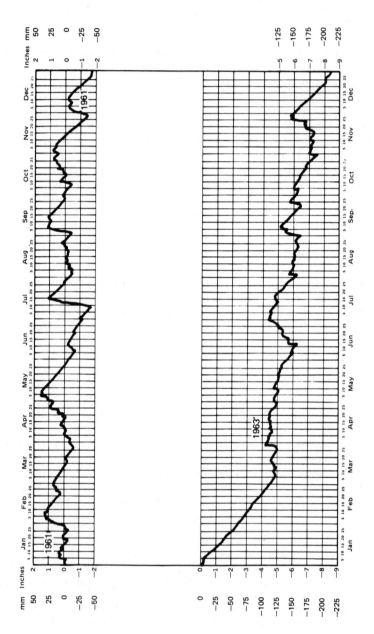

Fig. 10.19 Cumulative departure from mean daily rainfall.

fall (1961–68) for selected stations; it is a very good example of the output of the increased computational facilities afforded by large digital computers.

The evaluation of water resources requires the appraisal of the incidence of rainfall quantities over periods longer than daily sequences in order to determine storage capacities to meet demands. Hence in the UK more importance is attached to the study of monthly rainfall sequences and to the identification of rainfall deficiencies in the long term.

One of the most exhaustive statistical studies of long rainfall records was made by Ashmore (1944) for two stations in North Wales. He produced the following statistics, among many others, on sequences of dry months for Pentrebychan (Table 10.7). Comparable data have recently been compiled for England and Wales (Murray, 1977). The point values for the Welsh rain gauge station are representative of conditions in the rain shadow of the Welsh Mountains, whereas the England and Wales values pertain to averages calculated to represent conditions over the two countries.

Table 10.7   Driest Periods of Consecutive Months

Months	Pentrebychan 1880–1942, total fall (mm)	England & Wales 1820–1976, total fall (mm)
3	50 Feb.–Apr. 1929	56 Feb.–Apr. 1938
6	149 Feb.–July 1921	179 Feb.–July 1921
12	496 Mar. 1933 – Feb. 1934	570 Sept. 1975 – Aug. 1976
18	831 Arp. 1933 – Sept. 1934	908 Mar. 1975 – Aug. 1976
24	1201 Dec. 1932 – Nov. 1934	1439 Oct. 1853 – Sept. 1855

It is from such information that water engineers can assess the potential safe yield of upland catchment areas; a typical assumption is that about 80% of normal annual rainfall will arrive over the driest three consecutive years. (The driest 36 consecutive months at Pentrebychan yielded 2050 mm, 80.4% of the annual average.)

More refined analyses of sequences of dry periods using monthly rainfall records have come from Australia. Foley (1957) adapted the residual-mass technique, taking departures from monthly averages, reducing them to proportions of the average annual rainfall to overcome seasonal variability, and accumulating the residual proportions. He computed an index of drought severity that was related to the steepness of the fall of the residual mass curves. Further studies (Maher, 1966) endeavoured to define regions of Australia that had similar drought severities by cross correlation of long rainfall records.

The use of cumulative residuals or departures from average was extended in South Africa (Herbst *et al.*, 1966) to confine the definition of droughts more specifically to periods in which rainfall deficits were in excess of *average* deficits. Thus sequences of months with extremely dry conditions are identified beyond the shortfalls in monthly rainfall amounts that are normally experienced in some months of each year. This definition of a drought is given in Fig. 10.20. In addition, an average monthly drought *intensity* over

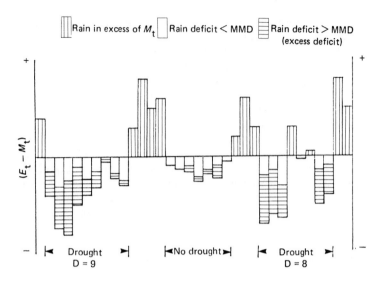

Fig. 10.20   Beginning and ending of droughts. (Reproduced from P.H. Herbst, D.B. Bredenkamp & H.M.G. Barker (1966) *J. Hydrol.*, **4**, 264–272, by permission of Elsevier Publishing Company.)

the period of a drought can be calculated from the following expression:

$$Y = \frac{\sum_{t=1}^{D}\left[(E_t - M_t) - (MMD)_t\right]}{\sum_{t=1}^{D}(MMD)_t}$$

where E_t = effective rainfall = $(R_{t-1} - M_{t-1})W_t + R_t$ (M_t is the monthly mean rainfall, W_t is a carry-over factor, ($t = 1...12$) and R_t are the monthly rainfalls), $(MMD)_t$ = monthly mean deficit ($t = 1, 2, ..., 12$) and D = drought duration in months.

$$MMD_t = \sum_{t=1}^{n}\frac{(M_t - R_t)}{n}$$

A *severity* index is given by $Y \times D$.

The Herbst method has been applied to the long-rainfall record at Kew Observatory 1607 – 1967 (Shaw, 1979) and the top ten droughts ranked according to their severity index are shown in Table 10.8. This study was made after the severe drought over England and Wales from May 1975 to August 1976. The drought *intensity* is the highest at Kew over the 15 month period June 1975 – August 1976, but it ranks only fifth in severity index value of the droughts identified in the long record. The high intensity reflects the low percentage, 53%, of the average rainfall for the 15-month period.

The regional variations of droughts in England and Wales were investigated by applying the Herbst method to 72 rain gauge stations with records from 1911. The most severe widespread drought occurred over an average 25-month period centred on 1933. The most seriously affected areas were East Anglia and Southeast England, but a narrow belt in the Midlands also had high severities (Fig. 10.21) (Shaw, 1979).

This technique for defining droughts from rainfall records is extremely useful, since it is quite objective. It can be used in most climatic regimes, the main requirement being homogeneous series of monthly rainfalls, and these are often more readily available than comparable runs of river flow records.

For hydrological applications, there is a limit to the information on extreme events that can be gleaned from precipitation records alone. The consequences of excess or deficient rainfalls depend so much on the nature

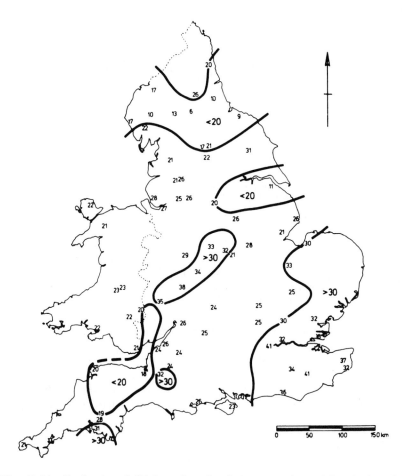

Fig. 10.21 England and Wales selected raingauge stations. Map 1 drought severity. Average 25.0.

Table 10.8   The Ten Maximum Droughts at Kew

Period	Intensity	Duration (months)	Severity	% of mean rain for period
Oct. 1931 – Nov. 1934	1.07	38	40.8	76.1
Aug. 1920 – Mar. 1924	0.87	44	38.3	81.3
Sept. 1971 – Aug. 1974	1.01	36	36.3	75.9
July 1947 – Sept. 1949	1.14	27	30.8	68.9
June 1975 – Aug. 1976	1.95	15	29.3	53.2
Dec. 1741 – Aug. 1744	0.85	33	29.0	76.7
Nov. 1703 – Sept. 1705	1.23	23	28.4	67.5
Nov. 1862 – July 1865	0.83	33	27.5	79.8
Aug. 1713 – Feb. 1715	1.34	19	25.5	57.5
July 1746 – Nov. 1748	0.87	29	25.2	83.2

and antecedent condition of the ground that other hydrological processes must be investigated. The solving of engineering problems caused by extreme events only begins with the analysis of the precipitation records.

References

Ashmore, S.E. (1944). 'The rainfall of the Wrexham District.' *Quart. Jl Roy. met soc.*, **70**, 241–269.

Bell, F.C. (1969). 'Generalised rainfall-duration-frequency relationships.' *Proc. ASCE*, **95**, HY1, 311–327.

Bilham, E.G. (1936). 'Classification of heavy falls in short periods.' *British Rainfall 1935*, 262–280.

Bleasdale, A. (1963). 'The distribution of exceptionally heavy daily falls of rain in the United Kingdom, 1863 – 1960.' *J. Inst. Water Eng.*, **17**, 45–55.

Brooks, C.E.P. and Carruthers, N. (1953). *Handbook of Statistical Methods in Meteorology*. HMSO, M.O. 538.

Court, A. (1961). 'Area-depth rainfall formulas.' *J. Geophys. Res.*, **66** (6), 1823 – 1831.

Dillon, E.C. (1954). 'Analysis of 35-year automatic recordings of rainfall at Cork.' *Trans. Inst. Civ. Eng. Ireland*, **80**, 191–283.

Foley, J.C. (1957). *Droughts in Australia*. Bureau of Meteorology Bulletin No. 43.

Herbst, P.H., Bredenkamp, D.B. and Barker, H.M.G. (1966). 'A technique for the evaluation of drought from rainfall data.' *J. Hydrol.*, **4**, 264–272.

Hershfield, D.M. (1961). 'Estimating the probable maximum precipitation.' *Proc. ASCE*, **87**, HY5, 99–116.

Hershfield, D.M. and Wilson, W.T. (1960). 'A comparison of extreme rainfall depths from Tropical and Non-tropical storms.' *J. Geophys. Res.*, **65** (3), 959–982.

Holland, D.J. (1967). 'Rain intensity frequency relationships in Britain.' *British Rainfall 1961*, Part III, 43–51.

Institute of Hydrology (IOH).(1968). *Research 1968*, 48 pp.

Institution of Water Engineers (IWE).(1937). 'Report of Joint Committee to consider methods of determining general rainfall over any area.' *Trans. Inst. Wat. Eng.*, **XLII**, 231–259.

Jennings, A.H. (1950). 'World's greatest observed point rainfalls.' *Mon. Weath. Rev.*, **78** (1), 4–5.

Jones, A.E. (1973). *Surface Fitting Techniques Applied to Rainfall Depth Area-Duration Analysis*. M.Sc. dissertation Part 2, Imperial College, London.

Maher, J.V. (1966). *Drought Assessment by Statistical Analysis of Rainfall*. WMO Seminar on Agrometeorology, Melbourne.

Murray, R. (1977). 'The 1975/76 drought over the United Kingdom hydrometeorological aspects.' *Meteorol. Mag.*, **106**, 1258, 129–145.

National Environment Research Council NERC (1975). *Flood Studies Report*, Vol. II.

Paulhus, J.L.H. (1965). 'Indian Ocean and Taiwan rainfalls set new records.' *Mon. Weath. Rev.*, **93** (5), 331–335.

Pettersen, S. (1958). *Introduction to Meteorology*, 2nd ed. McGraw-Hill.

Pullen, R.A., Wiederhold, J.F.A. and Midgley, D.C. (1966). 'Storm Studies in South Africa: Large area storms, depth-area-duration analysis by digital computer.' *The Civil Engineer in South Africa*, **8** (6), 173–187.

Rodda, J.C. (1967). 'A country-wide study of intense rainfall for the United Kingdom.' *J. Hydrol.*, **5**, 58–69.

Rodda, J.C. (1973). 'A study of magnitude, frequency and distribution of intense rainfall in the United Kingdom.' *British Rainfall 1966*, Part III, 204–215.

Shaw, E.M. (1979). 'The 1975/76 drought in England and Wales in perspective.' *Disasters*, **3** (1), 103–110.

Shaw, E.M. and Lynn, P.P. (1972). 'Area rainfall evaluation using two surface fitting techniques. *Bull. IAHS*, **XVII** (4), 419–433.

Stidd, C.K. (1953). 'Cube-root-normal precipitation distributions.' *Trans. AGU*, **34**, 31–35.

Thiessen, A.H. (1911). 'Precipitation for large areas.' *Mon. Weath. Rev.*, **39**, 1082–1084.

Wiesner, C.J. (1970). *Hydrometeorology*. Chapman and Hall.

Wild, J. and Hall, J.K. (1982). 'Aspects of hydrology in the province of North Sumatra, Indonesia.' *Proc. Inst. Civ. Eng.*, **73** (2), 85–108.

World Meteorological Organization (WMO). (1965). *Guide to Hydrometeorological Practices*. WMO No. 168, TP82, A5.

World Meteorological Organization (WMO). (1969). *Manual for Depth-Area-Duration Analysis of Storm Precipitation*. WMO No. 237, TP 129.

11

Evaporation Calculations

The factors governing evaporation from open water and from a vegetated surface are described in Chapter 4. In that chapter, methods of measurement of water losses are outlined. The practical approach of those methods considers the evaporation process from the liquid phase as the loss rate from the water body with the final measurements being given in the form of an equivalent depth of water lost over a selected time period. The evaluation of evaporation into the gaseous phase considers the gain of water vapour by the air above the open water or vegetation, i.e. the absorption of the water vapour by the air measured over a period of time. The instrumentation required to measure the relevant properties of the air is described in that earlier chapter. Here the related analysis and computations using the measured meteorological variables are set out.

11.1 Calculation of E_o

There are two major approaches that may be adopted in calculating evaporation from open water, E_o. The *mass transfer* method, sometimes called the *vapour flux* method, calculates the upward flux of water vapour from the evaporating surface. The second or *energy budget* method considers the heat sources and sinks of the water body and air and isolates the energy required for the evaporating process. A third method uses a combination of the two physical approaches.

11.1.1 *Mass Transfer Methods*

The vapour flux calculations can be subdivided according to the siting of the meteorological sensors. In Fig. 11.1, three sets of measurements are shown diagrammatically and each gives rise to a separate method of calculation of E_o. It is customary to let e represent the saturated vapour pressure of the air, whence e_d is the saturated vapour pressure of the air at T_d, the dew point, and is equivalent to the vapour pressure at T_a (see Fig. 1.3).

The most straightforward method uses the bulk aerodynamic equation originating with Dalton in the early 19th century:

$$E = f(u)(e_s - e_d) \tag{11.1}$$

where E is the evaporation rate, $f(u)$ is a function of wind speed, and $(e_s - e_d)$ is the saturation deficit. Thus the evaporation is related to the wind speed and is proportional to the vapour pressure deficit, the difference between the

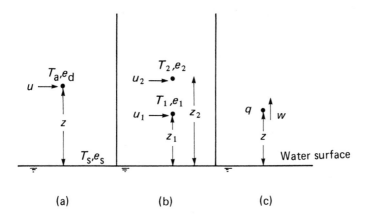

Fig. 11.1 Measurements for mass transfer. (i) and (ii) T_s = temperature of the water surface, T_a = air temperature at height z (T_1 and T_2 at two heights z_1 and z_2), e_s = saturated vapour pressure of air at water surface, e_d = vapour pressure of air (e_1 and e_2 at heights z_1 and z_2), and u = horizontal wind velocity at height z (u_1 and u_2 at heights z_1 and z_2). (iii) w = vertical wind velocity at height z, q = specific humidity at height z.

saturated vapour pressure at the temperature of the water surface and the actual vapour pressure of the air above.

$f(u)$ may take two forms: $a(b + u)$ and Nu, where a, b and N are empirical mass transfer coefficients. Many investigations have produced values of a, b and N for various conditions and measurements. Special care is needed in noting the height of the anemometer and thermometers above the surface. In the UK, the detailed studies of Penman (1948) using the first form of $f(u)$, resulted in:

$$E_0 = 0.35 (0.5 + u_2/100) (e_s - e_d) \qquad (11.2)$$

for which the air measurements are made at 2 m above the surface. The vapour pressures are in mm of mercury, wind speed in miles per day and E_0 is in mm day^{-1}.

In the USA and Australia, the second form of the wind term, Nu, is more commonly used, and again values of the mass transfer coefficient N are dependent on the height and units of the air measurements. Harbeck and Meyers (1970) give values of N from three famous studies. Thus in the general equation:

$$E_0 = N u_2 (e_s - e_d) \qquad (11.3)$$

with wind speed at 2 m in m s^{-1}, vapour pressures in mb and E_0 in cm day^{-1}. Harbeck and Meyers (1970) found:

$N = 0.0120$ for Lake Hefner
$N = 0.0118$ for Lake Mead and
$N = 0.01054$ for Falcon Reservoir.

The last value of N was obtained from detailed measurements made over 2 years by equating the aerodynamic E_o with E_o values calculated by the energy budget method (to be described later). From a study of numerous reservoirs of different sizes up to 12 000 hectares, the further factor of surface area can be incorporated into the equation to determine evaporation loss from a reservoir (Harbeck, 1962). Thus:

$$E_o = 0.291 A^{-0.05} u_2 (e_s - e_d) \text{ mm day}^{-1} \tag{11.4}$$

with A in m^2, u_2 in m s^{-1} at height 2 m, and e_s and e_d in mb.

Example.   Calculate the annual water loss from a 5 km^2 reservoir, when u_2 is 10.3 km h^{-1} and e_s and e_d are 14.2 and 11.0 mm of mercury, respectively.

$$A = 5 \text{ km}^2 = 5 \times 1000^2 m^2$$
$$u_2 = 10.3 \text{ km h}^{-1} = \frac{10.3 \times 1000}{60 \times 60} = 2.86 \text{ m s}^{-1}$$
$$e_s = 14.2 \text{ mm of mercury} = 14.2 \times 1.33 = 18.9 \text{ mb}$$
$$e_d = 11.0 \text{ mm of mercury} = 11.0 \times 1.33 = 14.6 \text{ mb}$$
$$\therefore E_o = 0.291 (5 \times 1000^2)^{-0.05} \times 2.86 (18.9 - 14.6) \text{ mm day}^{-1}$$
$$= \frac{0.291 \times 2.86 \times 4.3}{2.16}$$
$$= 1.66 \text{ mm day}^{-1}$$
$$= 606 \text{ mm year}^{-1} \text{ (assuming constant loss rate)}$$

Total annual water loss from reservoir = $0.606 \times 5 \times 1000^2 = 3.03 \text{ Mm}^3$.

The second vapour flux method of calculating evaporation uses air measurements at two fairly close levels above the water surface and considers the turbulent transfer of water vapour through the small height difference. The equation for E_o due to Thornthwaite and Holtzman (1939) takes the form:

$$E_o = \frac{0.623 K^2 \rho (u_2 - u_1)(e_1 - e_2)}{p (\ln z_2/z_1)^2} \text{ cm s}^{-1} \tag{11.5}$$

where e_1 and e_2 are vapour pressures (mb) at heights z_1 and z_2 (cm), u_1 and u_2 are wind speeds (cm s^{-1}) at heights z_1 and z_2 (cm), p is atmospheric pressure (mb), ρ is the density of air (g cm^{-3}), and K is von Karman's constant = 0.41. This equation is valid only for neutral conditions when the lapse rate of fall of temperature with height is small and the mechanism of the vapour transfer is by frictional turbulence. With greater heating of the ground and an increased lapse rate, the vapour flow is affected by convective turbulence. Under such conditions, and also in temperature inversions, the logarithmic relationship of wind speed with height used in the Thornthwaite – Holtzman equation does not hold. Much research has been undertaken by meteorologists on this approach to evaporation but owing to the very varied atmospheric conditions, the method remains too complex for general application by the engineer.

The direct eddy-flux or eddy-transfer method calculates energy for evaporation from measurements of vertical wind velocity and vapour content of the air at a single point above the evaporating surface:

$$E_o = L \, \overline{(\rho w)' \, q'} \, (\text{mW cm}^{-2}) \qquad (11.6)$$

where ρ is the air density (g cm^{-3}), w is the vertical wind speed (m s^{-1}), q is the specific humidity and L is the latent heat of vaporization of water $(2.47 \times 10^6 \, \text{J kg}^{-1})$.

The prime denotes the departure of an instantaneous value from a mean value and the bar signifies a mean value over a specific time period. Thus $(\rho w)'$ represents an instantaneous fluctuation in the rate of upward air flow at the point of measurement and q' is the associated moisture content fluctuation. The evaporation rate is the mean of the products of such measurements during a given time period. The theory and methods of measurement for the eddy-flux method have been developed in Australia, and a single instrument to give evaporation directly has been designed. This semi-automatic instrument, the Evapotron, has small rapid-response sensing elements and electronic apparatus to effect the computations (Dyer & Maher, 1965). Although the instrument has performed well under test conditions, its use has not become widespread owing to the complexity of the apparatus and the limitations in the response time of the wet bulb sensors for the humidity. Research and development on instruments to measure eddy-flux evaporation continue and the Institute of Hydrology's Hydra has been well tested under UK and tropical conditions and over a variety of surfaces. This robust battery-powered compact instrument employs an ultrasonic vertical windspeed sensor and infrared hygrometers. A digital microprocessor analyses the measurements in the field.

11.1.2   Energy Budget Method

Evaporation from a lake or reservoir may be calculated on a weekly or monthly basis by taking into consideration the heat or energy required to effect the evaporation. A heat balance following the principle of the conservation of energy is evaluated from incoming, outgoing and stored energy. The elements of the heat balance are shown in Fig. 11.2.

In the diagram, Q_s is the short-wave solar radiation, Q_{rs} is the reflected short-wave radiation, Q_1 is the long-wave radiation from the water body, Q_c is the sensible heat transfer to the air, Q_g is the change in stored energy and Q_v is the energy transfer between water and bed. Then Q_{E_o}, the energy required for evaporation, can be calculated as follows:

$$Q_{E_o} = Q_s - Q_{rs} - Q_1 - Q_c \pm Q_g \pm Q_v \qquad (11.7)$$

Care must be taken to see that all the terms are in the same energy units, W m^{-2}. Then E_o, the evaporation from open water is given by:

$$E_o = \frac{Q_{E_o}}{\rho L} \, \text{mm s}^{-1}$$

where L is the latent heat of vaporization of water. Some energy budget

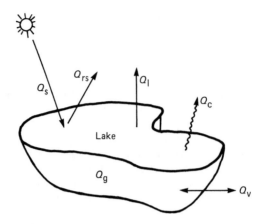

Fig. 11.2   Energy budget measurements.

equations include a separate term for incoming long-wave radiation, but this contributes a comparatively small amount of energy. All the radiation terms can be accounted for by a net radiometer. The measurements required to implement this formula are numerous, involving temperature of the air, of the water surface and at various depths in addition to the net radiation, and the data processing involved is consequently extensive and time consuming. Modern sensors and recording equipment, although simplifying the work, make it an expensive method to use. However, the energy budget method is reliable and can be used over a suitable period of time for a specific reservoir until satisfactory mass transfer coefficients have been determined. Then empirical mass transfer calculations can give the required reservoir losses while the expensive energy budget equipment is used at other locations.

11.1.3   Penman Formula (Combination Method)

In a classical study of natural evaporation, Penman (1948) developed a formula for calculating open water evaporation based on fundamental physical principles, with some empirical concepts incorporated, to enable standard meteorological observations to be used. This latter facility resulted in the Penman formula being enthusiastically acclaimed and applied the world over, especially by practising engineers seeking an answer to the question of water loss.

The physical principles combine the two previous approaches to evaporation calculation, the mass transfer (or aerodynamic) method and the energy budget method. The basic equations are modified and rearranged to use meteorological constants and measurements of variables made regularly at climatological stations.

In a simplified energy balance equation:

$$H = E_o + Q \tag{11.8}$$

where H is the available heat, E_o is energy for evaporation and Q is energy for heating the air.

The values of E_o and Q can be defined by the aerodynamic equations (see Equation 11.1):

$$E_o = f(u)(e_s - e_d) \qquad (11.9)$$

and

$$Q = \gamma f_1(u)(T_s - T_a) \qquad (11.10)$$

γ is the hygrometric constant (0.27 mm of mercury/°F) to keep units consistent. It is generally assumed that $f(u) = f_1(u)$. If the aerodynamic equation (Equation 11.9) is based on the *air humidity* using the air temperature T_a, then:

$$E_a = f(u)(e_a - e_d) \qquad (11.11)$$

where e_a is the saturated vapour pressure at air temperature T_a, and thus $(e_a - e_d)$ is the saturation deficit (e_d, the vapour pressure of the air, is the saturated vapour pressure at the dew point, T_d). The temperature, T_a, is easily measured, whence e_a is easily obtained, whereas e_s in Equation 11.9 is difficult to evaluate.

If Δ represents the slope of the curve of saturated vapour pressure plotted against temperature, then:

$$\Delta = \frac{de}{dT} \simeq \frac{e_s - e_d}{T_s - T_d} \simeq \frac{e_a - e_d}{T_a - T_d} \quad \text{(if gradients are small)}$$

Then from Equation 11.10:

$$Q = \gamma f(u)\left[(T_s - T_d) - (T_a - T_d)\right]$$
$$= \gamma f(u)\left[\frac{(e_s - e_d)}{\Delta} - \frac{(e_a - e_d)}{\Delta}\right]$$
$$= \frac{\gamma E_o}{\Delta} - \frac{\gamma E_a}{\Delta}$$

Substituting for Q in the energy balance equation (Equation 11.8):

$$E_o = H - \frac{\gamma E_o}{\Delta} + \frac{\gamma E_a}{\Delta}$$

$$\Delta E_o + \gamma E_o = \Delta H + \gamma E_a$$

$$E_o = \frac{\Delta H}{\Delta + \gamma} + \frac{\gamma E_a}{\Delta + \gamma}$$
$$= \frac{\Delta}{\gamma} H + E_a \left/ \left(\frac{\Delta}{\gamma} + 1\right)\right. \qquad (11.12)$$

This final equation is the basic Penman formula for open water evaporation. It requires values of H and E_a as well as Δ for its application.

If net radiation measurements are available, then H, the available heat may be obtained directly. More often, H is calculated from incoming (R_I) and outgoing (R_O) radiation determined from sunshine records, temperature and humidity, using:

$$H = R_{\mathrm{I}}(1 - r) - R_{\mathrm{O}} \tag{11.13}$$

where r is the albedo and equals 0.05 for water. R_{I} is a function of R_{a}, the solar radiation (fixed by latitude and season) modulated by a function of the ratio, n/N, of measured to maximum possible sunshine duration. Using $r = 0.05$ gives:

$$R_{\mathrm{I}}(1 - r) = 0.95 \, R_{\mathrm{a}} f_{\mathrm{a}}(n/N) \tag{11.14}$$

Penman used $f_{\mathrm{a}}(n/N) = 0.18 + 0.55 \, n/N$ in the original work, but later studies have shown that the function $f_{\mathrm{a}}(n/N)$ depends on the clarity of the atmosphere and latitude (MAFF, 1967).

The term R_{O} in Equation 11.13 is given by:

$$R_{\mathrm{O}} = \sigma \, T_{\mathrm{a}}^4 (0.56 - 0.09 \sqrt{e_{\mathrm{d}}})(0.10 + 0.90 \, n/N) \tag{11.15}$$

where σT_{a}^4 is the theoretical black body radiation at T_{a} which is then modified by functions of the humidity of the air (e_{d}) and the cloudiness (n/N).

Thus H in Equation 11.12 is obtained from values found via Equations 11.14 and 11.15 inserted into Equation 11.13.

Next, E_{a} in Equation 11.12 is found using the coefficients derived by experiment for open water:

$$E_{\mathrm{a}} = 0.35 \, (0.5 + u_2/100)(e_{\mathrm{a}} - e_{\mathrm{d}}) \tag{11.16}$$

Finally a value for Δ is found from the curve of saturated vapour pressure against temperature corresponding to the air temperature, T_{a}.

The equations given are those originally published by Penman and subsequently used by Wales-Smith (1971) and McCulloch (1965). The four measurements required to calculate the open water evaporation are thus:

T_{a} mean air temperature for a week, 10 days or a month, °F or °C
e_{d} mean vapour pressure for the same period, mm of mercury
n bright sunshine over the same period, h day^{-1}
u_2 mean wind speed at 2 m above the surface, miles day^{-1}

R_{a} and N are obtained from standard meteorological tables (Appendix 11.1).

With meteorological observations made in various units and the tendency to work now in SI units, care is needed in converting measurements into the appropriate units for the formula. The evaporation from open water E_{o} is finally in mm/day.

Example. A calculation of E_{o} using the original formula and empirical equations is to be made from data assembled at the University of Keele climatological station, latitude 53° N for the month of June 1972. Temperature and vapour pressure values come from thermometer measurements in a Stevenson screen, wind speed from an anemometer at 10 m above the ground and sunshine hours from a Campbell – Stokes recorder.

T_{a} 10.7 °C (51 °F) giving $e_{\mathrm{a}} = 9.56$ mm of mercury — (Appendix 11.2.8)
e_{d} 7.3 mm of mercury (5.5 mb)
u_{10} 6.6 miles/h $\equiv u_2$ 5.15 miles h^{-1} = 124 miles day^{-1}
n 142.2 h = 4.74 h day^{-1}

From meteorological tables, R_a = 16.54 (in mm of water day^{-1}) (Appendix 11.1.1) and N = 16.7 h day^{-1} (Appendix 11.1.2). Thus n/N = 4.74/16.7 = 0.28 and $(e_a - e_d)$ = 9.56 – 7.3 = 2.26 mm of mercury.

To find H from the incoming radiation, $R_I (1 - r)$ and outgoing energy R_O:

$$R_I (1 - r) = 0.95\, R_a\, (0.18 + 0.55\, n/N)$$
$$= 0.95 \times 16.54 \times (0.18 + 0.55 \times 0.28)$$
$$= 15.713 \times 0.334$$
$$= 5.25 \text{ mm day}^{-1}$$

$$R_O = \sigma T_a^4 (0.56 - 0.09 \sqrt{e_d}) (0.10 + 0.90\, n/N)$$

Given the Stefan – Boltzmann constant:

$$\sigma = 5.67 \times 10^{-8} \text{ Wm}^{-2}\text{ K}^{-4} \text{ then } T_a \text{ must be converted to K}$$

$$\sigma T_a^4 = 5.67 \times 10^{-8} \times 283.7^4 \text{ Wm}^{-2}$$
$$= 5.67 \times 2.837^4 \times 2.388 \times 10^{-5} \times 1440 \text{ mm day}^{-1} \text{ (to convert to equivalent mm day}^{-1} \text{ of water)}$$
$$= 12.63 \text{ mm day}^{-1}$$
$$R_O = 12.63\, (0.56 - 0.09\, \sqrt{7.3})\, (0.10 + 0.90 \times 0.28)$$
$$= 12.63 \times 0.3168 \times 0.352$$
$$= 1.41 \text{ mm day}^{-1}$$

Thus

$$H = R_r (1 - r) - R_O = 5.25 - 1.41 = 3.84 \text{ mm day}^{-1}$$

Next

$$E_a = 0.35\, (0.5 + u_2/100)\, (e_a - e_d)$$
$$= 0.35\, (0.5 + 124/100)\, 2.26$$
$$= 0.35 \times 1.74 \times 2.26$$
$$= 1.38 \text{ mm day}^{-1}$$

Finally, Δ at 51 °F = 0.354 and therefore Δ/γ = 1.31. Then:

$$E_o = \frac{\Delta}{\gamma}\, H + E_a \left/ \left(\frac{\Delta}{\gamma} + 1 \right) \right.$$
$$= \frac{1.31 \times 3.84 + 1.38}{2.31}$$
$$= 2.78 \text{ mm day}^{-1}$$

and therefore an estimate of E_o for the month of June 1972 = 83 mm.

11.2   Calculation of E_t

As outlined in Chapter 4, the evaluation of actual water loss from a vegetated land surface by evaporation plus transpiration (E_t) adds further complexities to the processes involved in the evaporation from an open water surface (E_O). Instead of adding directly to the body of water, some of the rainfall (or snowfall) is intercepted by the vegetation, and, from the various wetted surfaces, moisture is readily returned to the atmosphere by direct evaporation. However, much of the precipitation eventually reaches the ground, where it is absorbed by the soil or runs off impervious surfaces. The assessment of transpiration loss must take into account the water available to the plants in the rooting zones of the soil. The mechanism of transpiration, i.e. the regulation of the passage of water through the plant pores, is well understood, but difficult to quantify or present in mathematical terms. However, increasing use is being made of the Penman-Monteith formula which is based on the physical principles applied by Penman with extended physical representation of the water loss from vegetation (Shuttleworth, 1979).

The evapotranspiration rate is given by:

$$E_t = \frac{\Delta R_n + \rho c_p (e_a - e_d)/r_a}{\lambda (\Delta + \gamma(r_s + r_a)/r_a)} \text{ g m}^{-2} \text{ s}^{-1}$$

where R_n = net radiation (W m^{-2})
   ρ = density of air
   c_p = specific heat of air
   r_s = net resistance to diffusion through the surfaces of the leaves and soil (sm^{-1})
   r_a = net resistance to diffusion through the air from surfaces to height of measuring instruments (sm^{-1})

and other variables are previously defined. Vapour pressures need to be in units of mb and the properties of the air in SI units. Mean values of r_s for grasses during daylight hours and with water unlimited are in the range 40-50 sm$_{-1}$ with r_a 20-40 sm^{-1} in low wind speeds 1-5 ms^{-1}.

Calculations of evaporation from catchment areas have tended to adopt the meteorological approach and attempt to assess, at the atmospheric receiving end of the process, the amount of water vapour taken up from the ground and vegetation.

The mass transfer methods used to evaluate E_o may also be applied to measurements of saturation deficit and wind speed made over land surfaces to estimate values of E_t. A great deal of research by many workers has shown that there can be wide variations in those estimates. These variations are due to the many changes in surface roughness influencing the wind pattern existing over an area of natural vegetation. Even in special detailed studies over fields of oats and wheat, the mechanism of evaporation and transpiration loss has been shown to vary according to the degree of turbulence of the wind (which changes with the weather conditions) and the state of the crop (e.g. Rider, 1954). Thus the vapour flow methods are suitable only for measurements over short time periods and over extensive areas with a uniform surface such as is provided by low growing crops or by an even forest canopy. It is to overcome the problems of the variable wind profile that great expectations are held for the Evapotron and now the Hydra which give records of continuous changes in evaporation rate. These instruments would readily give values of E_t when installed at selected sites representative of the major vegetational areas of a catchment.

In current practice, recourse is sometimes made to one of the several empirical formulae for a general estimate of catchment loss.

11.2.1   Empirical Formulae

Turc.   A widely used formula to estimate annual values of E_t for catchment areas was published by Turc (1954, 1955). Taking catchment data from 254 drainage basins, representing all the different climates in Europe, Africa, America and the East Indies, he used the water balance equation to evaluate E_t from P and Q, the precipitation and runoff.

He then established the formula:

$$E_t = \frac{P}{(0.9 + (P/L)^2)^{\frac{1}{2}}} \text{ mm per annum} \tag{11.17}$$

where P is the mean annual precipitation (mm), $L = 300 + 25T + 0.05T^3$ (mm) and T is the mean air temperature (°C).

Turc showed that the formula could be applied in humid and arid climates, either hot or cold. He considered that in obtaining by this method the best estimate of the mean annual evaporation, his results demonstrated that precipitation and temperature could be the dominant factors in evaporation.

Thornthwaite.   In the 1920s and 1930s, several climatologists studied the relationship of precipitation and temperature in attempting to classify different climates. The effectiveness of precipitation was expressed in terms of the precipitation/evaporation ratio. Thornthwaite (1948) used data for the summer months in the arid parts of the USA to establish the following relationship between monthly P/E and T:

$$\frac{P}{E} = 11.5 \left(\frac{P}{(T - 10)}\right)^{10/9} \tag{11.18}$$

where P and E are in inches and T is in °F. This formula was modified by Crowe (1971) to:

$$\frac{P}{E} = \frac{9P}{T - 10}$$

since monthly temperature and rainfall data have always some inherent errors. From the Crowe formula, a very simple direct evaporation/temperature relationship of $E = (T - 10)/9$ can be inferred. The validity of this too-simple equation is very questionable at the lower and upper limits, but for a rough estimate of E_t in the months of the growing season in moderate climates, the Thornthwaite – Crowe formulae may be useful when only precipitation and temperature data are available.

11.3   Calculation of Potential Evaporation

A value of the actual evapotranspiration (E_t) over a catchment is more often obtained by first calculating the potential evaporation plus transpiration (PE), i.e. assuming an unrestricted availability of water, and then modifying the answer by accounting for the actual soil moisture content (see also Soil Moisture Deficit, Section 11.4). This is the method developed from Penman's studies and recommended for general use in the UK by the Meteorological Office. There are several formulae for calculating potential evaporation. Their derivation has been largely due to the need to assess irrigation demands, but their application, particularly that of the Penman formula, has served more widely to help provide a numerical evaluation of the moisture content of a catchment as well as a first stage in the calculation of E_t.

11.3.1   Penman

Penman 1.   In the early stages of his work, Penman (1950a) related potential evaporation from a vegetated land surface to the evaporation from open water calculated via Equation 11.12 and proposed:

$$PE = f.E_o \tag{11.19}$$

By comparison with catchment data in England and Wales and in neighbouring European countries, he obtained the following values of f considered applicable in the climate of Western Europe:

Summer $f = 0.8$ (May, June, July, August)
Winter $f = 0.6$ (November, December, January,
                                February)
Equinoctial months $f = 0.7$ (March, April, September, October)

Thus, having calculated E_o for a particular month using Equation 11.12 and its defined components, the potential evaporation is obtained by applying the appropriate factor.

Penman 2.   Resulting from later experience, the E_o formula was modified (MAFF, 1967) to allow for the conditions under which evaporation plus transpiration takes place from a vegetated surface.
   The basic equation becomes one for PE directly:

$$PE = \frac{(\Delta/\gamma)\, H_T + E_{at}}{(\Delta/\gamma) + 1} \tag{11.20}$$

where the extra subscript t signifies inclusion of transpiration effects. Then $H_T = 0.75\, R_I - R_O$ (see Equation 11.13) since the reflective coefficient for incident radiation, the albedo, from a short grassed surface is 0.25. (The Penman equation was established to use standard meteorological observations at stations where the instruments are set conventionally in well kept lawns).
   The term E_{at} is very similar to E_a in Equation 11.16, the coefficient 0.5 being replaced by 1 to allow for extra roughness in the wind speed function:

$$E_{at} = 0.35\, (1 + u_2/100)\, (e_a - e_d)$$

In response to general demand, particularly among agriculturalists, for the simplification of the computations to facilitate application, the Meteorological Office prepared sets of tables that were published by the Ministry of Agriculture, Fisheries and Food (MAFF, 1967). The tables are primarily for use in the UK. The empirical equations for the incoming and outgoing radiation in the energy term H_T are different from those given previously. In addition to the albedo modification in Equation 11.14:

$$R_I(1 - r) = 0.75 \, R_a f_a(n/N),$$

$f_a(n/N)$ takes several forms:

1. $f_a(n/N) = (0.135 + 0.68 \, n/N)$ for smoky areas, $n/N \leqslant 0.40$
2. $f_a(n/N) = (0.16 + 0.62 \, n/N)$ for latitudes south of $54\frac{1}{2}°$ N
3. $f_a(n/N) = (0.155 + 0.69 \, n/N)$ for latitudes $54\frac{1}{2}°$ to $56°$ N

For latitudes north of $56°$ N, 0.01 is added to $f_a(n/N)$ for each extra degree up to the maximum 0.04 at $60°$ N.

The empirical equation for the outgoing reduction takes the form:

$$R_O = \sigma \, T_a^4 \, (0.47 - 0.075 \, \sqrt{e_d}) \, (0.17 + 0.83 \, n/N)$$

Selected abstracts from the tables are given in Appendix 11.2 at the end of the chapter. As before, care is needed with units, and provided the basic measurements are converted to the required form, the tables give the necessary values in equivalent mm of water.

Since the publication of the tables, the Meteorological Office have reverted to using the original coefficients given in Equations 11.14 and 11.15, but taking account of the increased albedo for vegetation and introducing the multiplying factor 0.95 to σT_a^4 since vegetation does not radiate as a perfect black body. In calculating PE, the following equation for H is now in operation:

$$H = 0.75 \, R_a \, (0.18 + 0.55 \, n/N) - 0.95 \, \sigma T_a^4 \, (0.10 + 0.90 \, n/N) \\ (0.56 - 0.092 \, \sqrt{e_d})$$

Example.   The PE for the month of June 1972 at the University of Keele Climatological Station, $53°$ N, is to be calculated from the same observational data as used for E_O in the previous example, but the form of the Penman equation used is that given in *Potential Transpiration* (MAFF, 1967). Values for the several expressions are taken from the Tables in Appendix 11.2; slight differences may be observed in values taken from Tables in Appendix 11.1:

$T_a = 10.7 \, °C \, (51 \, °F)$ thus $e_a = 9.56$ mm of mercury.

(Appendix 11.2.8)

$e_d = 7.3$ mm of mercury
$u_{10} = 6.6$ miles h^{-1}
$n = 142.2 \, h = 4.74$ h day^{-1}
$e_a - e_d = 9.56 - 7.3 = 2.26$ mm of mercury
$N = 16.75$ h day^{-1}

Thus $n/N = 4.74/16.75 = 0.28$.

The net heat H_T is given by $R_I(1 - r) - R_O$. For a short grass surface, the reflection coefficient r is 0.25, and using the appropriate $f_a(n/N)$:

$$R_I(1 - r) = 0.75 R_a (0.16 + 0.62\, n/N) \quad \text{(Appendix 11.2.1 and 11.2.3)}$$
$$= 12.24 \times 0.33$$
$$= 4.04 \text{ mm day}^{-1}$$

$$R_O = \sigma T_a^4 (0.47 - 0.075\, \sqrt{e_d})(0.17 + 0.83\, n/N) \quad \text{(Appendix 11.2.4,}$$
$$= 13.0 \times 0.27 \times 0.40 \qquad\qquad\qquad 11.2.5 \text{ and } 11.2.6)$$
$$= 1.40 \text{ mm day}^{-1}$$

$$E_{at} = 0.35\,(1 + u_2/100)(e_a - e_d) \qquad\qquad \text{(Appendix 11.2.7)}$$
$$= 0.79 \times 2.26$$
$$= 1.79 \text{ mm day}^{-1}$$

Weighting factor at 51 °F Δ/γ = 1.31. Then: (appendix 11.2.9)

$$PE = \frac{(\Delta/\gamma)\, H_T + E_{at}}{(\Delta/\gamma) + 1}$$
$$= \frac{1.31\,(4.04 - 1.40) + 1.79}{1.31 + 1}$$
$$= \frac{1.31 \times 2.64 + 1.79}{2.31}$$
$$= 2.27 \text{ mm day}^{-1}$$

Then PE for the month of June 1972 = 68 mm. Comparing the answers for E_o (83 mm) and PE (68 mm), the results conform to Equation 11.19 with $f = 0.8$ for the summer months.

11.3.2   Thornthwaite

An empirical formula for E_t developed by Thornthwaite was given in the previous section. Further work was directed towards finding an expression for potential evapotranspiration to serve the needs of irrigation engineers. The resulting formula is based mainly on temperature with an adjustment being made for the number of daylight hours. An estimate of the potential evapotranspiration, PE_m, calculated on a monthly basis, is given by:

$$PE_m = 16\, N_m \left(\frac{10\, \overline{T}_m}{I} \right)^a \text{ mm}$$

where m is the months 1, 2, 3...12, N_m is the monthly adjustment factor related to hours of daylight, $\overline{T}_m$ is the monthly mean temperature °C, I is the heat index for the year, given by:

$$I = \Sigma i_m = \Sigma \left(\frac{\overline{T}_m}{5} \right)^{1.5} \qquad \text{for } m = 1...12$$

and:

$$a = 6.7 \times 10^{-7} I^3 - 7.7 \times 10^{-5} I^2 + 1.8 \times 10^{-2} I + 0.49 \text{ (to 2 significant figures)}$$

Given the monthly mean temperatures from the measurements at a climatological station, an estimate of the potential evaporation for each month of the year can be calculated. The method has been used widely throughout the world, but strictly it is not valid for climates other than those similar to that of the area where it was developed, the eastern USA. Compared with the estimates from the Penman formula, Thornthwaite values tend to exaggerate the potential evaporation, as will be seen in the following example. This is particularly marked in the summer months with the high temperatures having a dominant effect in the Thornthwaite computation, whereas the Penman estimate takes into consideration other meteorological factors.

Example.   The potential evaporation using the Thornthwaite formula is to be calculated for each month using the monthly mean temperatures derived from the 18 years of observations at the University of Keele, 1953–69. The stages in the computations are shown in Table 11.1, in which E_m is the unadjusted evaporation:

Table 11.1   Computation of Thornthwaite Potential Evaporation

Months	Mean temp. (°C)	i_m	E_m (mm)	N_m	PE_m (mm)	Penman PE (mm)
J	2.4	0.33	12.97	0.68	8.8	– 0.2
F	2.7	0.40	14.49	0.82	11.9	8.6
M	5.1	1.03	26.35	0.98	25.8	28.2
A	7.6	1.87	38.33	1.15	44.1	47.0
M	10.6	3.09	52.41	1.31	68.7	70.6
J	13.4	4.39	65.33	1.39	90.8	82.0
J	14.7	5.04	71.27	1.36	96.9	75.4
A	14.5	4.94	70.36	1.23	86.5	60.2
S	12.8	4.10	62.58	1.06	66.3	37.8
O	9.8	2.74	48.68	0.88	42.8	16.0
N	5.6	1.19	29.77	0.72	20.7	3.1
D	3.5	0.59	18.49	0.63	11.6	– 2.5
Year		29.71			574.9	426.2

$I = \Sigma i = 30$

$a = 6.7 \times 10^{-7} (30)^3 - 7.7 \times 10^{-5} (30)^2 + 1.8 \times 10^{-2} (30) - 0.49$

$\quad = 0.018 - 0.069 + 0.54 + 0.49$

$\quad = 0.94$

Calculating the unadjusted evaporation (E_m) uses this value of a in:

$$E_m = 16 \left(\frac{10 \overline{T}_m}{I} \right)^{0.94}$$

while the daylight factors (N_m) are obtained from Appendix 11.1.2 by dividing the possible sunshine hours for the appropriate latitude (Keele, 53° N) by 12.

The Penman estimates for comparison have been calculated for each month of the years 1956–1969 and the monthly means for the 14 years are given in the last column of Table 11.1.

11.3.3   Turc

Turc also extended his empirical method for calculating the annual evapotranspiration (E_t) from annual values of precipitation and temperature to produce a formula for potential evaporation (PE) (with non-limiting water) over a shorter period of time. This approach again was particularly directed towards the needs of the agronomists for irrigation schemes.

The Turc short term formula for potential evaporation over 10 days is:

$$PE = \frac{P + a + 70}{\left[1 + \left(\frac{P + a}{L} + \frac{70}{2L} \right)^2 \right]^{\frac{1}{2}}} \quad mm$$

where P is the precipitation in a 10 day period (mm), a is the estimated evaporation in the 10-day period from the bare soil when there has been no precipitation, (1 mm $\leqslant a \leqslant$ 10 mm) and L is the 'evaporation capacity' of the air given by:

$$L = \frac{(T + 2) \; Q_s^{\frac{1}{2}}}{16}$$

with T the mean air temperature over the 10 days (°C) and Q_s the mean short-wave radiation (cal cm^{-2} day^{-1}). Although the formula takes into account soil moisture (via the parameter a) and a crop factor (the mean value of 70 reflecting growth with unlimited water), the estimate of PE depends mostly on L.

11.4   Soil Moisture Deficit

The calculation of potential evaporation (PE) from readily available meteorological data is seen to be a much simpler operation than the computation or measurement of actual evapotranspiration (E_t) from a vegetated surface. However, water loss from a catchment area does not always proceed at the potential rate, since this is dependent on a continuous water supply. When the vegetation is unable to abstract water from the soil, then the actual evaporation becomes less than potential. Thus the relationship between E_t and PE depends upon the soil moisture content.

When the soil is saturated, it will hold no more water. After rainfall ceases, saturated soil relinquishes water and becomes unsaturated until it can just hold a certain amount against the forces of gravity; it is then said to be at 'field capacity'. In this range of conditions, E_t = PE; evapotranspiration occurs at the maximum possible rate determined by the meteorological conditions. If there is no rain to replenish the water supply, the soil moisture gradually becomes depleted by the demands of the vegetation to produce a soil moisture deficit (SMD), viz the amount of water required to restore the soil to field capacity. As SMD increases, E_t becomes increasingly less than PE. The values of SMD and E_t vary with soil type and vegetation, and the relative changes in E_t with increasing SMD have been the subject of considerable study by botanists and soil physicists. Penman (1950) introduced the concept of a '*root constant*' (RC) that defines the amount of soil moisture (mm depth) that can be extracted from a soil without difficulty by a given vegetation. Table 11.2 lists some typical root constants. It is assumed that E_t = PE for a particular type of vegetation until the SMD reaches the appropriate root constant plus a further 25 mm approximately, which is added to allow for extraction from the soil immediately below the root zone. Thereafter, E_t becomes less than PE as moisture is extracted with greater difficulty. As the SMD increases further, the vegetation wilts and E_t becomes very small or negligible. Before the onset of wilting, vegetation will recover if the soil moisture is replenished, but there is a maximum SMD for each plant type at a 'permanent wilting point' from which the vegetation cannot recover and dies.

Table 11.2   Root Constants in Soil Moisture Depths

Vegetation Type	mm
Permanent grassland	75
Root crops, e.g. potatoes	100
Cereals, e.g. wheat, oats	140
Woodland	200

Fig. 11.3 shows diagrammatically the average sequences of soil moisture content related to the rainfall and potential evaporation through the annual seasonal cycle in the UK. The soil moisture state is given at each stage for three types of vegetation: short-rooted grass, medium-rooted shrubs and deep-rooted trees. The actual evaporation (E_t) is given in terms of PE for each vegetational type at each stage. During the spring, when potential evaporation exceeds rainfall, the soil moisture deficit begins first in the surface soil layers and then moves downwards into the lower layers as the water in the soil is used up, until in the summer months there could be soil moisture deficits in all the rooting zones of the soil. When rainfall totals begin to exceed potential evaporation in the autumn, the soil moisture stores are gradually replenished from the top soil layer downwards, until all the soil layers reach field capacity again in the winter.

A soil moisture budget can be made on a monthly basis for various types of vegetation classified according to their root constants. Permanent grass, RC 75 mm, can extract a further 25 mm from the soil with difficulty, and it is considered that a maximum SMD of 125 mm could be reached before there would be permanent wilting. For deep-rooted woodlands with a RC of 200 mm, a maximum SMD of 250 mm has been suggested (Grindley, 1970).

To evaluate soil moisture deficit and actual evaporation over a catchment area, the proportions of the different types of vegetation covering the catchment must be known. This entails a land-use survey and a classification of the vegetation for allocation of RCs before water budgeting may be carried out. To simplify the procedure, the Meteorological Office for some years adopted the model proposed by Penman (1950b) whereby 0.2 of a catchment area, the riparian land bordering all the streams, is considered to have a

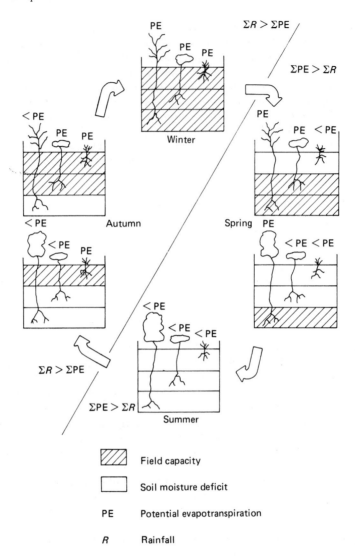

Fig. 11.3   Idealized annual soil moisture cycle for three vegetational types.

water table sufficiently close to the root zone that the vegetation will always transpire at the potential rate. A half of the catchment was assumed to be covered by permanent grass and thus have a RC of 75 mm whereas the remaining 0.3 of the catchment was woodland with a RC of 200 mm. This simple model was applied on an operational basis, and the Meteorological Office published fortnightly maps of soil moisture deficits for the whole of Great Britain. The information provided on these maps was a great aid to agriculturalists concerned with the irrigation of crops and to river engineers

Table 11.3 Calculation of Soil Moisture Deficit and Actual Evaporation For a Catchment (mm)

Month	Rain	PE	Rain – PE	Potent. SMD	RC 200 mm SMD	E_t	RC 75 mm SMD	E_t	Catchment SMD	E_t
April	48	51	– 3	3	3	51	3	51	2.4	51
May	20	86	– 66	69	69	86	69	86	55	86
June	31	102	– 71	140	140	102	112	74	98	88
July	84	86	– 2	142	142	86	112	84	99	85
August	69	76	– 7	149	149	76	113	70	101	73
September	9	46	– 37	186	186	46	116	12	114	29
October	3	25	– 22	208	208	25	118	5	121	15
November	81	5	+ 76	132	132	5	42	5	61	5
December	51	4	+ 47	85	85	4	0	4	26	4

responsible for flood forecasting on the rivers. The original model was later expanded to include more detailed land-use and vegetational groups when applied to specific catchments. With a good coverage of rainfall and potential evaporation data, one of the final products of the more detailed model is now a map of a hydrometric area or major river catchment with isopleths of actual evaporation (E_t) (Grindley, 1970).

Example. In Table 11.3, the computation of catchment soil moisture deficit and actual evaporation is demonstrated using the earlier simple model (sometimes referred to as the Grindley model) on a series of rainfall and potential evaporation data, beginning at the start of a growing season. The calculations are shown on a monthly basis, but the budgeting on a shorter time period (preferably daily or weekly) is normally carried out to avoid anomalies caused by persistent dry spells and the irregular incidence of rainfall.

The first three columns of numbers are self-explanatory. Potential SMD is the soil moisture deficit that would result if the potential evaporation was always fulfilled. It is the aggregate of the rainfall – potential evaporation considered as a deficit, and is assumed to apply to the riparian lands at or above field capacity. The calculations for the woodland zone (RC 200 mm) are in the next two columns; the SMD follows the potential SMD until it reaches the woodland zone maximum of 250 mm. In this example, the values are all equal to potential SMD and the actual evaporation E_t = PE through to December. For the grassland areas (RC 75 mm), E_t proceeds at the PE rate until 96 mm of SMD has developed and thereafter it is less than the PE. Then values of the actual SMD are taken for the potential – actual relationship of the vegetational type RC 75 mm from Appendix 11.3.1. This table for the two vegetational types RC 75 mm and RC 200 mm is adapted from Meteorological Office Hydrological Memorandum No. 38 (Grindley, 1969). The actual evaporation from the grassland zone equals PE for April and May and then is given by the change in SMD plus the rainfall, e.g. for June:

$$E_t = \text{SMD June} - \text{SMD May} + \text{June rain}$$
$$= 112 - 69 + 31$$
$$= 74 \text{ mm}$$

The catchment values in the last two columns are obtained from:

SMD $= 0.3 \text{ SMD}_{200} + 0.5 \text{ SMD}_{75}$ with the riparian areas remaining at field capacity

$$E_t = 0.2 \text{ PE} + 0.3 E_{t,200} + 0.5 E_{t,75}$$

11.5 Meteorological Office Rainfall and Evaporation Calculating System (MORECS)

The advent of faster computers with greatly increased data storage facilities has meant that computations of actual evaporation and soil moisture deficit can be made with greater attention to the processes involved and can also incorporate a wider variety of land uses in a budgeting model. Additionally, research in the past 10 years has indicated that modifications in the albedo used in the Penman PE formula can account more realistically for variations in available radiation caused by seasonal and land-use changes. In its increased attention to the growing hydrological needs in agriculture and water engineering, the UK Meteorological Office has been in the forefront of these developments and has produced a comprehensive computer package, MORECS, designed to serve requirements for evaporation, effective rainfall and soil moisture deficits (Meteorological Office, 1981). The calculations are of areal values at an average height above sea level for each of 188 40 × 40 km grid squares covering Great Britain. A flow chart of the computer system is shown in Fig. 11.4.

The meteorological variables required for the calculation of PE by the Penman formula, viz mean air temperature (T_a), average vapour pressure (e_d), duration of bright sunshine (n) and wind speed at 2 m (u_2) (using the same notation as before), are assembled in computer storage from all the reporting climatological stations in the country. After quality control of the basic data, the areal values for the grid squares are calculated using techniques developed in the Meteorological Office and then the PE for each 1600 km² grid area is obtained for 7-day periods. The Penman evaporation computer subroutine in the system can easily be updated to include any improvements that may be recommended by further research.

The water budgeting model incorporated into MORECS is a development from the Grindley model. It is called the soil moisture extraction model (SMEM) and for each vegetation type it deals with the soil moisture in two layers. The top layer, assumed to hold 40% of the soil moisture, contains all the densely packed shallow roots that enable the plants to evaporate freely at the potential rate. From the bottom layer, the remaining 60% of the moisture is extracted with increasing difficulty by the more aparsely distributed deep roots. Thus the moisture content of the soil layers varies with the vegetation type. Allowances are also made for changing plant water use after the harvesting of crops. The MORECS system contains many options including the necessary particulars of albedo and plant water use relationships for some 16 different land uses. The model can also take into account interception by the vegetation with differences being made for seasonal changes in deciduous broadleaved trees. For example, large trees in full leaf are assumed to intercept up to 2 mm of rain from each daily rainfall. Any surplus rain is assumed to be immediately making good any soil moisture deficiency or becoming effective rainfall. The daily calculations are made for each type of land use represented in a grid square. The SMD for individual crops may be obtained, but the main results are the grid square averages calculated from the different land use proportions. The routine output values are shown in Fig. 11.4 and monthly values for each of the 188 grid squares are available from the Meteorological Office for each month for the period 1961–85.

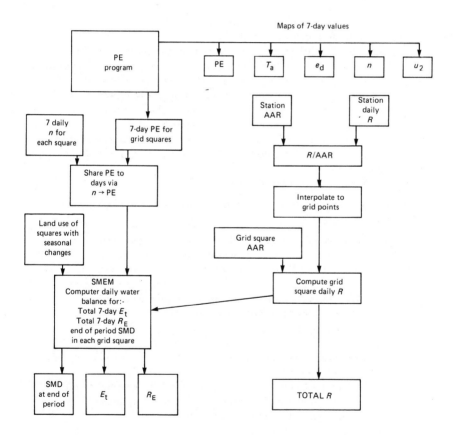

Fig. 11.4 Flowchart of program to calculate soil moisture deficit, actual evaporation and effective rainfall. R = rainfall, R_E = effective rainfall, AAR = average annual rainfall. (Reproduced from B. G. Wales-Smith (1975) In *Engineering Hydrology Today* (Ed. M. Monro) by permission of the Institution of Civil Engineers.)

Although MORECS is fully operational on a regular basis, the system is being thoroughly tested, and research workers are comparing MORECS results with measurements of evaporation and soil moisture deficits made in field experiments. (Discrepancies in the various answers may in some measure be due to comparing areal values (over 1600 km²) with much smaller areal or even point values.) The MORECS model as currently implemented assumes a uniform standard soil, and further improvements may be made in adjusting the moisture content of the soil layers to differing soil characteristics as well as to vegetation type. Future developments of great potential value to hydrologists could also include the linking of MORECS, which provides an evaluation of effective rainfall, to deterministic catchment models and to hydrogeological models.

11.6   Crop Water Requirements

In designing irrigation schemes and assessing the quantity of water to be supplied, the hydrologist must make estimates of crop water requirements. The needs vary for different crops and during the periods of growth. The several computational methods use a reference crop evapotranspiration, which is the rate of evapotranspiration from an extensive uniformly covered grass surface which is never short of water. This is essentially the potential evaporation (PE) of the Penman formula. Two other methods to derive the reference crop evapotranspiration have widespread usage: the Blaney-Criddle formula, developed and applied successfully by American irrigation engineers, and the measurements of E_0 from evaporation pans (Doorenbos and Pruitt, 1977).

There are two main components in the calculations:

(1) Determining the reference crop evapotranspiration ($ET_0 = PE$)

  (a) Results of the Penman calculations in Section 11.3.1 using monthly mean data provide ET_0 in mm d^{-1} but these are subject to an adjustment factor dependent on relative humidity, incoming radiation and the ratio of daytime to night-time mean wind speeds.

  (b) The Blaney-Criddle formula requires only mean daily temperatures (T°C) over each month.

$$ET_0 = p(0.46T + 8) \text{ mm d}^{-1}$$

  with p the mean daily percentage (for the month) of total annual daytime hours.

  An adjustment factor is applied similarly for relative humidity, sunshine hours and daytime wind speed estimates.

  (c) From evaporation pan measurements

$$ET_0 = k_p E_{pan} \text{ mm d}^{-1} \text{ (mean daily value)}$$

  Pan coefficients (k_p) are available for the US Class A pan (see Fig. 4.1b) according to different ground cover round the pan, mean relative humidity and the daily run of wind.

(2) The second stage in estimating crop water requirements is the selection of the crop coefficient (k_c) according to the cropping pattern during a production season and the growth characteristics of the crop. Then $ET_{crop} = k_c ET_0$ calculated for each of the 30 or 10 day periods through the growing season, depending on the chosen budgeting period for the application of water to supplement any rainfall.

Irrigation is of prime importance in hot climates especially in those regions with a dry season. The seasonal values of ET_{crop} for a selection of irrigated crops are shown in Table 11.4, and sample values for k_c are given in Table 11.5.

Table 11.4   Ranges of Seasonal Totals of ET_{crop} (mm)

Alfalfa	600–1500	Bananas	700–1700
Maize	400–750	Coffee	800–1200
Rice	500–950	Cotton	550–950
Sorghum	300–850	Oranges	600–950
Soyabeans	450–825	Sugar cane	1000–1500
Vegetables	250–500	Vines	450–900

Table 11.5   Sample k_c values

	J	F	M	A	M	J	J	A	S	O	N	D
ORANGES Mature tree ground clean	.75	.75	.7	.7	.7	.65	.65	.65	.65	.7	.7	.7
GRAPES Mature vine hot areas	–	–	.25	.45	.6	.7	.7	.65	.55	.45	.35	–
RICE (Asia) Wet season						1.1	1.1	1.05	1.05	1.05	.95	
Dry season	1.1	1.25	1.25	1.25	1.0							1.1
COTTON Semi-arid hot				.3	.6	1.0	1.12	1.15	.06			

In evaluating the total water requirements for an irrigation scheme, many more factors need consideration. The ET_{crop} is variable over time and area and can be affected by changing local conditions. The quality of the soil and method of application together with agricultural practices, all have to be assessed in calculating the total water needs.

Appendix 11.1.1 R_a Expressed in Equivalent Evaporation (mm day^{-1})

(Reproduced from J.S.G. McCulloch (1965) *E. African Agric. Forest. J.*, **xxx** (3) 286–295, by permission.)

	Lat.	Jan	Feb	Mar	Apr	May	June	July	Aug	Sept	Oct	Nov	Dec
		NORTHERN			HEMISPHERE								
	60	1.4	3.6	7.0	11.1	14.6	16.4	15.6	12.6	8.5	4.7	2.0	0.9
	50	3.7	6.0	9.2	12.7	15.5	16.6	16.1	13.7	10.4	7.1	4.4	3.1
	40	6.2	8.4	11.1	13.8	15.9	16.7	16.3	14.7	12.1	9.3	6.8	5.6
	30	8.1	10.5	12.8	14.7	16.1	16.5	16.2	15.2	13.5	11.2	9.1	7.9
	20	10.8	12.4	14.0	15.2	15.7	15.8	15.8	15.4	14.4	12.9	11.3	10.4
	10	12.8	13.9	14.8	15.2	15.0	14.8	14.9	15.0	14.8	14.2	13.1	12.5
Equator	0	14.6	15.0	15.2	14.7	13.9	13.4	13.6	14.3	14.9	15.0	14.6	14.3
	10	15.9	15.7	15.1	13.9	12.5	11.7	12.0	13.1	14.4	15.4	15.7	15.8
	20	16.8	16.0	14.5	12.5	10.7	9.7	10.1	11.6	13.6	15.3	16.4	16.9
	30	17.2	15.8	13.5	10.9	8.6	7.5	7.9	9.7	12.3	14.8	16.7	17.5
	40	17.3	15.1	12.2	8.9	6.4	5.2	5.6	7.6	10.7	13.8	16.5	17.8
	50	16.9	14.1	10.4	6.7	4.1	2.9	3.4	5.4	8.7	12.5	16.0	17.6
	60	16.5	12.6	8.3	4.3	1.8	0.9	1.3	3.1	6.5	10.8	15.1	17.5
		SOUTHERN		HEMISPHERE									

Appendix 11.1.2   Mean Daily Duration of Maximum Possible Sunshine Hours (*N*)

(United Nations (FAO) (1977) *Crop Water Requirements*; and Ministry of Agriculture, Fisheries and Food (1967) *Potential Transpiration*. Reproduced by permission of the Controller, Her Majesty's Stationery Office, © Crown copyright.)

North Lats.	Jan.	Feb.	Mar.	Apr.	May	June	July	Aug.	Sept.	Oct.	Nov.	Dec.
South Lats.	July	Aug.	Sept.	Oct.	Nov.	Dec.	Jan.	Feb.	Mar.	Apr.	May	June
60	6.7	9.0	11.7	14.5	17.1	18.6	17.9	15.5	12.9	10.1	7.5	5.9
58	7.2	9.3	11.7	14.3	16.6	17.9	17.3	15.3	12.8	10.3	7.9	6.5
56	7.6	9.5	11.7	14.1	16.2	17.4	16.9	15.0	12.7	10.4	8.3	7.0
54	7.9	9.7	11.7	13.9	15.9	16.9	16.5	14.8	12.7	10.5	8.5	7.4
52	8.3	9.9	11.8	13.8	15.6	16.5	16.1	14.6	12.7	10.6	8.8	7.8
50	8.5	10.0	11.8	13.7	15.3	16.3	15.9	14.4	12.6	10.7	9.0	8.1
48	8.8	10.2	11.8	13.6	15.2	16.0	15.6	14.3	12.6	10.9	9.3	8.3
46	9.1	10.4	11.9	13.5	14.9	15.7	15.4	14.2	12.6	10.9	9.5	8.7
44	9.3	10.5	11.9	13.4	14.7	15.4	15.2	14.0	12.6	11.0	9.7	8.9
42	9.4	10.6	11.9	13.4	14.6	15.2	14.9	13.9	12.6	11.1	9.8	9.1
40	9.6	10.7	11.9	13.3	14.4	15.0	14.7	13.7	12.5	11.2	10.0	9.3
35	10.1	11.0	11.9	13.1	14.0	14.5	14.3	13.5	12.4	11.3	10.3	9.8
30	10.4	11.1	12.0	12.9	13.6	14.0	13.9	13.2	12.4	11.5	10.6	10.2
25	10.7	11.3	12.0	12.7	13.3	13.7	13.5	13.0	12.3	11.6	10.9	10.6
20	11.0	11.5	12.0	12.6	13.1	13.3	13.2	12.8	12.3	11.7	11.2	10.9
15	11.3	11.6	12.0	12.5	12.8	13.0	12.9	12.6	12.2	11.8	11.4	11.2
10	11.6	11.8	12.0	12.3	12.6	12.7	12.6	12.4	12.1	11.8	11.6	11.5
5	11.8	11.9	12.0	12.2	12.3	12.4	12.3	12.3	12.1	12.0	11.9	11.8
Equator 0	12.0	12.0	12.0	12.0	12.0	12.0	12.0	12.0	12.0	12.0	12.0	12.0

Appendix 11.2.1   0.75 R_A (mm of water)

(Ministry of Agriculture, Fisheries and Food (1967) *Potential Transpiration.*
Reproduced by permission of the Controller, Her Majesty's Stationery
Office, © Crown copyright.)

Month	50	52	54	56	58	60
January	2.73	2.38	2.04	1.69	1.34	0.98
February	4.48	4.13	3.76	3.38	3.00	2.55
March	6.91	6.58	6.24	5.89	5.54	5.20
April	9.51	9.30	9.06	8.81	8.59	8.35
May	11.47	11.35	11.25	11.14	11.03	10.92
June	12.30	12.26	12.22	12.18	12.14	12.09
July	12.00	11.92	11.83	11.73	11.63	11.52
August	10.20	10.02	9.84	9.64	9.45	9.25
September	7.71	7.44	7.13	6.83	6.53	6.25
October	5.15	4.80	4.46	4.11	3.77	3.40
November	3.13	2.80	2.45	2.10	1.75	1.40
December	2.32	1.98	1.64	1.31	0.96	0.63

Appendix 11.2.2   Length of day, N (h)

(Ministry of Agriculture, Fisheries and Food (1967) *Potential Transpiration.*
Reproduced by permission of the Controller, Her Majesty's Stationery
Office, © Crown copyright.)

| Month | Latitude °N | | | | | |
	50	52	54	56	58	60
January	8.55	8.25	7.95	7.60	7.20	6.70
February	10.00	9.85	9.65	9.50	9.25	9.00
March	11.80	11.80	11.75	11.75	11.70	11.70
April	13.70	13.80	13.95	14.10	14.30	14.50
May	15.35	15.60	15.90	16.20	16.60	17.05
June	16.25	16.55	16.95	17.40	17.95	18.60
July	15.85	16.10	16.45	16.85	17.35	17.90
August	14.40	14.60	14.80	15.00	15.25	15.55
September	12.60	12.65	12.70	12.75	12.80	12.85
October	10.70	10.60	10.50	10.40	10.25	10.10
November	9.00	8.80	8.55	8.25	7.90	7.50
December	8.05	7.80	7.40	7.00	6.50	5.95

Appendix 11.2.3   Sunshine Function

(Ministry of Agriculture, Fisheries and Food (1967) *Potential Transpiration.* Reproduced by permission of the Controller, Her Majesty's Stationery Office, © Crown copyright.)

n/N	0.00	0.01	0.02	0.03	0.04	0.05	0.06	0.07	0.08	0.09
$f_a(n/N) = (0.135 + 0.68\,n/N)$ (Smoky areas)										
0.10	0.20	0.21	0.22	0.22	0.23	0.24	0.24	0.25	0.26	0.26
0.20	0.27	0.28	0.28	0.29	0.30	0.30	0.31	0.32	0.33	0.33
0.30	0.34	0.35	0.35	0.36	0.37	0.37	0.38	0.39	0.39	0.40
0.40	0.41	0.42	0.42	0.43	0.43	0.44	0.45	0.45	0.46	0.46
0.50	0.47	0.48	0.48	0.49	0.49	0.50	0.51	0.51	0.52	0.53
0.60	0.53	0.54	0.54	0.55	0.56	0.56	0.57	0.57	0.58	0.59
$f_a(n/N) = (0.16 + 0.62\,n/N)$ (areas south of $54\frac{1}{2}°$ N)										
0.10	0.22	0.23	0.23	0.24	0.25	0.25	0.26	0.27	0.27	0.28
0.20	0.28	0.29	0.30	0.30	0.31	0.32	0.32	0.33	0.33	0.34
0.30	0.35	0.35	0.36	0.36	0.37	0.38	0.39	0.38	0.40	0.40
0.40	0.41	0.41	0.42	0.43	0.43	0.44	0.45	0.45	0.46	0.46
0.50	0.47	0.48	0.48	0.49	0.49	0.50	0.51	0.51	0.52	0.53
0.60	0.53	0.54	0.54	0.55	0.56	0.56	0.57	0.57	0.58	0.59
$f_a(n/N) = (0.155 + 0.69\,n/N)$ (areas $54\frac{1}{2}°$ N – 56° N)										
0.10	0.22	0.23	0.24	0.24	0.25	0.26	0.27	0.27	0.28	0.29
0.20	0.29	0.30	0.31	0.31	0.32	0.33	0.33	0.34	0.35	0.36
0.30	0.36	0.37	0.38	0.38	0.39	0.40	0.40	0.41	0.42	0.42
0.40	0.43	0.44	0.44	0.45	0.46	0.47	0.47	0.48	0.49	0.49
0.50	0.50	0.51	0.51	0.52	0.53	0.53	0.54	0.55	0.56	0.56
0.60	0.57	0.57	0.58	0.59	0.60	0.60	0.61	0.62	0.62	0.63

Appendix 11.2.4   Values of σT^4 (mm of water)

(Ministry of Agriculture, Fisheries and Food (1967) *Potential Transpiration.*
Reproduced by permission of the Controller, Her Majesty's Stationery
Office, © Crown copyright.)

°F	0	1	2	3	4	5	6	7	8	9
30	11.0	11.1	11.2	11.3	11.4	11.5	11.6	11.6	11.7	11.8
40	11.9	12.0	12.1	12.2	12.3	12.4	12.5	12.6	12.7	12.8
50	12.9	13.0	13.1	13.2	13.3	13.4	13.5	13.6	13.7	13.9
60	14.0	14.1	14.2	14.3	14.4	14.5	14.6	14.7	14.8	14.9
°C										
− 0	11.2	11.0								
0	11.2	11.4	11.5	11.7	11.9	12.0	12.2	12.3	12.5	12.7
10	12.9	13.1	13.3	13.5	13.7	13.9	14.0	14.2	14.4	14.6
20	14.8	15.0								

Appendix 11.2.5   Humidity Function $f(e_d) = (0.47 - 0.075 \sqrt{e_d})$ (e_d in mm)

(Ministry of Agriculture, Fisheries and Food (1967) *Potential Transpiration.*
Reproduced by permission of the Controller, Her Majesty's Stationery
Office, © Crown copyright.)

Vapour pressure (mm)	0.0	0.2	0.4	0.6	0.8
4	0.32	0.32	0.32	0.31	0.31
5	0.30	0.30	0.30	0.29	0.29
6	0.29	0.28	0.28	0.28	0.27
7	0.27	0.27	0.27	0.26	0.26
8	0.26	0.25	0.25	0.25	0.25
9	0.24	0.24	0.24	0.24	0.23
10	0.23	0.23	0.23	0.22	0.22
11	0.22	0.22	0.22	0.21	0.21
12	0.21	0.21	0.21	0.20	0.20
13	0.20	0.20	0.19	0.19	0.19
14	0.19	0.19	0.19	0.18	0.18

Appendix 11.2.6   Sunshine Function f_b $(n/N) = (0.17 + 0.83n/N)$ (All Areas)

(Ministry of Agriculture, Fisheries and Food (1967) *Potential Transpiration*. Reproduced by permission of the Controller, Her Majesty's Stationery Office, © Crown copyright.)

n/N	0.00	0.01	0.02	0.03	0.04	0.05	0.06	0.07	0.08	0.09
0.10	0.25	0.26	0.27	0.28	0.29	0.29	0.30	0.31	0.32	0.33
0.20	0.34	0.34	0.35	0.36	0.37	0.38	0.39	0.39	0.40	0.41
0.30	0.42	0.43	0.44	0.44	0.45	0.46	0.47	0.48	0.49	0.49
0.40	0.50	0.51	0.52	0.53	0.54	0.54	0.55	0.56	0.57	0.58
0.50	0.59	0.59	0.60	0.61	0.62	0.63	0.63	0.64	0.65	0.66
0.60	0.67	0.68	0.68	0.69	0.70	0.71	0.72	0.72	0.73	0.74

Appendix 11.2.7   Wind Speed Function 0.35 $(1 + u_2/100)$ from wind speed at 10 m

(Ministry of Agriculture, Fisheries and Food (1967) *Potential Transpiration*. Reproduced by permission of the Controller, Her Majesty's Stationery Office, © Crown copyright.)

Miles h^{-1}	0.0	0.2	0.4	0.6	0.8	Knots	0.0	0.2	0.4	0.6	0.8
3	0.55	0.56	0.58	0.59	0.60	3	0.58	0.59	0.61	0.62	0.64
4	0.61	0.62	0.64	0.65	0.67	4	0.65	0.66	0.68	0,69	0.71
5	0.68	0.69	0.70	0.72	0.73	5	0.72	0.74	0.76	0.77	0.79
6	0.75	0.76	0.77	0.79	0.80	6	0.80	0.82	0.84	0.85	0.87
7	0.81	0.82	0.84	0.85	0.86	7	0.88	0.90	0.91	0.92	0.94
8	0.88	0.89	0.90	0.92	0.93	8	0.95	0.97	0.99	1.00	1.02
9	0.94	0.95	0.96	0.98	0.99	9	1.03	1.05	1.06	1.08	1.09
10	1.00	1.02	1.03	1.05	1.06	10	1.10	1.12	1.14	1.15	1.17
11	1.07	1.08	1.10	1.11	1.12	11	1.18	1.20	1.21	1.23	1.24
12	1.13	1.15	1.16	1.18	1.19	12	1.25	1.27	1.28	1.30	1.32
13	1.20	1.21	1.23	1.24	1.26	13	1.33	1.35	1.36	1.38	1.39
14	1.27	1.28	1.30	1.31	1.32	14	1.40	1.42	1.44	1.45	1.47
15	1.33	1.35	1.36	1.37	1.38	15	1.48	1.50	1.51	1.53	1.54
16	1.40	1.41	1.43	1.44	1.45	16	1.55				
17	1.46	1.48	1.49	1.50	1.52	17	1.63				
18	1.53										
19	1.59										
20	1.66										

Appendix 11.2.8 Air Temperature and Saturation Vapour Pressure (mm)

(Ministry of Agriculture, Fisheries and Food (1967) *Potential Transpiration*. Reproduced by permission of the Controller, Her Majesty's Stationery Office, © Crown copyright.)

Air temp. (°F)	30 +	40 +	50 +	60 +	Air temp. (°C)	0 +	10 +	20 +
0.0	4.20	6.29	9.21	13.26	− 0.5	4.40		
0.5	4.30	6.42	9.38	13.49	0.0	4.58	9.21	17.53
1.0	4.38	6.54	9.56	13.73	0.5	4.75	9.52	
1.5	4.48	6.67	9.74	13.98	1.0	4.93	9.84	18.68
2.0	4.58	6.80	9.92	14.23	1.5	5.11	10.18	
2.5	4.67	6.94	10.10	14.48	2.0	5.30	10.52	
3.0	4.77	7.07	10.29	14.73	2.5	5.49	10.87	
3.5	4.87	7.21	10.48	15.00	3.0	5.69	11.23	
4.0	4.97	7.35	10.67	15.27	3.5	5.89	11.61	
4.5	5.07	7.50	10.87	15.54	4.0	6.10	11.99	
5.0	5.17	7.64	11.07	15.81	4.5	6.32	12.38	
5.5	5.27	7.79	11.28	16.08	5.0	6.54	12.79	
6.0	5.38	7.93	11.48	16.36	5.5	6.77	13.13	
6.5	5.49	8.09	11.69	16.65	6.0	7.02	13.63	
7.0	5.60	8.24	11.90	16.95	6.5	7.26	14.08	
7.5	5.71	8.40	12.12		7.0	7.52	14.53	
8.0	5.82	8.55	12.34	17.53	7.5	7.79	15.00	
8.5	5.94	8.71	12.56		8.0	8.05	15.49	
9.0	6.05	8.88	12.79	18.16	8.5	8.33	15.97	
9.5	6.17	9.05	13.02		9.0	8.62	16.47	
					9.5	8.91		

Appendix 11.2.9 Weighting Factor Δ/γ and Temperature

(Ministry of Agriculture, Fisheries and Food (1967) *Potential Transpiration.* Reproduced by permission of the Controller, Her Majesty's Stationery Office, © Crown copyright.)

°F	30	40	50	60	°C	0	10	20
0	0.66	0.91	1.26	1.75	0	0.69	1.26	2.23
0.5	0.66	0.92	1.28	1.77	0.5	0.71	1.30	
1.0	0.67	0.94	1.31	1.80	1.0	0.73	1.34	
1.5	0.68	0.96	1.33	1.83	1.5	0.76	1.38	
2.0	0.69	0.97	1.35	1.86	2.0	0.78	1.42	
2.5	0.70	0.99	1.37	1.88	2.5	0.80	1.47	
3.0	0.71	1.01	1.40	1.91	3.0	0.83	1.51	
3.5	0.72	1.02	1.42	1.94	3.5	0.85	1.56	
4.0	0.74	1.04	1.44	1.97	4.0	0.88	1.60	
4.5	0.75	1.06	1.47	2.00	4.5	0.91	1.65	
5.0	0.76	1.07	1.49	2.03	5.0	0.94	1.69	
5.5	0.77	1.09	1.52	2.06	5.5	0.97	1.74	
6.0	0.79	1.11	1.54	2.10	6.0	1.00	1.79	
6.5	0.80	1.13	1.57	2.13	6.5	1.03	1.84	
7.0	0.82	1.15	1.59	2.16	7.0	1.06	1.89	
7.5	0.83	1.17	1.62	2.20	7.5	1.09	1.94	
8.0	0.84	1.19	1.64	2.23	8.0	1.13	1.99	
8.5	0.86	1.20	1.67	2.27	8.5	1.16	2.04	
9.0	0.87	1.22	1.69	2.32	9.0	1.19	2.11	
9.5	0.89	1.24	1.72		9.5	1.23	2.17	

Appendix 11.3.1   Derivation of Actual Soil Moisture Deficit From Potential Soil Moisture Deficit

Potential (mm)	Actual (mm)	Potential (mm)	Actual (mm)	Potential (mm)	Actual (mm)
RC 75 mm					
96	96	116	106	170	114
98	97	118	107	180	115
100	98	120	108	190	116
102	100	124	109	200	117
104	101	128	110	210	118
106	102	132	111	220	119
108	103	136	112	240	120
110	104	140	112	260	122
112	105	150	113	280	124
114	106	160	113	300	125
RC 200 mm					
224	224	244	234	290	241
226	225	246	234	300	242
228	226	248	235	320	243
230	227	250	236	340	245
232	228	254	237	360	247
234	229	258	238	380	248
236	230	262	239	400	250
238	231	266	239		
240	232	270	240		
242	233	280	240		

References

Crowe, P.R. (1971). *Concepts in Climatology*. Longman, P. 98.

Doorenbos, J. and Pruitt, W.O. (1977). *Crop Water Requirements FAO Irrigation and Drainage*. Paper 24, 144 pp.

Dyer, A.J. and Maher, F.J. (1965). 'Automatic eddy-flux measurement with the evapotron.' *J. appl. Met.*, **4** (5), 622–625.

Grindley, J. (1969). *The Calculation of Actual Evaporation and Soil Moisture Deficit over Specified Catchment Areas*. Hydrological Memorandum No. 38, Meteorological Office.

Grindley, J. (1970). 'Estimation and mapping of evaporation.' IASH Pub. No. 92. *World Water Balance*, Vol. 1, pp. 200–213.

Harbeck, G.E. (1962). *A Practical Field Technique for Measuring Reservoir Evaporation Utilizing Mass Transfer Theory*. US Geol. Surv. Prof. Paper 272-E.

Harbeck, G.E. and Meyers, J.S. (1970). 'Present day evaporation measurement techniques.' *Proc. ASCE*, **HY7**, 1381–1389.

McCulloch, J.S.G. (1965). 'Tables for the rapid computation of the Penman estimate of evaporation.' *E. African Agric. Forest. J.*, **XXX** (3), 286–295.

Meteorological Office (1981). *The Meteorological Office Rainfall and Evaporation Calculation System*. MORECS, Hydrological Memorandum No. 45.

Ministry of Agriculture, Fisheries & Food (MAFF).(1967). *Potential Transpiration*. Tech. Bull. 16., HMSO.

Penman, H.L. (1948). 'Natural evaporation from open water, bare soil and grass.' *Proc. Roy. Soc. Lond.*, **193**, 120–145.

Penman, H.L. (1950a). 'Evaporation over the British Isles.' *Quart. Jl. Roy. Met. Soc.*, **LXXVI**, 330, 372–383.

Penman, H.L. (1950b). 'The water balance of the Stour catchment area.' *J. Inst. Water Eng.*, **4**, 457–469.

Rider, N.E. (1954). 'Evaporation from an oat field.' *Quart. Jl. Roy. Met. Soc.*, **LXXX**, 198–211.

Shuttleworth, W.J. (1979). *Evaporation*. Institute of Hydrology Report No. 56, 61 pp.

Thornthwaite, C.W. (1948). 'An approach towards a rational classification of climate.' *Geog. Rev.*, **38**, 55–94.

Thornthwaite, C.W. and Holzman, B. (1939). 'The determination of evaporation from land and water surfaces.' *Mon. Weath. Rev.*, **67** (1), 4–11.

Turc, L. (1954). 'Calcul du bilan de l'eau evaluation en fonction des precipitations et des temperatures.' IASH Rome Symp. 111 Pub No. 38, 188–202.

Turc, L. (1955). 'Le bilan d'eau des sols. Relations entre les precipitations, l'evaporation et l'ecoulement.' *Ann. Agron.*, **6**, 5–131.

United Nations (FAO).(1977). *Crop Water Requirements*. Food and Agricultural Organisation, Irrigation and Drainage Paper 24. 144 pp.

Wales-Smith, B.G. (1971). *The Use of the Penman Formula in Hydrology*. Hydrological Memorandum No. 39, Meteorological Office.

Wales-Smith, B.G. (1975). *The Estimation of Irrigation Needs*. In: *Engineering Hydrology Today*, pp. 45–54.

Further Reading

Hounam, C.E. (1973). *Comparison Between Pan and Lake Evaporation*. WMO Tech. Note No. 126.

McIlroy, I.C. (1966). Evaporation and its measurement Parts A & B. *Agric. Met. Proc. WMO. Seminar*, 243–265 and 409–431.

Thom, A.S. and Oliver, H.R. (1977). 'On Penman's equation for estimating regional evaporation.' *Quart. Jl. Roy. Met. Soc.*, **C111**, 345–357.

Webb, E.K. (1966). 'A pan-lake evaporation relationship'. *J. Hydrol.*, **4**, 1–11.

Webb, E.K. (1975). 'Evaporation from catchments'. In *Prediction in Catchment Hydrology*, pp. 203–236. Aust. Acad. Sci.

12

River Flow Analysis

The ultimate aim of many computational techniques in engineering hydrology is the derivation of river discharges, and it might appear that once these are obtained the hydrologist's work is done. However, whether they are gained indirectly from considerations of other hydrological variables (to be described in following chapters) or directly from river discharge measurements, the discharge data are only samples in time of the behaviour of the river. The hydrologist then must assess the worth of the data and their representativeness over the period for which the information is required, usually the expected life of a water engineering project.

From river stages and stage – discharge relationships derived from complementary velocity – area measurements or from discharges obtained with a calibrated structure (Chapter 6), processed and quality controlled (Chapter 9), the basic data readily available are daily mean discharges and instantaneous peak discharges. For greater detail, for example, for small catchments in the UK, the engineering hydrologist may obtain the 15-min interval stage recordings and equivalent discharges.

An example of a year's record of daily mean discharges is shown in Fig. 12.1a from the gauging station at Broken Scar on the River Tees in the Northumbrian Water Authority area (note that the discharge scale is logarithmic). A more meaningful plot in Fig. 12.1b relates the same year's daily record (1973) to the highest (MAX) and lowest (MIN) discharges on corresponding days in the period of the record (1956 to 1972). Although 1973 has the appearance of an average year, the rainfall, usually dominant in the winter, has been well distributed throughout the twelve months, with extra large discharges occurring in the summer months even when evaporation losses are at their maximum. The great irregularity shown by the sequence of daily mean flows is indicative of a catchment that responds rapidly to rainfall. The occurrence of extremely low daily mean flows (0.02 m^3s^{-1}), which are seen in the minimum values in October, emphasizes the lack of storage in the drainage basin. A contrasting record is seen in Fig. 12.2a. The daily mean discharges from the Theale gauging station on the River Kennet for the same year, 1973, show a much more even pattern. Intense summer rainstorms have produced sharp peaks on the hydrograph, but the low flows are sustained at about 4 m^3s^{-1}. The lack of variation in the daily flows is a notable feature of a catchment with a large storage. The Kennet drains an area predominantly composed of chalk, which dampens the effects of minor irregularities in daily rainfalls.

The records that are shown here are gauged daily mean discharges calculated directly from the measurements made at a gauging station. However, some rivers may be affected by large abstractions upstream of the

(a)

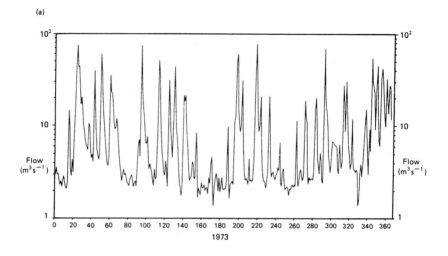

(b)

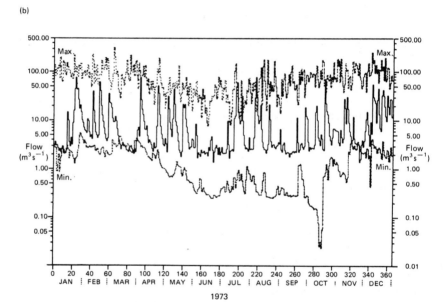

Fig. 12.1   025001 River Tees at Broken Scar. Daily mean flows. Drainage area 818 km². (Reproduced from Department of the Environment (1978) *Surface Water: United Kingdom 1971–73*, by permission of the Controller, Her Majesty's Stationery Office. © Crown copyright.)

gauging station or the river flow may be controlled totally by regulated reservoir releases. The hydrologist must be alive to such modifications as shown in Figure 12.2b. For certain studies, the abstractions must be added to

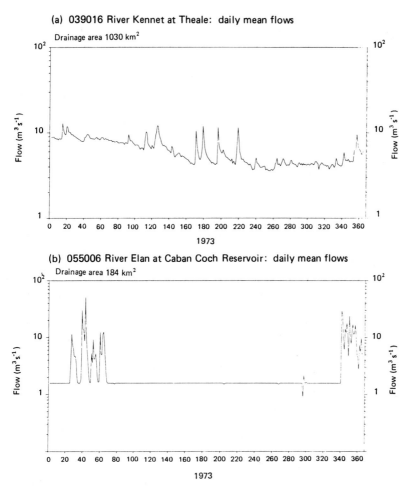

(a) 039016 River Kennet at Theale: daily mean flows

(b) 055006 River Elan at Caban Coch Reservoir: daily mean flows

Fig. 12.2   (Reproduced from Department of the Environment (1978) *Surface Water: United Kingdom 1971–73*, by permission of the Controller, Her Majesty's Stationery Office. © Crown copyright.)

the gauged flows to give the naturalized or gross flows. The discharges in both the Tees and the Kennet are modified by man, but the changes are small in proportion to the total daily mean flows and are not readily discernible on the daily mean flow plots.

Abstractions from a river, although taken regularly each day for domestic or other water supplies, are usually quantified on a monthly basis. Hence a very useful statistic of river flow is the average of the daily mean flows over a month (the monthly mean flow). Table 12.1 shows the monthly mean gauged and naturalized discharges for the River Thames at Teddington for 1973. Teddington Weir is the tidal limit of the Thames, and thus the difference between the gauged and naturalized flows gives a measure of the demands of London on the freshwater resources of the river.

Table 12.1   Monthly Mean Gauged and Naturalized Discharges (m^3s^{-1}): River Thames at Teddington 1973 (Drainage Area 9870 km^2)

Figures rounded off to the first from three decimal places. (Reproduced from Department of the Environment (1978) *Surface Water: United Kingdom 1971–73*, by permission of the Controller, Her Majesty's Stationery Office. © Crown copyright.)

	J	F	M	A	M	J	J	A	S	O	N	D
Gauged	50.1	42.9	25.4	22.0	33.7	21.7	22.6	12.7	11.6	9.8	10.0	20.3
Naturalized	69.5	60.3	43.5	42.1	50.7	40.4	41.3	29.6	27.5	28.2	27.3	41.4
Difference	19.4	17.4	18.1	20.1	27.0	18.7	18.7	16.9	15.9	18.4	17.3	21.1

Further monthly statistics for the River Tees given in Table 12.2 are the highest instantaneous peaks recorded (which are essential for assessing regulation requirements) and the maximum and minimum daily mean discharges (indicating the range of water availability). The maxima and minima can readily be identified on Fig. 12.1 plots. The value of these statistics is enhanced with each year of record: the longer the record at a gauging station, the more reliable can be the evaluation of water resources and the estimation of extreme events either of dangerous floods or of harmful droughts.

Monthly data may be used more directly for evaluating the amount of storage required in a reservoir to guarantee a given demand or supply rate. To determine the reservoir capacity needed to ensure supplies over a low flow period, the cumulative volumes over that period are evaluated. In Fig. 12.3, the mean monthly flows for a catchment of 150 km^2 for the 5 years 1973–77 have been converted into volumes of water and the cumulative sums plotted. The slope of the straight line OA represents an average catchment flow rate of 74 Ml day^{-1} over the period from O to A. To sustain this flow rate as a steady abstraction from a reservoir full at O, a storage of 13 000 Ml would be required at the point of maximum deficit, X. Over this period, there are two minor dry spells, but from A a more severe dry spell develops. The more gentle slope of the line AB represents only 53 Ml day^{-1} average inflow rate over the period A to B, and if this rate is to be maintained as a steady abstraction a storage of 12 000 Ml is necessary, as seen at the point of maximum deficit Y. On the evidence of what is known to be an exceptionally dry period 1975–76 (AB), a reliable yield of 53 Ml day^{-1} would be assured with a storage of 12 000 Ml. This method gives a good first estimate of yields and required reservoir capacities, but gives little information on probabilities of reservoir failure to meet demand.

12.1   Peak Discharges

High river discharges are caused by various combinations of extreme conditions. Heavy rainfalls over short durations, deep snow cover melted by

Table 12.2 Monthly Statistics of Gauged Flows (m^{3}s^{-1})

Instantaneous peaks, highest and lowest daily means. River Tees at Broken Scar, 1973. Drainage area 818 km^2. (Reproduced from Department of the Environment (1978) *Surface Water: United Kingdom 1971–73*, by permission of the Controller, Her Majesty's Stationery Office. © Crown copyright.)

	J	F	M	A	M	J	J	A	S	O	N	D
Peaks	111.8	77.0	65.9	164.4	60.6	12.6	170.4	275.9	32.8	149.6	59.5	130.7
Max. daily mean	75.5	60.9	35.6	75.3	44.2	8.5	60.8	78.1	18.8	69.3	30.5	55.3
Min. daily mean	2.1	3.7	2.2	2.3	1.8	1.4	1.7	2.1	1.8	2.6	1.4	3.0

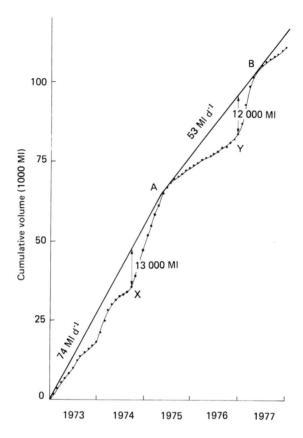

Fig. 12.3 Cumulative monthly volumes.

warming rain and moderate rainfalls on frozen ground or saturated soil, can all contribute to a rapid and large runoff. Flood conditions are of great concern, and notable events are always studied in detail. With the expansion of the river gauging network in the UK, many more floods are now being measured, but when the river has overtopped the gauging control and where a river is ungauged, only estimates of the peak discharges can be made by hydraulic calculations.

Peak discharges from major runoff events in the UK have been assembled by engineers in relation to reservoir practice (ICE, 1933) and a revised analysis with records updated (ICE, 1960). More recently (NERC, 1975), with wider terms of reference, a more comprehensive study of floods has been made involving the collation of nationwide records for the whole of the British Isles. From a combination of flood discharge measurements and post-event discharge calculations taken from the ICE and NERC published records, a graph of peak discharge against catchment area has been drawn (Fig. 12.4). Twenty peak discharges for areas ranging from 4 km² to nearly 1800 km² are plotted, and an enveloping curve is drawn. Most of the points represented by catchments less than 25 km² pertain to the tributary flows contributing to the disastrous Lynmouth flood in August 1952, where the peak flow was estimated at 650 m³s⁻¹ from 101 km² (Dobbie & Wolf, 1953). The highest measured flow in the UK is the 2402 m³s⁻¹ on the River Findhorn at Forres in Scotland in 1969 from a catchment area of 782 km². There are other historic records of flood levels on many rivers in the country, and some of these could possibly provide peak discharges above the curve in Fig. 12.4. However changes in the river profiles and cross-sections over the ensuing years makes the conversion of level data into discharges unreliable. Comparable data for the USA are also shown in Fig. 12.4. The drainage areas range from 2 km² to over 8000 km², with corresponding peak discharges from 144 m³s⁻¹ to 21 000 m³s⁻¹ (Crippen & Bue, 1977). These extreme events in the USA come from a wide range of different climatic regions, some of which experience tropical rainfall intensities. From the enveloping curves of the two plots, a 5 km² catchment somewhere in the USA may well produce a peak of 600 m³s⁻¹, but in the UK only a 150 m³s⁻¹ peak may be expected from the same drainage area. (The Findhorn record in Scotland would not be worth plotting on the USA graph!) Thus, when making a general appraisal of peak records, such a relationship of peak discharge to catchment area should be made according to climatic region. The USA has been divided into 17 flood regions (Crippen & Bue, 1977), and such subdivisions in the UK have been made for more detailed studies of flood events (NERC, 1975).

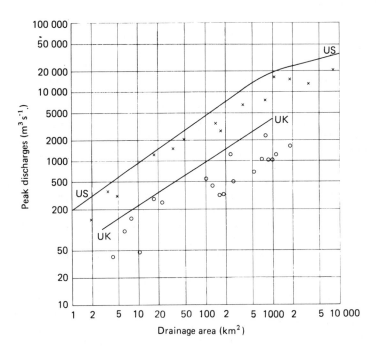

Fig. 12.4   Maximum flood discharges.

Table 12.3   Peak Discharges per Unit Area in The UK

River	Area km²	Peak discharge m³s⁻¹ km⁻²	
Hoaroak Water	8.06	18.44	(1952)
Hoaroak Water	17.0	16.80	(1952)
Lyn (Lynmouth)	101.	6.44	(1952)
Wenning	236.	5.28	(1966)
Findhorn	782.	3.07	(1969)

Some hydrologists prefer to convert the absolute peak discharges into relative values per unit area for plotting against catchment area, thus giving an inverse relationship and a declining curve with increased catchment area. Sample values from the UK plot are shown in Table 12.3.

Numerous empirical formulae given in the hydrological literature have been derived from the relationship between peak discharges and catchment areas, with the coefficients specifically determined for particular countries or climatic regions. In attempting to use such formulae to obtain peaks for ungauged catchments, the hydrologist must guard against applying them to inappropriate conditions and areas.

12.2   River Regimes

In Table 12.1, the monthly mean discharges for the River Thames in 1973 exhibit a distinctive seasonal pattern, with the highest values occurring in the winter months. The expected pattern of river flow during a year is known as the *river regime*. (This term may also be applied to the absolute range of flow in a river and has been used in outdated flow theories.) Flow records for 20 – 30 years are required to provide a representative pattern, since there may be considerable variations in the seasonal discharges from year to year. The averages of the monthly mean discharges over the years of record calculated for each month, January to December, give the general or expected pattern, the flow regime of the river.

The river regime is the direct consequence of the climatic factors influencing the catchment runoff, and can be estimated from a knowledge of the climate of a region. The eminent French geographer, Pardé, has identified and classified distinctive river regimes and this can be a great help to engineers faced with unfamiliar conditions and sparse data (Rochefort, 1963). The classification is based on an understanding of the role of the main climatic features, temperature and rainfall, in causing river runoff. An illustration of *simple river regimes* resulting from a single dominant factor is given in Fig. 12.5a. For each river, the monthly mean discharges from January to December are represented as proportions of the mean of the 12 monthly values. In this way, the graphs are comparable and independent of the absolute values of the monthly mean discharges and catchment areas.

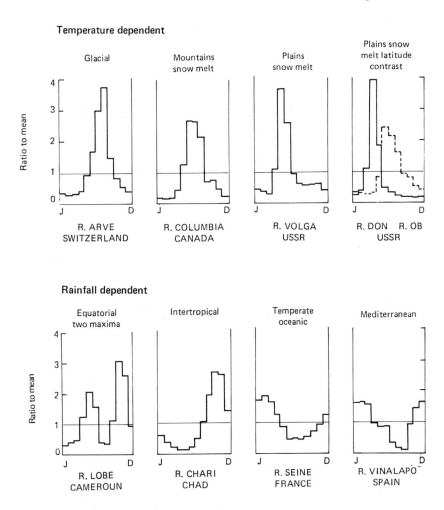

Fig. 12.5   (a) Simple river regimes (Pardé).

12.2.1   Temperature-Dependent Regimes

Rivers with a dominant single source of supply, initially in the solid state (snow or ice), produce a simple maximum and minimum in the pattern of monthly mean discharges according to the seasonal temperatures.

Glacial.   When the catchment area is over 25 – 30% covered by ice, the river flow is dominated by the melting conditions. Such rivers are found in the high mountain areas of the temperate regions. There is little variation in the pattern from year to year, but in the main melting season, July and August, there are great diurnal variations in the melt water flows.

Mountain snow melt.   The seasonal peak from snow melt is lower and earlier

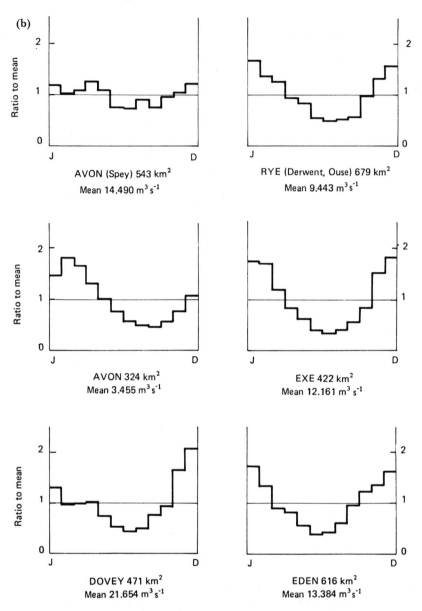

Fig. 12.5   (b) River régimes in the UK.

than in a glacial stream, but the pattern is also regular each year providing there has been adequate winter snowfall. The low winter flows are caused by freezing conditions.

Plains snow melt.   The regular winter snow cover of the interior regions of the large continents in temperate and sub-Polar latitudes melts quickly to give a short 3-month season of high river flows. The timing of the peak month depends on latitude, with the more southerly rivers (e.g. Don) having rapid flood peaks in April, whereas further north there is a slower melt and the lower peak occurring in June is more prolonged.

12.2.2   Rainfall-Dependent Regimes

In the equatorial and tropical regions of the world with no high mountains, the seasonal rainfall variations are the direct cause of the river regimes. Temperature effects in these areas are mostly related to evaporation losses, but with these being dependent on rainfall, the overall effect of evaporation is of secondary importance in influencing the river flow pattern.

Equatorial.   Drainage basins wholly within the equatorial belt experience two rainfall seasons with the annual migration of the intertropical convergence zone, and these are reflected directly in the river regime (e.g. R. Lobe).

Tropical.   Within the tropics, there are usually marked wet and dry seasons that vary in length according to latitude, and the flow ratios vary according to rainfall quantities. The R. Chari in Chad is a clear example. Some catchments have double peaks, as in the equatorial region, but only separated by 1 or 2 months rather than 6 months.

Temperate oceanic.   The R. Seine demonstrates the characteristic regime of these regions. Rainfall occurs all the year round, but the summer evaporation provides the relatively small variation in the seasonal flow pattern, which is however annually irregular.

Mediterranean.   The regime resembles that of the temperate oceanic regions, but is more extreme. The dry summers result in very low flows; along the desert margins, rivers dry up completely. Most of the flow results from the winter rains, but occasional very heavy summer storms may produce flood flows. The river regime is also very variable from one year to another, in parallel with the irregular incidence of rainfall.

The flow regime of a river draining a catchment area within a single climatic region may be readily estimated by considering the features demonstrated by the simple regimes. It must be remembered that away from the equatorial region, the patterns will be reversed in the southern hemisphere. However, the conditions in the high mountains in any of the regions will result in modifications to the expected pattern. Rivers with two or more sources of supply have a *mixed regime*. For example, a spring maximum may be identified as snow melt, whereas early winter rains give a second peak.

More *complex regimes* result from the overlapping of different causes. These are usually characteristic of large rivers, especially those flowing through several climatic zones. The major rivers of the world, the Congo, Nile, Mississippi and the Amazon come into this category. With mixed or complex regimes, the range between the extreme months is usually small, and the annual variability decreases with increase in catchment size.

Examples of river regimes within the UK are given in Fig. 12.5b. The general pattern resembles that for temperate oceanic regions with a relatively small range between the wettest and driest months. The Scottish Avon, a tributary of the Spey, shows the least seasonal variation, the result of persistent rainfall throughout the year. Further south, the effect of summer evaporation losses becomes more apparent, and the Rye in east Yorkshire reflects this tendency. The slight differences in the other four diagrams stem from varying location and catchment characteristics. There are sustained later spring flows above normal from the chalk of the south coast Avon, and the exposure of the Dovey catchment in central Wales causes high flows from the moutains in early winter. These are average conditions calculated from the years of record available for each station, but there are great variations from year to year that result from the irregular incidence of rainfall.

12.3   Flow Frequency

For many problems in water engineering, the hydrologist is asked for the frequency of occurrence of specific river flows or for the length of time for which particular river flows are expected to be exceeded. Thus frequency analysis forms one of the important skills required of a hydrologist. In attempting to provide any answer to the questions of frequency, good

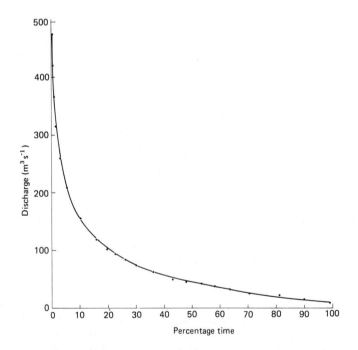

Fig. 12.6   Flow duration curve. River Thames at Teddington Weir.

Table 12.4   Flow Frequencies over a 4-Year Period: River Thames at Teddington

Daily mean discharge (m^3s^{-1})	Frequency	Cumulative frequency	Percentage cumulative frequency
Over 475	3	3	0.21
420–475	5	8	0.55
365–420	5	13	0.89
315–365	8	21	1.44
260–315	25	46	3.15
210–260	36	82	5.61
155–210	71	153	10.47
120–155	82	235	16.08
105–120	52	287	19.64
95–105	42	329	22.52
85–95	50	379	25.94
75–85	58	437	29.91
65–75	83	520	35.59
50–65	105	625	42.78
47–50	72	697	47.71
42–47	75	772	52.84
37–42	73	845	57.84
32–37	84	929	63.59
26–32	103	1032	70.64
21–26	152	1184	81.04
16–21	128	1312	89.80
11–16	141	1453	99.45
Below 11	8	1461	100.00
Total days	1461		

reliable hydrological records are essential, and these must if possible extend beyond the expected life of the engineering scheme being considered.

From the basic assemblage of river flow data comprising the daily mean discharges and the instantaneous peaks, analysis of the daily mean flows will be considered first. Taking the n years of flow records from a river gauging station, there are $365(6)n$ daily mean discharges. The frequencies of occurrence in selected discharge classes (groups) are compiled, starting with the highest values. The cumulative frequencies converted into percentages of the total number of days are then the basis for the *flow-duration curve*, which gives the percentage of time during which any selected discharge may be equalled or exceeded. An example is demonstrated in Table 12.4, in which the daily mean discharges for four years for the River Thames at Teddington Weir are analysed. The flow duration curve plotted on natural scales is seen in Fig. 12.6. The area under the curve is a measure of the total volume of water that has flowed past the gauging station in the total time considered. For the reliable assessment of water supply, the flow-duration curves for the wettest and driest years of the record should be derived and plotted. In the sharp angled section of the graph in the high flow region, it can be difficult to measure accurately the required area.

The representation of the flow duration curve is improved by plotting the cumulative discharge frequencies on log–probability paper (Fig. 12.7). (The abscissa scale is based on the normal probability distribution; if the logarithms of the daily mean discharges were normally distributed, they would plot as a straight line on the log–probability paper.) From the plot (Fig. 12.7), it can be readily seen for example that for 2% of the 4-year period, flows exceeded 290 m^3s^{-1}. At the other extreme, flows of less than 12 m^3s^{-1} occurred for the same proportion of the time. Alternatively it can be stated that for 96% of that 4-year period, the flow in the River Thames at Teddington is between 12 and 290 m^3s^{-1}. The 50% time point provides an average value, the median (45 m^3s^{-1}).

The shape of the flow-duration curve gives a good indication of a catchment's characteristic response to its average rainfall history. An initially steeply sloped curve results from a very variable discharge, usually from small catchments with little storage where the stream flow reflects directly the rainfall pattern. Flow-duration curves that have a very flat slope indicates little variation in flow regime, the resultant of the damping effects of large storages. Groundwater storages are provided naturally by extensive chalk or limestone aquifers, and large surface lakes or reservoirs may act as runoff regulators either naturally or controlled by man. Examples of different flow-duration curves are given in Fig. 12.8. The comparisons are simplified by plotting the logarithms of the daily mean discharges as percentages of the overall daily mean discharge. The Honddu, a tributary of the Usk drains a

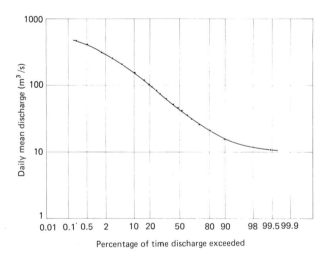

Fig. 12.7   River Thames at Teddington Weir.

AT. NO.: 056003   RIVER: HONDDU (Usk) S. Wales   STATION: THE FORGE BRECON
TA REGISTERED FOR CALENDER YEARS: 1800 – 1976
COMPLETE YEAR(S) USED IN ANALYSIS
AY MEANS USED

C.A. (km^2) =   62.1
MEAN FLOW   =   1.483
MAX FLOW   =   23.850
MIN FLOW   =   .345
ALL FLOWS IN CUMECS

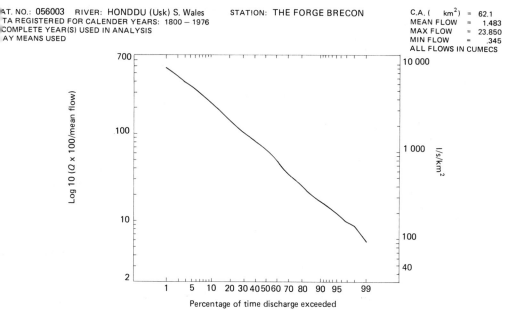

Fig. 12.8(a)

AT. NO.: 038003   RIVER: MIMRAM (Lee)   STATION: PANSHANGER PARK
TA REQUESTED FOR CALENDER YEARS: 1800 – 1976
COMPLETE YEAR(S) USED IN ANALYSIS
AY MEANS USED

C.A. (km^2) =   133.9
MEAN FLOW   =   .519
MAX FLOW   =   1.633
MIN FLOW   =   .135
ALL FLOWS IN CUMECS

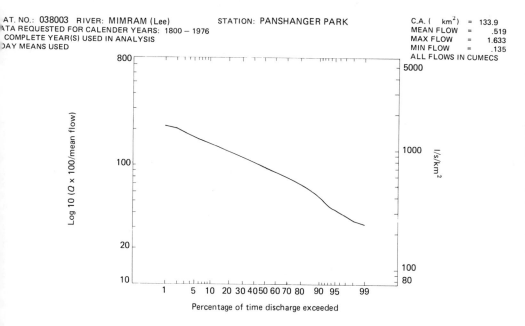

Fig. 12.8(b)

mountainous area of South Wales, whereas the catchment of the Mimram, a tributary of the Lee is nearly all chalk.

The comparison of flow-duration curves from different catchments can also be used to extend knowledge of the flow characteristics of a drainage area that has a very limited short record. One or two years of the latter's records are required overlapping with those of a long-term well established gauging station on a neighbouring river, whose flow duration curve for the whole length of its record may be taken to represent long-term flow conditions. For satisfactory results, the two catchments should be in the same hydrological region and should experience similar meteorological conditions. The method for supplementing the short-term record is to construct its long-term flow duration curve by relating the overlapping short-period flow duration curves of both catchments, as shown in Fig. 12.9. The available data are plotted and flow-duration curves (in full lines) are drawn in Fig. 12.9a and b. S_a and S_b are flow-duration curves for the short overlapping records; L_a is the flow-duration curve for the long-term neighbouring record. Selected percentage discharge values from the two short-period flow duration curves are plotted in Fig. 12.9c, and a straight-line relationship drawn. Time percentage discharge values for L_a are converted to corresponding percentage values for L_b via the S_a and S_b relation. The derived long-term duration curve for the short-period station, L_b, is shown by a broken line in

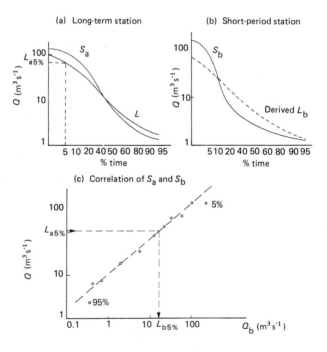

Fig. 12.9   Derivation of long-term flow duration curve for a short-period station. (After A. J. Raudkivi (1979) *Hydrology*, by permission of Pergamon Press.)

Fig. 12.9b. By these means, the variation in the flow characteristics embodied in the long-term flow duration curve has been translated to the short-period station.

Flow-duration curves from monthly mean and annual mean discharges can also be derived, but their usefulness is much less than those constructed from daily mean flows, since the extreme discharges are lost in the averaging. It should be stressed that no representation of the chronological sequence of events is portrayed or enumerated in flow-duration curves. For assesssing water resources, the frequency of sequences of wet or dry months can be evaluated as for monthly rainfall, given a suitably long record. The consideration of data series as time dependent sequences is reserved for a later chapter.

12.4 Flood Frequency

The measured instantaneous flood peak discharges abstracted from calibrated levels on autographic charts or automatic digital recorders constitute one of the most valuable data sets for the hydrologist. The longer a record continues, homogeneous and with no missing peaks, the more is its value enhanced. Even so, it is very rare to have a satisfactory record long enough to match the expected life of many engineering works required to be designed. As many peak flows as possible are needed in assessing flood frequencies.

The hydrologist defines two data series of peak flows: the annual maximum series and the partial duration series. These may be understood more readily from Fig. 12.10. *The Annual Maximum Series* takes the single maximum peak discharge in each year of record so that the number of data values equals the record length in years. For statistical purposes, it is necessary to ensure that the selected annual peaks are independent of one another. This is sometimes difficult, e.g. when an annual maximum flow in January may be related to an annual maximum flow in the previous December. For this reason, it is sometimes advisable to use the Water Year rather than the

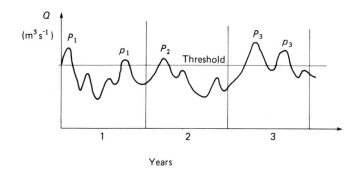

Fig. 12.10   Flood peak data series.

calendar year; the definition of the Water Year depends on the seasonal climatic and flow regimes. *The Partial Duration Series* takes all the peaks over a selected level of discharge, a threshold. Hence the series is often called the 'Peaks over Threshold' series (POT). There are generally more data values for analysis in this series than in the Annual Series, but there is more chance of the peaks being related and the assumption of true independence is less valid.

In Fig. 12.10, P_1, P_2 and P_3 form an annual series and P_1, p_1, P_2, P_3 and p_3 form a POT series. It will be noted that one of the peaks in the POT series, p_3, is higher than the maximum annual value in the second year, P_2. For sufficiently long records it may be prudent to consider all the major peaks and then the threshold is chosen so that there are N peaks in the N years of record, but not necessarily one in each year. This is called the *Annual Exceedance Series*, a special case of the POT series.

Flood frequency analysis entails the estimation of the peak discharge which is likely to be equalled or exceeded on average once in a specified period, T years. This is called the T-year event and the peak, Q_T, is said to have a *return period* or *recurrence interval* of T years. The return period, T years, is the long-term average of the intervals between successive exceedances of a flood magnitude, Q_T, but it should be emphasized that those intervals may vary considerably around the average value T. Thus a given record may show 25-year events, Q_{25}, occurring at intervals both much greater or much less than 25 years, even in successive years. Alternatively, it may be required to estimate the return period of a specified flood peak.

The annual series and the partial-duration series of peak flows form different probability distributions, but for return periods of 10 years and more, the differences are minimal and the annual maximum series is the one most usually analysed.

12.5   Flood Probabilities

When a series of annual maximum flows is subdivided by magnitude into discharge groups or classes of class interval ΔQ, the number of occurrences of the peak flows (f_i) in each class can be plotted against the discharge values to give a frequency diagram (Fig. 12.11a). It is convenient to transform the ordinates of the diagram in two ways: with respect both to the size of the discharge classes (ΔQ) and the total number, N, of events in the series. By plotting $f_i/(N.\Delta Q)$ as ordinates, it is seen that the panel *areas* of the diagram will each be given by f_i/N, and hence the sum of those areas will be unity, i.e.:

$$\sum_{i=1}^{n} \left(\frac{f_i}{N.\Delta Q} . \Delta Q \right) = \frac{N}{N} = 1$$

where n is the number of discharge classes.

If the data series is now imagined to be infinitely large in number and the

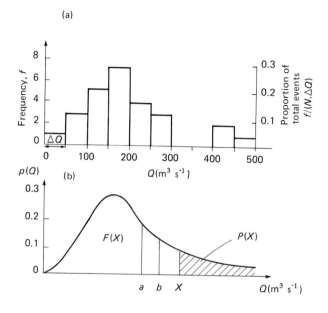

Fig. 12.11   Frequency diagram and probability distribution of annual maximum peak flows.

class intervals are made infinitessimally small, then in the limit, a smooth curve of the *probability density distribution* is obtained (Fig. 12.11b), the area under this curve being unity, i.e.:

$$\int_0^\infty p(Q)\,dQ = 1 \tag{12.1}$$

The probability that an annual maximum, Q, lies between two values, a and b, is given by:

$$P = \int_a^b p(Q)\,dQ$$

For any given magnitude, X, the probability that an annual maximum equals or exceeds X, i.e. that $Q \geqslant X$ is:

$$P(X) = \int_X^\infty p(Q)\,dQ$$

which is the area shaded under the probability curve (Fig. 12.11b).
    If $F(X)$ is the probability of $Q < X$:

$$F(X) \quad \int_0^X p(Q)\,dQ$$

and clearly:

$$P(X) = 1 - F(X) \tag{12.2}$$

$P(X)$ is the probability of an annual maximum equalling or exceeding X in any given year, since it is the relative proportion of the total number of annual maxima that have equalled or exceeded X. If X is equalled or exceeded r times in N years (N large), then $P(X) \rightarrow r/N$. The return period for X is, however, $T(X) = N/r$. Thus:

$$P(X) = \frac{1}{T(X)} \tag{12.3}$$

It follows that

$$T(X) = \frac{1}{P(X)} = \frac{1}{1 - F(X)}$$

and:

$$F(X) = \frac{T(X) - 1}{T(X)} \tag{12.4}$$

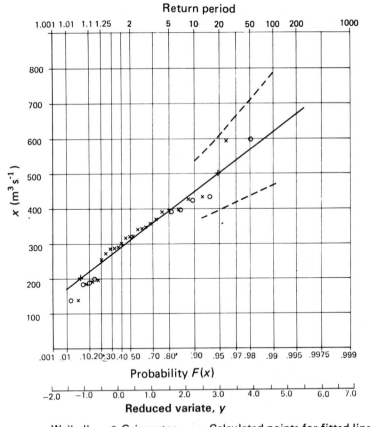

Fig. 12.12 Flood frequency (Gumbel plot).

Thus, if $T(X) = 100$ years, $P(X) = 0.01$ and $F(X) = 0.99$.

To facilitate flood frequency analysis for a data sample, special probability graph papers may be used. The probabilities, $F(X)$, are made the abscissa with X the ordinate. For any given probability distribution the values of $F(X)$ are transformed to a new scale such that the X versus $F(X)$ relationship is made linear for that distribution. An example of the Gumbel probability paper, designed for the Gumbel extreme value Type 1 probability distribution EVI, is seen in Fig. 12.12.

12.6   Analysis of an Annual Maximum Series

In Table 12.5, the procedure for analysing annual maximum flows is given using the 24 annual maximum discharges from a continuous homogeneous gauged river record. The peak flows (m^3s^{-1}) are arranged in decreasing order

Table 12.5   Flood Frequency Analysis (24 Years of an Annual Series)

X (m^3s^{-1})	Rank (r)	$P(X)$ (Gringorten)	$F(X)$	$P(X)$ (Weibull)	$F(X)$
594.7	1	.023	.977	.040	.960
434.7	2	.065	.935	.080	.920
430.4	3	.106	.894	.120	.880
402.1	4	.148	.852	.160	.840
395.9	5	.189	.811	.200	.800
390.8	6	.231	.769	.240	.760
369.5	7	.272	.728	.280	.720
356.8	8	.313	.687	.320	.680
346.9	9	.355	.645	.360	.640
342.9	10	.396	.604	.400	.600
342.6	11	.438	.562	.440	.560
321.4	12	.479	.521	.480	.520
318.0	13	.521	.479	.520	.480
317.2	14	.562	.438	.560	.440
300.2	15	.604	.396	.600	.400
290.0	16	.645	.355	.640	.360
284.6	17	.687	.313	.680	.320
283.2	18	.728	.272	.720	.280
273.3	19	.769	.231	.760	.240
256.3	20	.811	.189	.800	.200
196.8	21	.852	.148	.840	.160
190.3	22	.894	.106	.880	.120
185.5	23	.935	.065	.920	.080
138.8	24	.977	.023	.960	.040

$\overline{X} = 323.5$

$s_X = 97.1$

of magnitude with the second column showing the rank position. The probability of exceedence, $P(X)$, is then calculated for each value, X, according to a plotting position formula devised to overcome the fact that when N is not large, r/N is not a good estimator. Of the several formulae in use, the best is due to Gringorten:

$$P(X) = \frac{r - 0.44}{N + 0.12} \qquad (12.5)$$

where r is the rank of X and N is the total number of data values. The Weibull formula:

$$P(X) = \frac{r}{N + 1} \qquad (12.6)$$

is more widely used; it is easier to calculate than the Gringorten formula. Both formulae, which give very similar results, have been applied in the example and the two sets of values of $P(X)$ and $F(X)$ are shown in Table 12.5.

All the Weibull values of $P(X)$ have been plotted for their corresponding values of X (Fig. 12.12) but only the top 5 and bottom 4 Gringorten values have been plotted to show the location of their main divergence from the Weibull points. The distribution of the plotted points on the graph would be linear if the annual maxima came from a Gumbel distribution and N was large, but with small samples, departures from a straight line are always to be expected. It will be noted that the highest peak (594.7 m^{3}s^{-1}), which has the appearance of an outlier falls into line better in the Gringorten plotting position. Outliers commonly occur and arise from the inclusion in a short record of an event with a long return period.

Although the Gumbel distribution is used for the present example, many other probability distributions have been investigated for application to the extreme values produced by flood peak discharges. In the USA, after comparing six different distributions, the log Pearson Type III was selected, although many factors, other than statistical, governed the final choice (Benson, 1968). The log – Normal distribution, in which the logarithms of the peak flows conform to the Gaussian or Normal distribution, also takes the form of the probability curve with a positive skew shown in Fig. 12.11 and is often applied successfully to peak flow series.

12.6.1   Fitting the Extreme Value Distribution (EVI)

Although subjective graphical curve fitting by eye may be justifiably adequate in the analysis of inaccurately measured or even estimated flood discharges, objective methods of curve fitting are preferred by national agencies seeking uniformity of practice.

For example, the straight line drawn through the plotted points in Fig. 12.12 could well have been drawn by eye, especially using the Gringorten plotting positions, but in fact the line has been drawn by fitting the Gumbel distribution to the data by the *method of moments*.

The equation of the Gumbel extreme value distribution (Type 1) is given by

$$F(X) = \exp\left[-e^{-b(X-a)}\right] \qquad (12.7)$$

where $F(X)$ is the probability of an annual maximum $Q \leqslant X$ as defined previously, and a and b are two parameters related to the moments of the population of Q values. Defining the first moment (the mean) by μ_Q and the second moment (the variance) by σ_Q^2, the parameters a and b are given by the following expressions:

$$a = \mu_Q - \frac{\gamma}{b} \qquad (\gamma = 0.5772) \tag{12.8}$$

$$b = \frac{\pi}{\sigma_Q \sqrt{6}} \tag{12.9}$$

In Equations 12.8 and 12.9, μ_Q and σ_Q^2 pertain to the whole statistical population of floods at the station; with a finite sample they can only be estimated from the moments of the data sample. Thus:

$$\hat{\mu}_Q = \overline{Q} = \frac{1}{N} \sum_{i=1}^{N} Q_i \qquad \text{(the sample mean)}$$

$$\hat{\sigma}_Q^2 = s_Q^2 = \frac{1}{N-1} \sum_{i=1}^{N} (Q_i - \overline{Q})^2 \qquad \text{(the sample variance)}$$

with $\hat{\mu}_Q$ and $\hat{\sigma}_Q^2$ signifying sample estimates of μ_Q and σ_Q^2.

From Table 12.5, $\overline{X} = \overline{Q} = \hat{\mu}_Q = 323.5$ m³s⁻¹ and $s_x = s_Q = \hat{\sigma}_Q = 97.1$ m³s⁻¹ whence $b = 0.0132$ and $a = 279.8$. The straight line fitted to the data by the method of moments is thus given by

$$F(X) = \exp \left[- e^{-0.0132(X-279.8)} \right]$$

To evaluate two points for drawing the line:

for $X = 200, F(X) = 0.057$
        $X = 500, F(X) = 0.947$

The equation of the straight line can be used to provide an estimate of the return period of a given peak discharge, but extrapolations beyond the limits of the data must be treated with reserve. In the example of Fig. 12.12, the return periods of flows up to 500 m³s⁻¹ can be provided with some confidence by this analysis. The validity of the outlying peak needs further checking, and confidence limits to the fitted curve would be recommended before estimates are made of the return periods of higher discharges.

12.6.2   Estimating the T-Year Flood

The above formulation permits $F(X)$ to be found for a specified annual maximum X. Once $F(X)$ is known, $P(X) = 1 - F(X)$ is known, and therefore the return period $T(X) = 1/P(X)$ is known.

To reverse the procedure, the estimation of the annual maximum for a given return period can be obtained from combining Equations 12.4 and 12.7 for $F(X)$. Thus:

$$\exp \left[-e^{-b(X-a)} \right] = \frac{T(X) - 1}{T(X)}$$

Taking logarithms of both sides twice:

$$-b(X - a) = \ln\left[-\ln\frac{T(X) - 1}{T(X)} \right]$$

and rearranging:

$$X = a - \frac{1}{b}\ln\ln\left[\frac{T(X)}{T(X) - 1} \right] \tag{12.10}$$

Substituting for the parameters a and b with the sample mean $\overline{Q}$ and standard deviation s_Q as estimates of the population values μ_Q and σ_Q, then estimates of X may be obtained from:

$$\hat{X} = \overline{Q} - \frac{\gamma s_Q \sqrt{6}}{\pi} - \frac{s_Q\sqrt{6}}{\pi}\left(\ln\ln\left[\frac{T(X)}{T(X) - 1} \right] \right)$$

$$= \overline{Q} - \frac{\sqrt{6}}{\pi}\left(\gamma + \ln\ln\left[\frac{T(X)}{T(X) - 1} \right] \right)s_Q$$

$$= \overline{Q} + K(T)\,s_Q \tag{12.11}$$

where:

$$K(T) = -\frac{\sqrt{6}}{\pi}\left(\gamma + \ln\ln\left[\frac{T(X)}{T(X) - 1} \right] \right) \tag{12.12}$$

$K(T)$ is called the *frequency factor*. Although it is not dependent on the parameters of the probability distribution, $K(T)$ in Equation 12.12, a function only of the return period T, is specifically for the Gumbel Type 1 distribution (given in Table 12.6).

Thus if an estimate of the annual maximum discharge for a return period of 100 years is required, then $T(X) = 100$ years, $K(T) = 3.14$ and $\hat{Q}_{100} = \overline{Q} + 3.14\,s_Q$.

In the plotted example:

$$\hat{Q}_{100} = 323.5 + 3.14 \times 97.1$$
$$= 628.4 \text{ m}^3\text{s}^{-1}$$

With the mean and standard deviation of a sample of annual maximum flows and assuming the Gumbel distribution for the data, an estimate of the peak flow for any required return period can be obtained from Equation 12.11 using the appropriate $K(T)$ value from Table 12.6.

12.6.3   Confidence Limits for the Fitted Data

It is often advisable, particularly with a short data series, to construct confidence limits about the fitted straight-line relationship between the annual maxima and the linearized probability variable. A first step is the calculation of the standard error of estimate for a peak discharge (X) in terms of the return period (T). The expression for the standard error is dependent on the probability distribution used, and for the Gumbel distribution it is given by:

$$\text{SE}(\hat{X}) = \frac{s_Q}{\sqrt{N}}\left[1 + 1.14K(T) + 1.10(K(T))^2 \right]^{\frac{1}{2}} \tag{12.13}$$

Table 12.6   The $T - K(T)$ Relationship for the Gumbel Distribution

T	$K(T)$	T	$K(T)$	T	$K(T)$
1	$-\infty$	10	1.30	80	2.94
2	−0.16	15	1.64	90	3.07
3	0.25	20	1.86	100	3.14
4	0.52	25	2.04	200	3.68
5	0.72	30	2.20	400	4.08
6	0.88	40	2.40	600	4.52
7	1.01	50	2.61	800	4.76
8	1.12	60	2.73	1000	4.94
9	1.21	70	2.88		

where N is the number of annual maxima in the sample. Then the upper and lower confidence limits are calculated for a selection of return periods and estimated values of $\hat{X}$ from:

$$\hat{X} \pm t_{\alpha,\nu}\, SE(\hat{X})$$

where $t_{\alpha,\nu}$ are values of the t distribution obtained from standard statistical tables with α the probability limit required and ν the degree of freedom.

The calculations for the 95% confidence limits for the plotted example in Fig. 12.12 are set out in Table 12.7. The value of the t statistic is 2.06 for $\alpha = 100 - 95\% = 5\%$ and $\nu = n - 1 = 23$. The curves of the 95% confidence limits are plotted on Fig. 12.12 for the range of the selected T values.

12.7   Flood Prediction

The concept of the return period states that the T-year event, Q_T, is the *average* chance of exceedance once every T years *over a long record*. However, it is often required to know the actual probability of exceedance of the T year flood in a specific period of n years. $F(X)$ is the probability of X *not* being

Table 12.7   Calculation of 95% Confidence Limits (Gumbel Distribution)

T (Years)	10	20	30	50	100
$K(T)$	1.30	1.86	2.20	2.61	3.14
$\hat{X}$ (Equation 12.11) (m³s⁻¹)	449.7	504.1	537.1	576.9	628.4
$SE(\hat{X})$ (Equation 12.13) (m³s⁻¹)	41.3	52.2	58.9	67.1	77.8
$t_{5,23}.SE(\hat{X})$ (m³s⁻¹)	85.1	107.4	121.3	138.3	160.4
Upper $\hat{X}$ (m³s⁻¹)	534.8	611.5	658.4	715.2	788.8
Lower $\hat{X}$ (m³s⁻¹)	364.6	396.7	415.8	438.6	468.0

equalled or exceeded in any one year. With the assumption that annual maxima are independent, the probability of no annual maxima exceeding X in the whole n years is given by $(F(X))^n$. From Equation 12.2, the probability of exceedance of X at least once in n years is then $1 - (F(X))^n$, which may be signified by $P_n(X) = 1 - (1 - P(X))^n$. Substituting for $F(X)$ from Equation 12.4,

$$P(Q_T \geqslant X) \text{ in } n \text{ years } = P_n(X) = 1 - \left(\frac{T(X) - 1}{T(X)}\right)^n \qquad (12.14)$$

Thus for $T(X) = 100$ years and $n = 100$ years also:

$$P_n(X) = P(Q_{100} \geqslant X) = 1 - \left(\frac{99}{100}\right)^{100}$$

$$= 0.634$$

This means that the chance of the 100-year flood being equalled or exceeded at least once in 100-year is 63.4%, i.e. there is roughly a 2 in 3 chance that a 100-years event will be equalled or exceeded, once or more, in any given 100-year record, (and a 1 in 3 chance that it will not).

Rearranging Equation 12.14 gives

$$n = \frac{\log\left[1 - P_n(X)\right]}{\log\left[\dfrac{T(X) - 1}{T(X)}\right]} \qquad (12.15)$$

Thus for a required probability, $P_n(X)$, and return period, T, the length of record, n years, in which the T-year event has the probability $P_n(X)$ of being equalled or exceeded at least once, can be calculated. Such information is essential to engineers designing dams or flood protection works.

12.8   Droughts

In Chapter 10, droughts were considered as rainfall deficiencies; these have immediate effects on vegetation growth and evaporation losses. When there are subsequent deficiencies in river flows, it is the bulk users of water, namely industry and the large urban concentrations of domestic consumers, that begin to suffer. Water Authorities and Water Companies whose duty it is to maintain public supplies of water are always concerned for their resources during periods of low flows and by the increase in demand that dry spells generally stimulate. A thorough knowledge of the minimum flows in those rivers providing water supplies is therefore essential.

In the UK, all major rivers are perennial, but some of the headwaters and small tributary streams rising in limestone or chalk country are intermittent. Their upper courses dry up in summer, and in drought periods the lack of water in the streams may extend further down the valleys. Increasing attention is being paid to flow deficiencies, and in the perennial streams it is considered desirable to maintain a defined minimum discharge. The sustaining of a minimum discharge is particularly important in all rivers receiving waste water effluents in order to ensure required dilution of pollutants.

In the analysis of low flow conditions, it is preferable to have natural discharge records unaffected by major abstractions or sewage effluent discharge. In drought periods, hydrologists are encouraged to make extra gaugings on small tributary streams as well as at the established gauging stations. A catalogue of these dry-weather flows has been compiled by the central authority in the UK (Water Resources Board, 1970). Such records ($ls^{-1}km^{-2}$) provide more detailed information on the relationship of base flows to the geology of the catchment.

From the continuous river records, several features of the data sets can be abstracted or computed to give measures of the characteristics of low flows. These may be called *low flow indices*, and three groups of indices have been outlined (Beran & Gustard, 1977). The indices derive from the flow-duration curve, consideration of low-flow spells, and the frequency analysis of the annual minimum series of low flows.

The *flow-duration curve* (Fig. 12.7) gives the duration of occurrence of the whole range of flows in the river. Selected points on the lower end of the curve can give measures of flow deficiency. The flow that is exceeded 95% of the time, Q_{95}, and the percentage of time that a quarter of the average flow is exceeded, are two suggested indices. In Fig. 12.7, the 95% exceedence flow is 13.5 m^3s^{-1} and a quarter of the average flow (44 m^3s^{-1}) 11 m^3s^{-1}, is exceeded 99.45% of the time. To compare low-flow values between catchment areas, the Q_{95} value may be expressed as a runoff depth over the catchment in mm day^{-1}. Hence for the Thames curve, Q_{95}/A = 13.5 m^3s^{-1}/9870 km^2, which gives 0.118 mm day^{-1}.

Similarly for the other flow-duration curve examples (Fig. 12.8) where the flows are given as percentages of the average daily mean flows (ADF);

$$\text{Honddu } Q_{95}/A = \frac{13 \times 1.462}{100} \text{ m}^3\text{s}^{-1}/62.1 \text{ km}^2 = 0.264 \text{ mm day}^{-1}$$

and:

$$\text{Mimram } Q_{95}/A = \frac{42 \times 0.519}{100} \text{ m}^3\text{s}^{-1}/133.8 \text{ km}^2 = 0.141 \text{ mm day}^{-1}$$

The contrast between the Q_{95} values, 13% of the ADF for the Honddu and 42% of the ADF for the Mimram, reflects the geological differences between the two catchments, hard old rocks and chalk, respectively. Also the Honddu's 62.1 km^2 produces an ADF of 1.462 m^3s^{-1}, whereas the Mimram's 133.8 km^2 (twice the Honddu area) has an ADF of only 0.519 m^3s^{-1}, demonstrating the result of an annual rainfall difference of 1194 mm and 665 mm, respectively. The index Q_{95}/A can be correlated with various catchment characteristics and useful regional patterns of drought properties can be identified (IOH, 1980).

The study of *low flow spells* seeks to overcome the shortcomings of the flow-duration curve, which gives no indication of *sequences* of low flows. From the continuous record of daily mean discharges, the number of days for which a selected flow is not exceeded defines a low flow spell. A Q_{95} flow may not be exceeded for a sequence of 10 days, thus giving a low flow spell of 10 days' duration. The frequency or probability of occurrence of low flow spells of different durations may be abstracted and assessed from the record. Hence D days, the duration of spells of low flow $\leqslant Q_{95}$, can be plotted against the percentage of low flow spells $> D$. Catchments can again be compared by plotting individual results on the same graph.

The definition of a low flow spell depends on the flow chosen for non-exceedence, and there is no direct comparison of flows. The third group of low flow indices from the analysis of the annual minimum series of low flows repairs this omission, and is dealt with in the following section.

12.9   Frequency of Low Flows

Although the investigation of flood flows always attracts a great deal of research, the frequency analysis of low flows has not been neglected. In a study of low flows of rivers selected from all parts of the USA, Matalas investigated the suitability of four theoretical probability distributions for low flow data (Matalas, 1963). The principal requirements for annual minimum discharges are that the distribution should be skew, have a finite lower limit greater than or equal to 0 and be defined by a maximum of three parameters. The latter criterion needs to be considered, since the estimation of parameters is dependent on the data available, and record lengths are often not long enough for rigorous statistical analysis. The study demonstrated that the Gumbel distribution of the smallest value is one of the most reliable, and since it is relatively simple to compute, it is recommended for the assessment of the frequency of annual minimum flows.

The minimum daily mean flows for 10 years (Table 12.7) are used to

Table 12.7   Frequency of Low Flows (Annual Minimum Daily Mean Flows)

Annual minimum (m^3s^{-1})	Rank r	$P(X)$ Gringorten	$P(X)$ Weibull
0.408	1	.055	.091
0.351	2	.154	.182
0.315	3	.253	.273
0.256	4	.352	.364
0.238	5	.451	.455
0.222	6	.549	.545
0.210	7	.648	.636
0.187	8	.747	.727
0.152	9	.846	.818
0.074	10	.945	.909

$\overline{Q}_m = 0.241$
$s_m = 0.098$

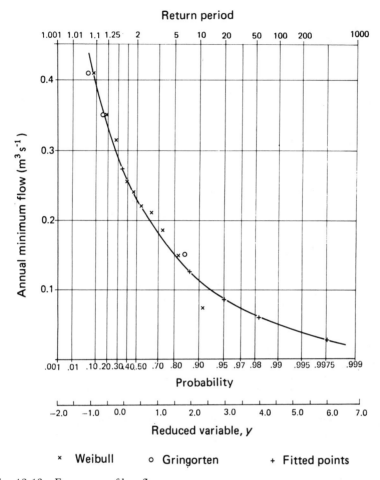

Fig. 12.13   Frequency of low flows.

demonstrate the method (IOH, 1980). In practice, at least 20 years should be obtained in order to achieve reliable results. The annual minimum values are ranked starting with the highest. The probabilities $P(X)$, although calculated from the Gringorten and Weibull plotting positions as before, have a different meaning:

$$P(X) = \text{Prob. (annual minimum} \leqslant X)$$

All the Weibull points and the two extremes of the Gringorten points are plotted in Fig. 12.13 on Gumbel extreme value paper. If a straight line were fitted by eye through the plotted points, it would give a probability of exceedence of zero flow of approximately 0.98 and thus apparently a probability of $Q < 0$ of 2%. However the form of Gumbel probability distribution of lowest values with a defined lower limit has a different form to the distribution used for maximum values with no defined upper limit (the so-called EVI distribution). The probability distribution applicable to low flows having a fixed lower limit is equivalent to the EVIII distribution in reverse and is given by:

$$P(X) = \exp(-\{(X - \epsilon)/\theta - \epsilon)\}^k) \qquad (12.16)$$

where ϵ is the minimum flow ≥ 0, and $X \geq \epsilon$, θ is called the characteristic drought, with $P(\theta) = e^{-1} = 0.368$ and k is the third parameter. If the minimum flow is assumed to be 0, then the formula reduces to a two-parameter distribution:

$$P(X) = \exp(-(X/\theta)^k) \qquad (12.17)$$

12.9.1   Fitting the EVIII Distribution

To linearize the abscissa of the Gumbel paper, the EVI Equation 12.7 is written:

$$\text{Prob. } (Q_T \leqslant X) = F(X) = \exp(-e^{-y})$$

where y, replacing $b(X - a)$, is known as the *reduced variate*.

For the EVIII distribution, $(X/\theta)^k$ is made equal to e^y so that Equation 12.17 becomes:

$$P(X) = \exp(-e^y) \qquad (12.18)$$

For the range of probability $0.001 \leqslant P \leqslant 0.999$, this variate y lies in the interval $2 > y > -7$. Thus the flood probability paper can be used for droughts by changing the sign of y (Gumbel, 1963).

To determine the parameters of the distribution, the sample estimates $(\overline{Q}_m$ and $s_m)$ of the population mean and standard deviation of minimum flows $(\mu_m$ and $\sigma_m)$ are required. The quotient μ_m/σ_m is a complex function of k, an estimate of $1/k$ can be read from Table 12.8 and an estimate of θ is given by:

$$\hat{\theta} = \frac{\overline{Q}_m}{\Gamma\left(1 + \dfrac{1}{k}\right)}$$

where $\Gamma(\ )$ is the gamma function available in standard tables. In the example, $\overline{Q}_m = 0.241$ and $s_m = 0.098$, whence $\overline{Q}_m/s_m = 2.459$, giving $1/k = 0.38$ (Table 12.8). Then:

$$\hat{\theta} = \frac{0.241}{\Gamma(1 + 0.38)} = 0.271$$

Taking logarithms of the equality:

$$e^y = \left(\frac{Q}{\theta}\right)^k$$

$$y = k\ln Q - k\ln \theta$$

and

$$\ln Q = \ln \theta + \frac{1}{k}y \qquad (12.19)$$

Table 12.8 μ/σ **as a Function of** $1/k$ **(Gumbel EVIII)**

(Reproduced from E.J. Gumbel (1963) *Bull. IASH* vol. VIII, no. 1, pp 5-23, by permission.)

Parameter $1/k$	μ/σ	Parameter $1/k$	μ/σ
0.01	78.534	0.10	8.312
0.02	39.543	0.20	4.366
0.03	26.542	0.30	3.024
0.04	20.039	0.40	2.337
0.05	16.135	0.50	1.913
0.06	13.531	0.60	1.621
0.07	11.669	0.70	1.408
0.08	10.271	0.80	1.242
0.09	9.183	0.90	1.109
0.10	8.312	1.00	1.000

This is the equation for the relationship Q versus y on Fig. 12.13, which becomes, after changing the sign of y:

$$\ln Q = -1.306 - 0.38\,y$$

Substituting selected values of y and obtaining corresponding values of Q, the non-linear fitted curve is drawn. (This would be a straight line on *logarithmic* Gumbel paper.) As with extrapolating flood peaks, it can be unsatisfactory to assess probabilities and return periods of low flows very far beyond the extent of the sample record, but forecasts ahead of twice the number of record years were found satisfactory for low flows on several American rivers (Gumbel, 1963).

The value of a frequency analysis using the annual minimum series is increased by selecting significant sequences of low flows, say 10 days, and taking the minimum 10-day average daily mean flow in each year of record.

After the UK drought of 1975-1976, in a study of low flows on a monthly basis, the log – Normal probability distribution was found to give acceptable answers to the questions of return periods of different durations of low monthly flows (Hamlin & Wright, 1978). Also following this event, the Institute of Hydrology made extensive and detailed studies of low flows from the records in their valuable data archive (IOH, 1980).

References

Allard, W., Glasspoole, J. and Wolf, P.O. (1960). 'Floods in the British Isles'. *Proc. Inst. Civ. Eng.*, **15**, 119-144.

Benson, M.A. (1968). 'Uniform flood frequency estimating methods for federal agencies.' *Water Resources Res.*, **4**(5), 891-908.

Beran, M.A. and Gustard, A. (1977). 'A study into the low flow characteristics of British rivers.' *J. Hydrol*, **35**, 147–157.

Crippen, J.R. and Bue, C.D. (1977). *Maximum Flood Flows in the Coterminus US*. US Geol. Surv., Water-Supply Paper 1887, 52 pp.

Department of the Environment (DOE) (1978). *Surface Water: United Kingdom* 1971–73. HMSO.

Dobbie, C.H. and Wolf, P.O. (1953). 'The Lynmouth flood of August 1952.' *Proc. Inst. Civ. Eng.*, Pt. III, 2, 522.

Gumbel, E.J. (1963). 'Statistical Forecast of Droughts.' *Bull. IASH.*, **1**, April, 5–23.

Hamlin, M.J. and Wright, C.E. (1978). 'The effects of drought on the river systems.' *Proc. Roy. Soc. Lond. A.*, **363**, 69–96.

Institute of Hydrology (IOH). (1980). *'Low Flow Studies.'* Research Report and addenda.

Institution of Civil Engineers (ICE). (1933). *Floods in Relation to Reservoir Practice*. Interim Report of the Committee on Floods.

Institution of Civil Engineers (ICE). (1960). *Floods in Relation to Reservoir Practice*. Reprint of 1933 Report with Additional Data, 66 pp.

Matalas, N.C. (1963). *Probability Distribution of Low Flows*. US Geol. Surv. Prof. Paper 434–A, 27 pp.

Natural Environment Research Council (NERC). (1975). *Flood Studies Report*, Vol. IV. Hydrological Data 541 pp.

Raudkivi, A.J. (1979). *Hydrology*. Pergamon Press, 479 pp.

Rochefort, M. (1963). *Les Fleuves*. Presses Universitaires de France, 128 pp.

Water Resources Board (1970). *Dry Weather Flows*. Technical Note 12.

13

Rainfall – Runoff Relationships

The derivation of relationships between the rainfall over a catchment area and the resulting flow in a river is a fundamental problem for the hydrologist. In most countries, there are usually plenty of rainfall records, but the more elaborate and expensive stream-flow measurements, which are what the engineer needs for the assessment of water resources or of damaging flood peaks, are often limited and are rarely available for a specific river under investigation. Evaluating river discharges from rainfall has stimulated the imagination and ingenuity of engineers for many years, and more recently has been the inspiration of many research workers.

To facilitate comparisons it is usual to express values for rainfall and river discharge in similar terms. The amount of precipitation (rain, snow, etc.) falling on a catchment area is normally expressed in millimetres (mm) depth, but may be converted into a total volume of water, cubic metres (m^3) falling on the catchment. Alternatively, the river discharge (flow rate), measured in cubic metres per second (m^3s^{-1} or cumecs) for a comparable time period may be converted into total volume (m^3) and expressed as an equivalent depth of water (in mm) over the catchment area. The discharge, often termed *runoff* for the defined period of time, is then easily compared with rainfall depths over the same time period.

Estimating runoff or discharge from rainfall measurements is very much dependent on the time scale being considered. For short durations (hours) the complex interrelationship between rainfall and runoff is not easily defined, but as the time period lengthens, the connection becomes simpler until, on an annual basis, a straight-line correlation may be obtained. The time interval used in the measurement of the two variables affects the derivation of any relationship, although with continuously recorded rainfall and stream discharge this constraint is removed and only the purpose of the study influences choice of time interval. Hence, relating a flood peak to a heavy storm requires continuous records, but determining water yield from a catchment can be accomplished satisfactorily using relationships between totals of monthly or annual rainfall and runoff.

Naturally, the size of the area being considered also affects the relationship. For very small areas of a homogeneous nature — a stretch of motorway, say — the derivation of the relationship could be fairly simple; for very large drainage basins on a national or even international scale (the River Danube, for example) and for long time periods, differences in catchment effects are smoothed out giving relatively simple rainfall – runoff relationships. However, in general and for short time periods, great complexities occur when spasmodic rainfall is unevenly distributed over an area of varied topography and geological composition. For catchments with many different surface characteristics affected by a single severe storm — say catchments up to about 200 – 300 km^2 in equable climates — the direct relationship between specific rainfalls and the resulting discharge or runoff is extremely complicated.

In the intermediate scale of both area and time, other physical and hydrological factors, such as evaporation, infiltration and groundwater flow are very significant, and thus any direct relationship between rainfall alone and runoff is not easily determined. The derivation of runoff from rainfall plus other measurable variables will be considered in the next chapter, but here the simpler methods relating rainfall alone to runoff will be given. Many of these methods have been derived by engineers for immediate practical use.

13.1   Rational Method

The first concern of a civil engineer engaged upon construction work on or near a river is to gain some idea of the flow regime of the river throughout the proposed life of the project. Briefly, this usually means that the designer wants to know the flood levels and the chance of occurrence of major flood discharges. Hence, the hydrologist may be called upon to estimate likely peak flows of the river at the point in question. In former times, the river engineer made all his own calculations using any local information he could lay hands on. If he could not find any records for a river gauging station, he looked for historical evidence of flood peaks and searched the records for rainfall observations.

The concept of the rational method for determining flood peak discharges from measurements of rainfall depths owes its origins to Mulvaney, an Irish engineer who was concerned with land drainage (Mulvaney, 1850). Some Americans attribute first mention of the formula to one of their engineers engaged upon sewer design (Kuichling, 1889). The use of the rational formula is sometimes referred to as the Lloyd-Davies method, since it was applied also to sewer design calculations in England (Lloyd-Davies, 1906). The formula to give the peak flow Q_p is:

$$Q_p = CiA \tag{13.1}$$

where C is the coefficient of runoff (dependent on catchment characteristics), i is the intensity of rainfall in time T_c and A is the area of catchment.

T_c is the *time of concentration*, the time required for rain falling at the farthest point of the catchment to flow to the measuring point of the river. Thus, after time T_c from the commencement of rain, the whole of the catchment is taken to be contributing to the flow. The value of i, the mean intensity, assumes that the rate of rainfall is constant during T_c, and that all the measured rainfall over the area contributes to the flow. The peak flow Q_p occurs after the period T_c.

The rational formula was devised in Imperial units. Thus, with i in in h^{-1} and A in acres, the value of Q_p is approximately in cusecs (ft^{3}s^{-1}). In metric units, with i in mm h^{-1} and A in km^2, the time conversion needs a factor of

0.278 to give the Q_p in cumecs (m^3s^{-1}). Thus:

$$Q_p\,(m^3s^{-1}) = 0.278\,C\,i\,(mm\ h^{-1})\,A\,(km^2) \tag{13.2}$$

Values of C vary from 0.05 for flat sandy areas to 0.95 for impervious urban surfaces, and considerable knowledge of the catchment is needed in order to estimate an acceptable value. The coefficient of runoff also varies for different storms on the same catchment, and thus, using an average value for C, only a crude estimate of Q_p is obtained, that may have wide margins of error. However, used with discretion, the formula can provide a rough first value, especially in small uniform urban areas. It has been used for many years as a basis for engineering design for small land drainage schemes and storm-water channels. However when the rational method leads to over design, modern engineers need to proceed to more precise and reliable methods in order to optimize designs and thereby reduce the construction costs of major schemes.

13.2   Time – Area Method

The time – area method of obtaining runoff or discharge from rainfall can be considered as an extension and improvement of the rational method. The peak discharge Q_p is the sum of flow-contributions from subdivisions of the catchment defined by time contours (called *isochrones*), which are lines of equal flow-time to the river section where Q_p is required. The method is illustrated in Fig. 13.1a. The flow from each contributing area bounded by two isochrones $(T - \Delta T, T)$ is obtained from the product of the mean intensity of effective rainfall (i) from time $T - \Delta T$ to time T and the area (ΔA). Thus Q_4, the flow at X at time 4 h is given by:

$$Q_4 = i_3\Delta A_1 + i_2\Delta A_2 + i_1\Delta A_3 + i_0\Delta A_4$$

i.e:

$$Q_T = \sum_{k=1}^{T} i_{(T-k)}\Delta A_{(k)} \tag{13.3}$$

As before, the whole catchment is taken to be contributing to the flow after T equals T_c.

Using the above nomenclature it is seen that the peak flow at X when the whole catchment is contributing to the flow, a period T_c after the commencement of rain, is:

$$Q_p = \sum_{k=1}^{n} i_{(n-k)}\Delta A_{(k)} \tag{13.4}$$

where n, the number of incremental areas between successive isochrones, is given by $T_c/\Delta T$, and k is a counter.

The unrealistic assumption made in the rational method of uniform rainfall intensity over the whole catchment and during the whole of T_c is

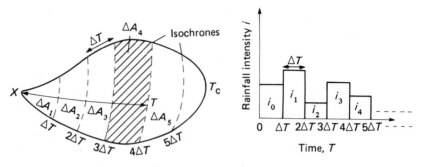

(a) Rainfall bar graph and catchment showing isochrones of travel time

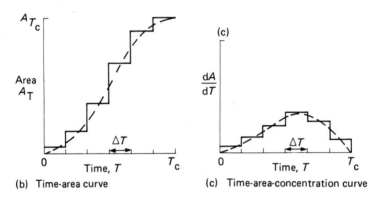

(b) Time-area curve

(c) Time-area-concentration curve

Fig. 13.1   Time – area method.

avoided in the time – area method, where the catchment contributions are subdivided in time. The varying intensities within a storm are averaged over discrete periods according to the isochrone time interval selected. Hence, in deriving a flood peak for design purposes, a design storm with a critical sequence of intensities can be used for the maximum intensities applied to the contributing areas of the catchment that have most rapid runoff. However, when such differences within a catchment are considered, there arises difficulty in determining T_c, the time after the commencement of the storm when, by definition, Q_p occurs.

For small natural catchments, a formula derived from data published by Kirpich for agricultural areas could be used to give T_c in hours (Kirpich, 1940):

$$T_c = 0.00025 \left(\frac{L}{\sqrt{S}} \right)^{0.80}$$

where L is the length of the catchment along the longest river channel (in m) and S is the overall catchment slope (in m m^{-1}).

There are many similar formulae for T_c, and further considerations of this critical measure will be given in Chapter 18 when urban catchments are discussed.

To fix the isochrones considerable knowledge of the catchment is required, so that the times of overland flow and flow in the river channels may be determined for wet or even saturated conditions, the worst conditions likely to cause a major flood flow. Isochrones for urban areas are more readily obtained by direct observation during storm periods and are, of course, more simply determined for small sewer-drained catchments.

The time – area method for calculating peak flows from rainfall has been elaborated and extended by many engineers and research workers since its inception in the 1920s. Many of the amendments have been related to specific catchment areas or conditions, and are available from the more advanced hydrological textbooks and original published papers (see Dooge, 1973). The method forms the basis for the Transport and Road Research Laboratory (TRRL) technique for designing storm-water sewers, which will be described in detail in Chapter 18.

The simple discrete form of the time – area concept can be generalized by making ΔT very small and considering increases in the contributing area to be continuous with increasing time. Thus, in Fig. 13.1b, the plot of catchment area against time is shown as a continuous line and this is known as the time – area curve. Its limits are the total area of the catchment and the time of concentration. For any value of T, the corresponding area A gives the maximum flow at the river outfall caused by a rainfall of duration T. The derivative of the time – area curve shown in Fig. 13.1c gives the rate of increase in contributing area with time, and is called the time – area – concentration curve, since the length of the time base is equal to the time of concentration of the catchment.

13.3   Hydrograph Analysis

Before trying to analyse a hydrograph, which describes the whole time history of the changing rate of flow from a catchment due to a rainfall event rather than just the peak flow, it is essential first to appreciate some of its simple components. In Fig. 13.2, rainfall intensity (i, in mm h^{-1}) is shown in discrete block intervals of time (T). The lower continuous curve of discharge (Q, in m^{3}s^{-1}) is the hydrograph resulting from the event. The discharge hydrograph is obtained from continuously recorded river stages and the stage – discharge relationship (Chapter 6) appropriate to the river gauging station.

The hydrograph of discharge against time has two main components, the area under the hump, labelled *surface runoff* (which is produced by a volume of water derived from the storm event), and the broad band near the time axis, representing *baseflow* contributed from groundwater.

At the beginning of the rainfall, the river level (and hence the discharge) is low and a period of time elapses before the river begins to rise. During this period the rainfall is being intercepted by vegetation or is soaking into the ground and making up soil-moisture deficits. The length of the delay before the river rises depends on the wetness of the catchment before the storm and

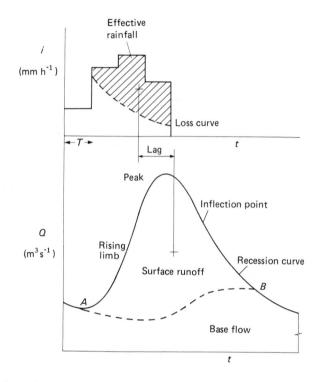

Fig. 13.2   Rainfall and a river hydrograph.

on the intensity of the rainfall itself.

When the rainfall has made up catchment deficits and when surfaces and soil are saturated, the rain begins to contribute to the stream flow. The proportion of rainfall that finds its way into a river is known as the *effective rainfall*, the rest being lost (to quick runoff) in evaporation, detention on the surface or retention in the soil. As the storm proceeds, the proportion of effective rainfall increases and that of lost rainfall decreases as shown by the loss curve (Fig. 13.2).

The volume of surface runoff, represented by the area under the hydrograph minus the baseflow, can be considered in two main subdivisions to simplify the complex water movements over the surface and in the ground. The effective rainfall makes the immediate contribution to the rising limb from A to the peak of the hydrograph and, even when the rainfall ceases, continues to contribute until the *inflection point* (Fig. 13.2). Beyond this point, it is generally considered that the flow comes from the water temporarily stored in the soil. This so-called *interflow* continues to provide the flow of the *recession curve* until the water from the whole of the effective rainfall is completely depleted at B (Fig. 13.2). One final term, lag or lag time requires explanation. There are many definitions of lag, which is a measure of the catchment response time, but here it is taken from the centre of gravity of the effective rainfall to the centre of gravity of the direct surface runoff.

The boundary between surface runoff and baseflow is difficult to define and depends very much on the geological structure and composition of the catchment. Permeable aquifers, such as limestone and sandstone strata, sustain high baseflow contributions, but impervious clays and built-up areas provide little or no baseflow to a river. The baseflow levels are also affected by the general climatic state of the area: they tend to be high after periods of wet weather and can be very low after prolonged drought. During the course of an individual rainfall event, the baseflow component of the hydrograph continues to fall even after river levels have begun to rise, and only when the storm rainfall has had time to percolate down to the water table does the baseflow division curve (shown schematically in Fig. 13.2) begin to rise. The baseflow component usually finishes at a higher level at the end of the storm surface runoff than at the rise of the hydrograph and thus there is enhanced river flow from groundwater storage after a significant rainfall event. Groundwater provides the total flow of the general recession curve until the next period of wet weather.

The main aims of the engineering hydrologist are to quantify the various components of the hyetogram and the hydrograph, by analysing past events, in order to relate effective rainfall to surface runoff, and thereby to be able to estimate and design for future events. As a result of the complexity of the processes that create stream flow from rainfall, many simplifications and assumptions have to be made.

13.4   The Unit Hydrograph

A major step forward in hydrological analysis was the concept of the *unit hydrograph* introduced by the American engineer Sherman in 1932. He defined the unit hydrograph as the hydrograph of surface runoff resulting from effective rainfall falling in a unit of time such as 1 hour or 1 day and produced uniformly in space and time over the total catchment area (Sherman, 1942).

In practice, a T hour unit hydrograph is defined as resulting from a unit depth of effective rainfall falling in T h over the catchment. The magnitude chosen for T depends on the size of the catchment and the response time to major rainfall events. The standard depth of effective rainfall was taken by Sherman to be 1 in, but with metrication, 1 mm or sometimes 1 cm is used. The definition of this rainfall – runoff relationship is shown in Fig. 13.3a, with 1 mm of uniform effective rainfall occurring over a time T producing the hydrograph labelled TUH. The units of the ordinates of the T-hour unit hydrograph are m^3s^{-1} per mm of rain. The volume of water in the surface runoff is given by the area under the hydrograph and is equivalent to the 1 mm depth of effective rainfall over the catchment area.

The unit hydrograph method makes several assumptions that give it simple properties assisting in its application.

(a) There is a direct proportional relationship between the effective rainfall and the surface runoff. Thus in Fig. 13.3b two units of effective rainfall falling in time T produce a surface runoff hydrograph that has its ordinates twice the TUH ordinates, and similarly for any propor-

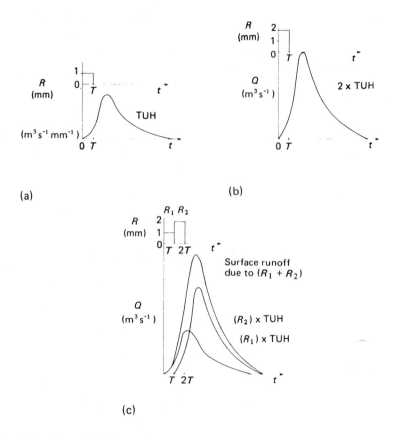

Fig. 13.3   The unit hydrograph.

tional value. For example, if 6.5 mm of effective rainfall fall on a catchment area in T h, then the hydrograph resulting from that effective rainfall is obtained by multiplying the ordinates of the TUH by 6.5.

(b) A second simple property, that of superposition, is demonstrated in Fig. 13.3c. If two successive amounts of effective rainfall, R_1 and R_2 each fall in T h, then the surface runoff hydrograph produced is the sum of the component hydrographs due to R_1 and R_2 separately (the latter being lagged by T h on the former). This property extends to any number of effective rainfall blocks in succession. Once a TUH is available, it can be used to estimate design flood hydrographs from design storms.

(c) A third property of the TUH assumes that the effective rainfall – surface runoff relationship does not change with time, i.e. that the same TUH always occurs whenever the unit of effective rainfall in T h is applied. Using this assumption of invariance, once a TUH has been derived for a catchment area, it could be used to represent the response of the catchment whenever required.

The assumptions of the unit hydrograph method must be borne in mind when applying it to natural catchments. In relating effective rainfall to surface runoff, the amount of effective rainfall depends on the state of the catchment before the storm event. If the ground is saturated or the catchment is impervious, then a high proportion of the rain becomes effective. In absorbing the rainfall, unsaturated ground will have a certain capacity before releasing effective rainfall to contribute to the surface runoff. Only when the ground deficiencies have been made up and the rainfall becomes fully effective will extra rainfall in the same time period produce proportionally more runoff. The first assumption of proportionality of response to effective rainfall conflicts with the observed non-proportional behaviour of river flow. In a second period of effective rain, the response of a catchment will be dependent on the effects of the first input, although the second assumption (Fig. 13.3c) makes the two component contributions independent. The third assumption of time invariance implies that whatever the state of the catchment, a unit of effective rainfall in T h will always produce the same TUH. However, the response hydrograph of a catchment must vary according to the season: the same amount of effective rainfall will be longer in appearing as surface runoff in the summer season when vegetation is at its maximum development and the hydraulic behaviour of the catchment will be 'rougher'. In those countries with no marked seasonal rainfall or temperature differences and constant catchment conditions throughout the year, then the unit hydrograph would be a much more consistent tool to use in deriving surface runoff from effective rainfall.

Another weakness of the unit hydrograph method is the assumption that the effective rainfall is produced uniformly both in the time T and over the area of the catchment. The areal distribution of rainfall within a storm is very rarely uniform. For small or medium sized catchments (say up to 500 km^2), a significant rainfall event may extend over the whole area, and if the catchment is homogeneous in composition, a fairly even distribution of effective rainfall may be produced. More usually, storms causing large river discharges vary in intensity in space as well as in time, and the consequent response is often affected by storm movement over the catchment area. However, rainfall variations are damped by the integrating reactions of the catchment, so the assumption of uniformity of effective rainfall over a selected period T is less serious than might be supposed at first. The effect of variable rainfall intensities can be reduced by making T smaller, and where a catchment is affected by major storms of different origins, separate TUHs can be derived for each storm type.

However, the unit hydrograph method has the advantage of great simplicity. Once a unit hydrograph of specified duration T has been derived for a catchment area (and/or specific storm type) then for any sequence of effective rainfalls in periods of T, an estimate of the surface runoff can be obtained by adopting the assumptions and applying the simple properties outlined above. The technique has been adopted and used world-wide over many years.

Examples of unit hydrographs for differing values of T are shown for two catchment areas in the Thames Basin in Fig. 13.4. These demonstrate the effect of catchment size and of the selection of T. For large catchments (over

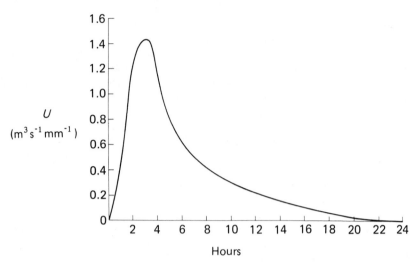

(a) 3h – UH River Mole at Horley Weir (89.9 km^2)

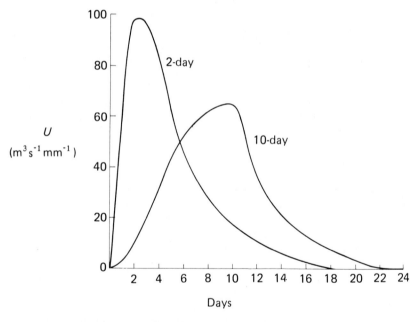

(b) 2-day and 10-day UH River Thames at Teddington Weir (9593 km^2)

Fig. 13.4   Unit hydrographs from the Thames Basin.

1000 km^2 for example) with longer response times, the unit hydrograph can be derived from weekly or even 10-day effective rainfalls, when the rainfall event causing the increased river flow constituted a widespread wet spell over the whole area (Andrews, 1962).

The main problems in the derivation of a unit hydrograph are the assessment of the effective rainfall from the measured rainfall and the separation of the resulting surface runoff from the total hydrograph. How these difficulties are tackled is outlined in the following sections.

13.4.1   Derivation of the Unit Hydrograph from Simple Storms

A selection of typical single-peaked hydrographs is taken from the continuous river gauge records and then the corresponding rainfall records, preferably from autographic rain gauges, are abstracted. If there is only one autographic rainfall record available along with some daily gauges, areal rainfall values for the storm in appropriate durations are obtained by proportioning daily totals (see Chapter 10). As an example, the areal rainfall (R) given in half-hour totals and the resulting hydrograph of discharge (Q) from a 1150 hectare catchment, are shown in Fig. 13.5. Such rainfall hyetograms with just single major blocks of rainfall are to be preferred.

The unit hydrograph derivation from one such storm event proceeds in the following stages:

(a) The first problem is the separation of the baseflow from the storm runoff. Fig. 13.6 shows several methods. The simplest separation would be given by a horizontal line to a from the start of the rise of the hydrograph. This would assume that the storm has no effect on or makes no contribution to groundwater. The more realistic separation is curve c, which shows a marked storm effect peaking some time after the peak of the stream flow. However, the drawing of this curve would be quite subjective, and various answers would be produced by different analysts. The straight line to point b on the recession curve can be singularly determined. After point b, the shape of the hydrograph recession is assumed to become exponential, and is fixed uniquely on the semi-logarithmic plot as shown. This is a satisfactory compromise and the method is straightforward and gives consistent results. It is seen applied in Fig. 13.5.

(b) The area above the baseflow separation line and enclosed by the hydrograph is then measured (by planimeter, or by using squared paper) to give the volume of surface runoff. The equivalent depth of runoff (mm) is evaluated for the catchment area. This is then by definition the effective rainfall from the storm (12 mm in Fig. 13.5).

(c) The determination of that part of the measured total rainfall that constitutes the effective rainfall is the next step. In Fig. 13.5, the effective rainfall (12 mm) is the hatched area deducted from the large 1-h central block of the areal rainfall bargraph. The duration, T, of the effective rainfall is then clearly indicated as 1 h in the example shown in Fig. 13.5.

(d) The time axis of the storm hydrograph is then usually subdivided into periods of duration T, beginning at the rise of the hydrograph, but any convenient interval can be chosen. The corresponding ordinates of surface runoff (Q_s) given by the total hydrograph discharge (Q) minus the corresponding baseflow, are then each divided by the effective rainfall to give the ordinates (U) of the unit hydrograph $(m^3s^{-1}mm^{-1})$.

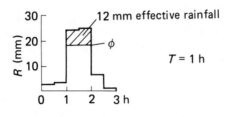

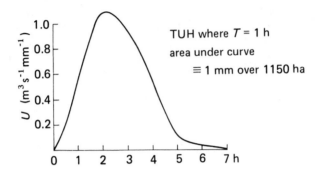

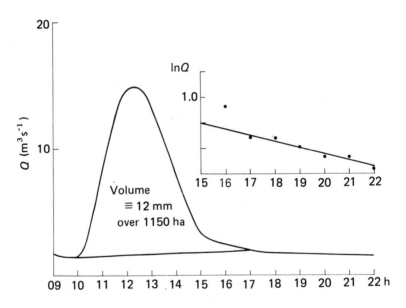

Fig. 13.5   Derivation of the unit hydrograph.

(e) The unit hydrograph for effective rainfall of duration T, the TUH, is
     then plotted, and the area under the curve is checked to see if the
     enclosed volume is equivalent to unit effective rainfall over the area of
     catchment.

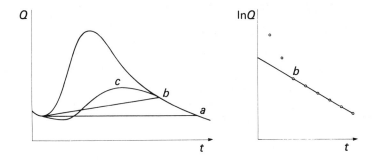

Fig. 13.6   Separation of base flow.

When all the single-peaked storms have been analysed and a corresponding number of unit hydrographs obtained, it will be noted that no two are identical, though they will all have the same general shape. Features of the unit hydrographs derived from eight storms on the same catchment are given in Table 13.1.

One way that an average unit hydrograph may be constructed is by taking the arithmetic means of the peak flows (U_p) and the times to peak (T_p), plotting the average peak at the appropriate mean value of T_p, and drawing the hydrograph to match the general shapes of the individual unit hydrographs as illustrated in Fig. 13.7. The resulting average unit hydrograph is

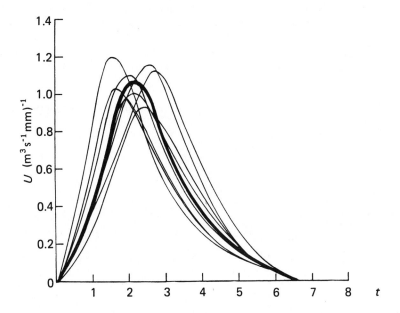

Fig. 13.7   Derivation of a unit hydrograph from eight storms.

Table 13.1   Unit Hydrograph Parameters from 8 Storms

Data for Storm 1 from Fig. 13.5.

Storm	1	2	3	4	5	6	7	8	Averages
U_p (m³s⁻¹mm⁻¹)	1.10	1.16	0.93	0.98	1.12	1.05	1.01	1.20	1.07
T_p (h)	2.1	2.6	2.4	1.9	2.7	1.7	2.2	1.5	2.1

then checked to ensure that the enclosed volume of runoff is equivalent to a unit of effective rainfall.

For rainfall bargraphs of a complex pattern, a more sophisticated rainfall separation procedure is needed. In Fig. 13.2, an idealized separation is shown by a curved loss-rate line. At the beginning of a storm there could be considerable interception of the rainfall and initial wetting of surfaces before the rainfall becomes 'effective' — that is, begins to form surface runoff.

The loss-rate is dependent on the state of the catchment before the storm and is difficult to assess quantitatively. Two simplified methods of determining the effective rainfall are given in Fig. 13.8. The ϕ index method assumes a constant loss rate of ϕ mm from the beginning of the rainfall event: this amount accounts for interception, evaporation loss and surface detention in pools and hollows. It could rightly be considered, however, that there is a period of time after the commencement of the storm before *any* of the rainfall becomes effective. In Fig. 13.8b, all the rainfall up to the time of rise of the hydrograph is considered lost, and there is a continuing loss-rate at some level afterwards. In both methods the rainfall separation line is positioned such that the hatched areas in Fig. 13.8 equate to the effective rainfall depth.

A choice between the two methods depends on knowledge of the catchment but, as the timing of the extent of initial loss is arbitrary, the fixing of the beginning of effective rainfall at the beginning of runoff in the stream

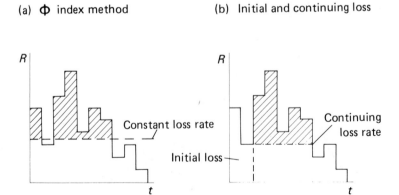

(a) Φ index method   (b) Initial and continuing loss

Fig. 13.8   Determination of effective rainfall.

neglects any lag time in the drainage process and is thus somewhat unrealistic. A constant loss-rate, the ϕ index, would therefore seem to be more
readily applicable.

13.4.2   Derivation of the Unit Hydrograph from Composite Storms

Very often, particularly on larger catchments, it is difficult to find in the
available records enough single-peaked storm events to provide a fair sample
for analysis. Multi-peaked sequences of rainfalls and the resultant hydrographs must then be used in the unit hydrograph derivation. If significant
peak flows can clearly be related to groups of higher rainfall values, it may be
possible to separate the records into distinctive individual events to be treated
as single storms. However, composite effects with overlapping storm
hydrographs call for more complex treatment.

In Fig. 13.9 a sequence of *effective* rainfalls (R) has been plotted for
durations of time T together with the resulting surface runoff hydrograph,
with ordinates (Q) indicated also at intervals T. In the centre of the diagram is
shown the TUH that commences under the rainfall block starting at time iT
for the general case. Now from the unit hydrograph superimposition
assumption shown in Fig. 13.3c, each rainfall block produces a component
part of the surface runoff at a later time jT, viz $R_i \times U_{j-i}$. Thus the surface

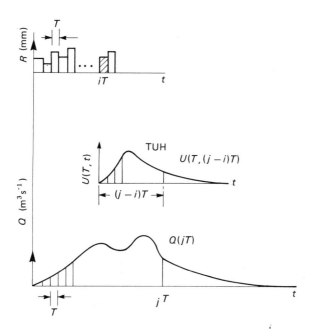

Fig. 13.9   The T-hour unit hydrograph (TUH). $Q(jT) = T \sum\limits_{i=0}^{j} R_i U\left(T, (j - i)\, T\right)$

with $j = 1,2,3 \ldots$ or $Q(j) = \sum\limits_{i=0}^{j} R_i U_{j-i}$ where R occurs in intervals of T h and

$R_0 = R\,(0 \rightarrow 1T)$.

runoff Q at time jT is given by summing all such components for all the rainfall blocks up to the block starting at time $(j - 1)T$:

$$Q_j = \sum_{i=0}^{j-1} R_i U_{j-i} \tag{13.5}$$

This shorthand notation represents a set of simultaneous equations for Q_j $(j = 1,2,3,...)$. If there are m blocks of effective rainfall $(R_0, R_1, R_2 ... R_{m-1})$ and n (non-zero) ordinates in the TUH $(U_1, U_2 ... U_n)$ then there will be $(m + n - 1)$ such equations, and hence $(m + n - 1)$ non-zero surface runoff ordinates:

$$
\begin{aligned}
Q_1 &= R_0 U_1 &+ &\ 0 &+ &\ 0 &+\ 0\ ... & &+\ 0\\
Q_2 &= R_1 U_1 &+ &R_0 U_2 &+ &\ 0 &+\ 0\ ... & &+\ 0\\
Q_3 &= R_2 U_1 &+ &R_1 U_2 &+ &R_0 U_3 &+\ 0\ ... & &+\ 0\\
&\ \vdots\\
Q_m &= R_{m-1} U_1 + R_{m-2} U_2 + R_{m-3} U_3 + ... + R_0 U_m + \ \ 0 \quad ... & &+\ 0\\
Q_{m+1} &= \ \ 0 \ + R_{m-1} U_2 + R_{m-2} U_3 + ... + R_1 U_m + R_0 U_{m+1} + ...\\
&\ \vdots\\
Q_{m+n-2} &= \ \ 0 \ + \ \ 0 \ + \ \ 0 \ + ... + \ \ 0 \ + \ \ 0 \ + ... + R_{m-1} U_{n-1} + R_{m-2} U_n\\
Q_{m+n-1} &= \ \ 0 \ + \ \ 0 \ + \ \ 0 \ + ... + \ \ 0 \ + \ \ 0 \ + ... + \ \ 0 \ + R_{m-1} U_n
\end{aligned}
$$

There are more equations than the n unknowns, U_1 to U_n. Expressing the equations first in matrix form followed by some matrix manipulation reduces the number of equations to n, as follows:

$$
\begin{bmatrix}
Q_1\\
Q_2\\
Q_3\\
\vdots\\
Q_m\\
Q_{m+1}\\
\vdots\\
Q_{m+n-2}\\
Q_{m+n-1}
\end{bmatrix}
=
\begin{bmatrix}
R_0 & 0 & & \cdots & & 0\\
R_1 & R_0 & 0 & \cdots & & 0\\
R_2 & R_1 & R_0 & 0 & \cdots & 0\\
\vdots & & & & & \vdots\\
R_{m-1} & R_{m-2} & R_{m-3} & \cdots & R_0\ 0 & \cdots & 0\\
0 & R_{m-1} & R_{m-2} & \cdots & R_1\ R_0\ 0 & \cdots & 0\\
\vdots & & & & & \vdots\\
0 & 0 & & \cdots & & R_{m-1} & R_{m-2}\\
0 & 0 & & \cdots & & 0 & R_{m-1}
\end{bmatrix}
\times
\begin{bmatrix}
U_1\\
U_2\\
U_3\\
\vdots\\
U_n
\end{bmatrix}
$$

Using conventional matrix notation, this becomes:

$$\underline{Q} = \underline{R} \cdot \underline{U} \tag{13.6}$$

In order to obtain the elements of $\underline{U}$ from this relationship, it is first necessary to make the rectangular $\underline{R}$ matrix into a square matrix. To make this transformation, $\underline{R}$ must be pre-multiplied by its transpose $\underline{R}^T$ and so, pre-multiplying each side of the equation by $\underline{R}^T$, gives

$$\underline{R}^T \cdot \underline{Q} = \underline{R}^T \cdot \underline{R} \cdot \underline{U}$$

and hence

$$\underline{U} = (\underline{R}^T \cdot \underline{R})^{-1} \cdot \underline{R}^T \cdot \underline{Q} \tag{13.7}$$

Thus given a series of effective rainfalls and corresponding surface runoffs, the ordinates of the TUH can be obtained using standard computer programs for matrix manipulation. The method automatically gives a best solution for the TUH in the least-squares sense. The matrix procedure often yields unreal oscillatory irregular shapes for the TUH. The method may be repeated with revised estimates of the effective rainfall and surface runoff data, and a smoothing process on the derived TUH ordinates employed until a smooth form of the TUH is obtained.

13.4.3   Changing the T of the TUH

Although a TUH may have been derived for a catchment from a selection of simple storms having the same duration of effective rainfall, or from a series of composite storms for which some given value of T has been used in the analysis, there is often a need to find an estimated pattern of runoff for a different duration of effective rainfall. It may be necessary then to derive a unit hydrograph for a new value of T. A comprehensive method which can be used for making the new T greater or less than the original T involves the construction of (or a tabular representation of) a standard S-curve. The S-curve is given by the sequential accumulation of the ordinates of the TUH; that is, a graph of cumulative U plotted against time. In effect, the S-curve based on a TUH represents the surface runoff hydrograph caused by an effective rainfall of intensity $1/T$ mm h^{-1} applied indefinitely (see Fig. 13.10a).

In the next stage, if a T_2 – UH is required, the same S-curve is plotted again (S_2) offset at a distance T_2 from the first S-curve (S_1), as shown graphically in Fig. 13.10b. The difference between the two S-curves displaced by T_2 then represents surface runoff from $(1/T_1) \times T_2$ mm of effective rainfall in T_2 h. Thus the differences ΔS_t between the ordinates of the two S-curves form the surface runoff hydrograph produced by an effective rainfall of T_2/T_1 mm in a T_2-h period. Therefore the new T_2 – hour unit hydrograph ordinates will be given by

$$\Delta S_t \div (T_2/T_1), \text{ i.e. } (T_1/T_2) \times \Delta S_t$$

The differences between the accumulated values of two S-curves are needed from the graphs at regular time intervals, but the operation is easier and more accurate if the working is done by means of a table. In Fig. 13.10c, unit hydrographs are shown for two values of T_2, $T_2 = \frac{1}{2}$ h and 2 h, derived from the 1 h – UH. The computations are carried out in Table 13.2. It will be noted that with $T_2 < T_1$ the new shorter period unit hydrograph peaks earlier and is an irregular shape because of the exaggeration of discrepancies in the data and the basic assumptions made in the method. For $T_2 > T_1$, variations in catchment response are damped down and the unit hydrograph is more generalized and therefore smoother. The new unit hydrographs should be checked to see that the enclosed area is equivalent to the unit of surface runoff and the final plotted curves amended if necessary. A 1-h UH and its S-curve are applicable to small and medium sized catchments that have response times measured in hours and for which the unit hydrograph method of analysis is realistic. For large catchments with slower responses

(a)

(b)

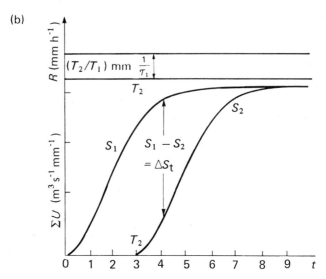

(c) Deriving ½ h–UH and 2 h–UH

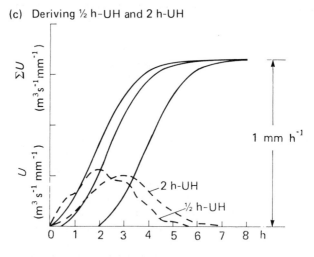

Fig. 13.10   Changing the T of the TUH.

and on which storms do not tend to be uniform in time or space, a longer standard period of 1 day could be used for changing T values. This latter operation would be rarely needed.

Table 13.2   Changing the T of the TUH

Time (h)	1 h UH	S-curve	S-curve	S-curve difference
Deriving the $\frac{1}{2}$ h UH			(offset $\frac{1}{2}$ h)	($\frac{1}{2}$ h UH)
0	0	0		0	0
0.5		0.25*	0	0.25	0.50
1	0.58	0.58	0.25	0.33	0.66
1.5		1.10	0.58	0.52	1.04
2	1.09	1.67	1.10	0.57	1.14
2.5		2.15	1.67	0.48	0.96
3	0.94	2.61	2.15	0.46	0.92
3.5		2.90	2.61	0.29	0.58
4	0.51	3.12	2.90	0.22	0.44
4.5		3.18	3.12	0.06	0.12
5	0.12	3.24	3.18	0.06	0.12
5.5		3.27	3.24	0.03	0.06
6	0.05	3.29	3.27	0.02	0.04
6.5		3.29	3.29	0	0
7	0	3.29	3.29	0	0
Deriving the 2 h UH			(offset 2 h)		(2 h UH)
0	0	0		0	0
1	0.58	0.58		0.58	0.29
2	1.09	1.67	0	1.67	0.83
3	0.94	2.61	0.58	2.03	1.02
4	0.51	3.12	1.67	1.45	0.72
5	0.12	3.24	2.61	0.63	0.32
6	0.05	3.29	3.12	0.17	0.08
7	0	3.29	3.24	0.05	0.03
8			3.29	0	0

* $\frac{1}{2}$ h values of S-curve taken from graph, Fig. 13.10c.

In carrying out analysis of effective rainfall and surface runoff records to obtain a unit hydrograph, it must be realized that some raw catchment data may not produce very good results, since sweeping assumptions are made in the theory. The simple linear properties of proportionality and superimposition do not hold for complex natural processes taking place in catchments of variable composition. There are catchments for which the method is not at all suitable, particularly those areas where the groundwater contribution to the flow may be high and where it is very difficult to carry out the baseflow separation. Many features of the flow within a catchment have non-linear characteristics and thus are not well modelled by the linear unit hydrograph procedure. Large catchments with varied geological structure and those experiencing variable storms over the area should be subdivided and more

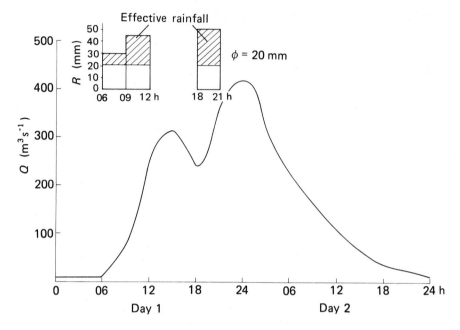

Fig. 13.11   A storm hydrograph from measured rainfall.

homogeneous subcatchments taken separately for the study of rainfall – runoff relationships.

13.4.4   Application of the TUH

Once a TUH has been determined satisfactorily for a catchment area, it can be used as a tool to obtain the storm hydrograph from measured rainfall amounts. Most often the interest of the engineer is concentrated on knowing the peak flow, but in some design projects it is necessary to know the total volume of flow or for how long a high flow rate over a critical level is likely to be sustained. The unit hydrograph method has the advantage of providing a continuous simulated record of discharge.

The computation of a storm hydrograph is demonstrated in Table 13.3. The 3 h UH is given at 3-h intervals with the U values in $m^3s^{-1}mm^{-1}$, and two periods of heavy rain occur, 75 mm between 0600-h and 1200 h and 50 mm between 1800 h and 2100 h (Fig. 13.11). A constant loss-rate of 20 mm is assumed. The three 3 h blocks of effective rainfall (10, 25 and 30 mm) are applied to the ordinates of the TUH at their appropriate starting times. An estimated series of baseflow values related to the initial value of flow before the storm is added. The contributions made by the effective rainfalls and the baseflow are summed along each row to provide the total flows in the last column. These are plotted on Fig. 13.11. When evaluating flood peaks, extreme rainfall intensities are used and the catchment is assumed to be saturated initially.

Table 13.3   Calculation of Discharge (m³s⁻¹) from Effective Rainfall (mm)

Hours	3 h UH	Time	10 × UH	25 × UH	30 × UH	Base flow	Total flow
				Effective rainfall × UH			
0	0	0600	0			10	10
3	6.0	0900	60	0		10	70
6	9.4	1200	94	150		9	253
9	7.1	1500	71	235		8	314
12	5.4	1800	54	177.5	0	8	239.5
15	4.0	2100	40	135	180	9	364
18	2.9	2400	29	100	282	10	421
21	1.8	0300	18	72.5	213	10	313.5
24	1.0	0600	10	45	162	11	228
27	0.4	0900	4	25	120	11	160
30	0	1200	0	10	87	12	109
		1500		0	54	12	66
		1800			30	12	42
		2100			12	12	24
		2400			0	12	12

As explained in the previous section, an estimate of the state of wetness of the catchment before a storm is required in order to assess the amount of effective rainfall that will form direct surface runoff. A general guide to the degree of moisture in the ground can be obtained from the *antecedent precipitation index (API)*. This is calculated on a daily basis and assumes that soil moisture declines exponentially when there is no rainfall. Thus:

$$\text{API}_t = k^t . \text{API}_0$$

i.e.

$$\text{API}_t = k . \text{API}_{t-1} \qquad (13.8)$$

where API_t is the index t days after the first day's API_0. The value of k is dependent on the potential loss of moisture (mainly through evapotranspiration) and varies seasonally usually between 0.85 and 0.98. If there is rainfall then this is added to the index and so a typical daily evaluation for the sixth day, for example, would be

$$\text{API}_6 = k . \text{API}_5 + R_5$$

where R_5 is the rainfall on the 5th day. At the beginning, an arbitrary value, e.g. 20 mm, may be assumed for API_0. As t increases, API_t becomes dominated by the recent rainfalls and the transient effect of API_0 disappears after 20 days or so. API_t thus gives a fair representation of the state of wetness of the catchment. The API can be related to the ϕ index by comparison with known events, and thus a good first estimate of the rainfall loss can be made.

13.5   The Instantaneous Unit Hydrograph

The general equation shown on Fig. 13.9:

$$Q_j = \sum_{i=0}^{j-1} R_i U_{j-i} \qquad j = 1, 2, 3, \ldots \qquad (13.5)$$

where R_0 is the effective rainfall in the first time interval T, describes the discrete summation operation, i.e. using depths of effective rainfall R, over finite intervals, T, to give individual Q ordinates also at discrete intervals T. If R is represented as a continuous rainfall intensity function $R(t)$ rather than as discrete depths R_i over finite intervals, then the rainfall term in the summation Equation 13.5 may be replaced by $\{R(t).dt\}$ for an infinitesimal time interval dt (Fig. 13.12). Considering the corresponding changes in the form of the TUH, as T becomes the infinitesimal dt, the unit depth of effective rainfall must be thought of as occurring at greater and greater intensity over shorter and shorter intervals, to the limiting case of being deposited instantaneously on the catchment. The resulting limiting form of the TUH is called the *instantaneous unit hydrograph* (IUH). Thus the ordinates U_{j-i} of the TUH at intervals T are replaced by $H(t-\tau)$ describing the continuous function of the IUH. The summation of the discrete Equation 13.5 becomes an integration of the product of the effective rainfall function and IUH to give a continuous function of surface runoff, $Q(t)$, instead of the discrete ordinates Q_j. Thus:

$$Q(t) = \int_0^t \{R(\tau).d\tau\}.H(t - \tau) = \int_0^t R(\tau).H(t - \tau).d\tau$$

with τ the dummy time variable of integration.

The relationship of a continuous rainfall with the corresponding surface runoff hydrograph via the IUH is shown in Fig. 13.12. The IUH is the impulse response of the catchment to an instantaneous unit input of effective rainfall. This theoretical concept, though physically impossible, has played a major role in UH theory and its development and has provided a fertile field for many research workers in hydrology (O'Donnell, 1966; Dooge, 1973).

A wide range of mathematical techniques surrounds the determination of $H(t)$, but the complexity of the methods has deterred practising engineers from applying them in day-to-day problems. For catchments that have no hydrometric records, techniques incorporating catchment parameters have had their attractions. However, these methods, though based on unit hydrograph theory, are more akin to mathematical models, and will be considered in the next chapter.

13.6   Rainfall – Runoff Relationships over Longer Periods

Water resources engineers are primarily concerned with catchment yields and usually study hydrometric records on a monthly basis. Most rainfall measurements are made daily, and a day's rain may be part of a continuous storm that can be related to resulting stream flow by means of the unit hydro-

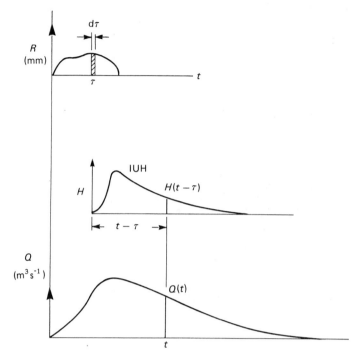

Fig. 13.12   Instantaneous unit hydrograph (IUH). $Q(t) = \int_0^t R(\tau).H(t-\tau).d\tau =$ $\int_0^t R(t-\tau).H(\tau)d\tau$ (convolution integral) or $Q(t) = R(t)*H(t)$.

graph method. With the summation of daily values to give a monthly total, periods of both wet and dry days are included, and the relationship of the rainfall with stream flow is indirect. Monthly mean discharges from a catchment area are converted to volumes of water produced and then to equivalent depths over the catchment area. Rainfall and runoff in the same units can then be compared.

The nature of the relationship of rainfall to runoff over longer periods again depends primarily on the structure and composition of the catchment area, but it can also be affected by the climate of the area. In the British Isles, rainfall occurs all year round and seasonal differences are small. Taking average monthly values over several years, regional variations in the rainfalls of the wettest and driest months can be identified. However, the range of variation of rainfall in any one month can be considerable, and in most places in the country any month can be the wettest or driest in any calendar year. Resultant effects on runoff are dependent on the time factor; the rainfall – runoff relationship is distinctly modified by the occurrence of sequences of wet or dry months.

The mean monthly rainfall pattern is shown in Fig. 13.13 for the catchment area of the River Severn rising in the mountains of mid-Wales and

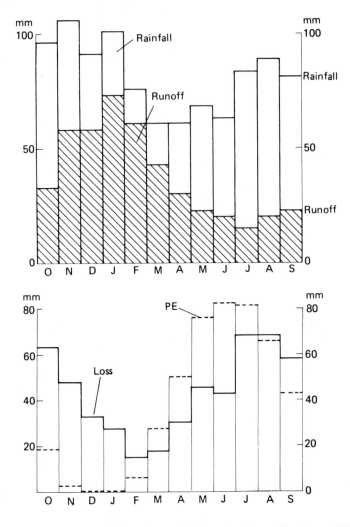

Fig. 13.13   Monthly rainfall and runoff. River Severn at Bewdley (1921–57). 4330 km². Loss = Rainfall – Runoff. PE = Average potential evaporation for Radnor.

flowing down to the river gauging station at Bewdley. The data are taken from the River Severn Basin Hydrological Survey (Ministry of Housing and Local Government, 1960).

The monthly runoff has been plotted on the same bar graph, and it must be noted that the 12 months are not shown in the ordinary sequence of a calendar year. In the UK the *water year* is usually taken to begin on October 1st at the end of the evaporation season when the groundwater storages are, on average, at a similar level each year. The bottom part of the diagram shows a value of loss given by the following relationship:

Loss = Rainfall – Runoff

Thus, given monthly rainfall totals, the monthly runoff should be obtained from:

Runoff = Rainfall – Loss

However, as can be seen from the plotted values of potential evaporation (MAFF, 1967) the water loss must have other components. In the months March to July, the loss is all due to evaporation, but where potential evaporation is less than the loss, (from September round to February), the extra losses are making up groundwater deficiencies.

The principle addition to a water balance equation on a monthly basis is the change in storage, and hence it is necessary to write:

$$RO = R - E - \Delta S$$

in which evaporation E plus ΔS (the change in storage), constitute the loss (the difference between rainfall and runoff). Thus, even over a period of one month, when the variables have been lumped together, there are still complexities in the rainfall – runoff relationship, and other factors must be considered. To obtain the runoff from rainfall for one month, a statistical appraisal can be made of the relationship for a specific month (say June), if many years of records for June can be analysed. Such an analysis generally needs to take account of the seasonal changes from month to month, and the study then becomes a problem in time series analysis. This will be considered in Chapter 15.

The problem is further complicated in those regions of the world that have distinctive rainy and dry seasons, and in such cases it is advisable to analyse the rainfall and runoff for the combined months of the wet season or seasons.

When the winter and summer seasons are considered for the British Isles by combining the winter months October to March and the summer months April to September, significant relationships between rainfall and runoff can be obtained. In Fig. 13.14, 42 years of winter and summer rainfall and runoff for the River Derwent catchment are shown. The data are from Law (1953), who showed that statistical correlation of the two variables could be used to estimate the reliable yield of a catchment. The regression lines have been fitted by 'least squares' to the scatter of points and the resulting equations give the best estimate of winter or summer runoff from a known or measured winter or summer rainfall. Further statistical properties of the estimates may be calculated if considered necessary. However, the correlation coefficients between runoff and rainfall are 0.87 for the winter period and 0.91 for the summer period, which indicate a satisfactory linear relationship. It is interesting to note that if the regression lines are extrapolated to intersect the rainfall axis, it could be inferred that 30 mm of rainfall are needed in the winter season before runoff occurs and 137 mm are needed in the summer season before there is any runoff. Although these figures, representing seasonal losses, are of the correct order, absolute values from such regression analyses should be treated with caution, since many other factors may also be operating.

In taking annual values for rainfall and runoff comparisons, seasonal

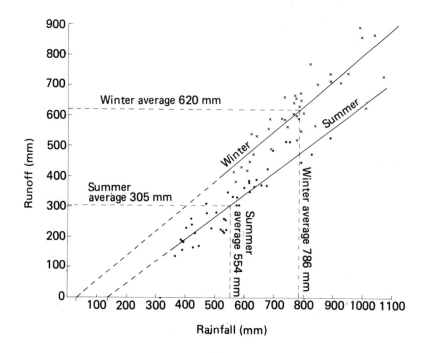

$\times$ Winter runoff = 0.82 rainfall − 25
• Summer runoff = 0.73 rainfall − 100

Fig. 13.14   Summer and winter rainfall and runoff. River Derwent catchment (Trent) 1906–47.

effects are removed and consideration of changes in groundwater storage can be neglected. Then, the straightforward linear relationship of Runoff = Rainfall − Loss (mainly evaporation) holds with even less scattering of points round the regression line.

In deriving relationships between rainfall and runoff for different time periods from hours up to years, straight-line approximations have been used with greater confidence as the time spans increase. It has been emphasized that the catchment response to inputs of rainfall is by no means linear, the superposition process and the time invariant property of the unit hydrograph method being a greatly simplified representation of the complex temporal interrelationship of rainfall and runoff in a drainage basin. More realistic non-linear relationships will be explored when considering catchment modelling.

References

Andrews, F.M. (1962). 'Some aspects of the hydrology of the Thames Basin.' *Proc. Inst. Civ. Eng.*, **21**, 55–90.

Dooge, J.C.I. (1973). *Linear Theory of Hydrologic Systems*. Tech. Bull. No. 1468, USDA Agriculture Research Service.

Kirpich, Z.P. (1940). 'Time of concentration of small agricultural watersheds.' *Civ. Eng.*, **10**(6), 362.

Kuichling, E. (1889). 'The relation between the rainfall and the discharge of sewers in populous districts.' ASCE *Trans.*, **20**, 1–56.

Lloyd-Davies, D.E. (1906). 'The elimination of storm water from sewerage systems.' *Proc. Inst. Civ. Eng.*, **164**, 41–67.

Law, F. (1953). 'The estimation of the reliable yield of a catchment by correlation of rainfall and runoff.' *J. Inst. Water Eng.*, **7**, 273–293.

O'Donnell, T. (1966). 'Methods of computation in hydrograph analysis and synthesis.' *Recent Trends in Hydrograph Synthesis Paper III*. TNO Proc. No. 13. The Hague.

Ministry of Agriculture, Fisheries & Food (MAFF). (1967). *Potential Transpiration*. Tec. Bull. No. 16. HMSO.

Ministry of Housing and Local Government (1960). *River Severn Basin*. Hydrological Survey, HMSO.

Mulvaney, T.J. (1850). 'On the use of self-registering rain and flood gauges.' *Trans. Inst. Civ. Eng. Ireland*, **4**(2), 1–8.

Sherman, L.K. (1932). 'Streamflow from rainfall by the unit-graph method.' *Eng. News Record*, **108**, 501–505.

Sherman, L.K. (1942). 'The unit hydrograph method,' Chapter X1E of *Hydrology*, ed. O.E. Meinzer, pp. 514–525.

14

Catchment Modelling

The advent of the high-speed digital computer with a large storage for data has stimulated research in many disciplines. Hydrologists are well to the fore in automating the application of existing analytical methods, but they have also used this modern tool to advantage in exploring new theories. Many advances and improvements in the development of rainfall–runoff relationships appeared in the 1950s and 1960s. However, it is only in the last 10 years or so that the practising engineer has had ready access to large computers, and therefore many of the advances in hydrological analysis have remained as research tools. Now, some of the new techniques are being applied more widely in solving engineering problems.

The principal techniques of hydrological modelling make use of the two powerful facilities of the digital computer, the ability to carry out vast numbers of iterative calculations and the ability to answer yes or no to specifically designed interrogations. Applying these facilities, mathematical models are built up by careful logical programming to describe the land phase of the hydrological cycle in space and time.

Hydrological models are divided broadly into two groups; the deterministic models seek to simulate the physical processes in the catchment involved in the transformation of rainfall to streamflow, whereas stochastic models describe the hydrological time series of the several measured variables such as rainfall, evaporation and streamflow involving distributions in probability. In providing information for the design engineer, a combination of the deterministic and stochastic approaches is proving to be the most successful. In this Chapter, deterministic models will be considered, and stochastic models are left to Chapter 15.

Firstly, the application of some of the linear models developed from unit hydrograph theory will be described. This will be followed by an introduction to non-linear models, much more difficult to use and still the basis of research. A section will be devoted to conceptual models in which the mathematical representations are inspired by considerations of the physical processes acting upon an input to produce an output. Many conceptual models are designed specifically to match the capabilities of the digital computer. The chapter will conclude with a mention of physically based component models that seek to describe realistically the several component processes of the hydrological cycle such as overland flow and infiltration.

14.1 The Nash Model

The derivation of a unit hydrograph for a river catchment described in the last chapter is dependent on the availability of some rainfall and runoff records. Very often, information on river flows is required for streams and rivers that are not gauged. For large projects where feasibility studies may occupy a few years, special temporary gauging stations may be established to provide the necessary data. However, design flow estimates may be required in a shorter time and, since the unit hydrograph represents the response of a catchment to a defined rainfall, many attempts have been made to synthesize the unit hydrograph; these methods can be found readily in the literature (ranging from Snyder, 1938, to NERC, 1975).

In formulating mathematical expressions for the shape of the unit hydro-graph, it is usually the instantaneous unit hydrograph (IUH) that is studied. Nash postulated that the transformation by a catchment of an effective rainfall into a surface runoff could be modelled by routing that rainfall down a cascade of equal linear reservoirs (Nash, 1957). The relationship between the storage, S, of a linear reservoir and the outflow is:

$$S = KQ \qquad (14.1)$$

where K is a time constant for the storage.

The resultant hydrographs developing after each reservoir, due to an instantaneous input of unit amount into the first reservoir, are shown in Fig. 14.1. After n reservoirs, the form of the IUH at time t, is given by

$$U(0,t) = \frac{1}{K(n-1)!}\left(\frac{t}{K}\right)^{n-1}e^{-t/K} \qquad (14.2)$$

Equation 14.2 has two parameters, n and K, needing to be identified for any given catchment. For *gauged* catchments, n and K may be found by means of simple relationships between the moments of the areas formed by the plotted values of the effective rainfall (R), and the surface runoff (Q) with time, and of the area of instantaneous unit hydrograph (IUH):

$$U_1' = Q_1' - R_1' \qquad \text{(moments about the origins)}$$
$$U_2 = Q_2 - R_2 \qquad \text{(moments about the centroid of the areas)}$$

For Equation 14.2, $U_1' = nK$ and $U_2 = nK^2$. The values of n and K can then be determined from the effective rainfall and surface runoff statistics:

$$n = \frac{(Q_1' - R_1')^2}{Q_2 - R_2}$$

and

$$K = \frac{Q_2 - R_2}{Q_1' - R_1'}$$

In practice, n is rarely an integer and therefore $(n-1)!$ in Equation 14.2 is replaced by $\Gamma(n)$, where Γ is the tabulated gamma function.

For *ungauged* catchments, values of Q and R are not available. To circumvent this, Nash carried out correlation studies between the moments of the Q and R values for gauged catchments in the UK and the significant physical characteristics of those catchments, such as catchment area, overland slope and river-length (Nash, 1960). Then from measurements of the same

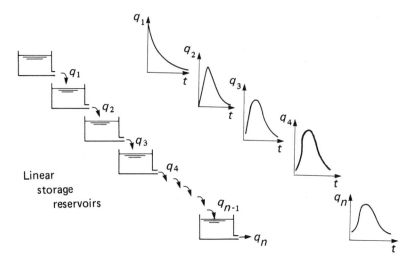

Fig. 14.1   Routing of instantaneous inflow through a series of linear storage reservoirs. (Reproduced from V.T. Chow (Ed.) (1964) *Handbook of Applied Hydrology*, by permission of McGraw Hill.)

physical characteristics of any ungauged catchment, using the correlation relationships, the parameters n and K can be found, and hence the instantaneous unit hydrograph, $U(0,t)$ synthesized from Equation 14.2. For reasons of statistical independence, a dimensionless measure of U_2 was used, namely $U_2/(U_1')^2 = m_2$ and writing $U_1' = m_1$, the correlation equations are given by:

$$m_1 = 36.7\,A^{0.3}.\,\text{OLS}^{-0.3} \tag{14.3}$$

and

$$m_1 = 23.1\,L^{0.3}.\,\text{EA}^{-0.33} \tag{14.4}$$

where A is the catchment area in km², L is the length of the longest stream to the catchment boundary, in km, EA is slope of main channel in parts per 10 000 and OLS is overland slope in parts per 10 000.

Very poor correlations betweem m_2 and any of the catchment characteristics were found, but the best regression result was selected for application:

$$m_2 = 0.39\,L^{-0.1} \tag{14.5}$$

Using these equations with measures of the characteristics from an ungauged catchment, values of m_1 and m_2 are found. Then the required parameters of the IUH, n and K are computed from:

$$K = m_1 m_2 \tag{14.6}$$

and

$$n = \frac{1}{m_2} \tag{14.7}$$

The Nash model with the parameters derived from the given regression equations is immediately applicable to small and medium sized ungauged catchments (up to about 2000 km^2) in Great Britain. To use the method in other climates and regions, new regression equations should be derived from a representative sample of gauged catchments.

A first estimate of the response of an ungauged catchment is obtained from this model, but even a few actual river and rainfall measurements would be preferable. However, the reader is referred to the original papers for a discussion of the reliability of the model results.

14.2   Further Developments of the Unit Hydrograph

Many research workers throughout the world have studied extensions of the unit hydrograph principles. One of the most searching and fundamental contributions was made by Dooge (Dooge, 1959). Concentrating on linear mechanisms, he suggested that the response of a catchment could be modelled by combining storage effects (used by Nash) with translation effects. The latter were defined by using the concept of a linear channel, i.e. a channel in which the rating curve is a straight-line relationship between discharge and cross-sectional area. Along any such river reach, an inflow hydrograph is translated without change of shape. Inflows for the Dooge model were obtained by the time – area method and used as distributed inputs to a generalized network of linear channels and reservoirs. This very general model yields a complicated formula for the IUH; its application has been limited even when the simplifying assumption of equal storages, needing only one value of K, has been made.

Further simplifications of the Dooge approach using linear theory were also made by Diskin, who modelled the catchment response with two series of equal linear reservoirs in parallel; in some respects these simulated rapid surface flow and slower interflow as components of the storm hydrograph (Diskin, 1964). Singh developed another linear catchment model by routing the time – area curve through two different linear storages in series, and showed that in practice it is possible to use simple geometrical forms in place of the real time – area curve (Chow, 1964). Kulandaiswamy produced a non-linear catchment response function using a non-linear storage expression and hence incorporated non-linear relationships that have long been recognized as being more realistic in the description of catchment behaviour (Chow, 1964).

14.3   The MIT Models

The unit hydrograph and its derivative models mostly have a single rainfall input usually an effective or excess rainfall after losses have been deducted. Sometimes the input value comes from the measurement of a single rain gauge, but more satisfactorily it is the areal rainfall over the catchment computed from several point rainfall measurements. In either event, it is a single value representing catchment rainfall and is called a *lumped* input. A team of workers led by Eagleson at the Massachusetts Institute of Technology (MIT) developed models using linear storages and linear channels, but with several simultaneous, but different, inputs at different points in the models. Such inputs clearly give a more realistic model of the different rainfalls occurring in any given time period over a catchment area. These models are called *distributed* models, since they seek to represent the areal rainfall distribution.

The simplest model, shown in Fig. 14.2, is a series of units comprising one linear reservoir, storage constant K and one linear channel module with τ, the translation time constant; values of K and τ remain the same for all the units. The lateral inputs R_1, R_2, . . . R_n . . . R_N are time varying.

The IUH for this sequence is given by the sum of the impulse responses for each reservoir and channel combination:

$$U_N(x,t) = \frac{1}{N}\sum_{n=1}^{N}\frac{1}{KTn}\left(\frac{t-n\tau}{K}\right)^{n-1}\exp[-(t-n\tau)/K] \qquad t > n\tau \quad (14.8)$$

where x represents a distance location on the catchment and N is the total number of linear units (NB, this is of the same form as Equation 14.2).

The use of this model for natural catchments is difficult, since the relation-

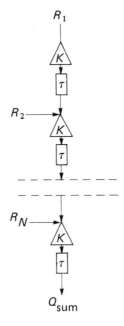

Fig. 14.2   Distributed linear reservoir model. (Reproduced with modifications from W.O. Maddaus & P.S. Eagleson (1969) *A Distributed Linear Representation of Surface Runoff*, MIT Department of Civil Engineering, Hydrodynamics Lab., Report No. 115, by permission.)

ship of the parameters n, K and τ to catchment characteristics is not straightforward. The reader is referred to the original reports for further explanation
of the experimental studies.

The principle of this MIT model was expanded further by incorporating
the open channel flow equations and the physical dimensions of the channel.
With this development, there is a connection with the more stringent mathematical component models considered later.

14.4   The Runoff-Routing Model

A different approach to determining runoff from a catchment was initiated
by Laurenson in Australia (Laurenson, 1964). He also was tackling the
problem of estimating floods from ungauged catchments; the runoff-routing
method he adopted gives the complete hydrograph, not just the peak flow as
in the empirical formulations. In this respect it is similar to the unit hydrograph method; also the runoff-routing method does not contain the base flow
contribution.

The Laurenson model is based on the routing of excess or effective rainfall
through non-linear catchment storages described by:

$$S = K(Q)Q \qquad (14.9)$$

where S is the stored volume in the 'reservoir' and the coefficient $K(Q)$ is a
function of Q, the outflow rate from the 'reservoir'. This relationship implies
that the storage delay time is dependent on Q, which of course varies during a
storm. Thus the flow through the various storage processes of interception,
surface detention and soil moisture detention, is made essentially non-linear.

In the original form of the model, the catchment area was divided into
contributing areas by sequential isochrones from the outfall, and a relative
time – area diagram constructed. Beginning at the head of the catchment, the
effective rainfalls over the top subarea were converted into an inflow hydrograph, $I(t)$, for increments of time, t. The hydrograph, $I(t)$, was then routed
through the non-linear storage equation (Equation 14.9) using the function
of K:

$$K = aQ^{-b} \qquad (14.10)$$

which had been derived from average values of lag and discharge from catchment flood records. To the resulting values of $Q(t)$, were added the excess
rainfall inputs, $I(t)$, from the next subarea, and the routing procedure
repeated. The sequence of inputs and routings was continued to give the final
discharges at the outfall. To apply the routing procedure, the non-linear
storage equation (Equation 14.9) is expanded over one time period (Δt) for
the first contributing area by substituting into the continuity equation
expressed in finite difference form:

$$S_2 - S_1 = \left[\frac{(I_1 + I_2)}{2} - \frac{(Q_1 + Q_2)}{2} \right] \Delta t \qquad (14.11)$$

whence

$$Q_2 = \frac{I_1\Delta t + I_2\Delta t + (2K_1 - \Delta t)Q_1}{(2K_2 + \Delta t)} \tag{14.12}$$

where K_1 and K_2 take values from Equation 14.10.

In the first instances K_2 is made equal to K_1 and an approximate value of Q_2 so obtained is used to find an improved K_2: by repeating, a final value of Q_2 is obtained by iteration (usually only 2 iterations needed). Q_2 then becomes Q_1 for the next routing.

The operation of the runoff-routing model over a catchment divided into, say 5 subareas, beginning with the hyetograph of excess rainfall converted to an inflow hydrograph, $I_1(t)$ follows the sequence:

$$I_1(t) \rightarrow Q_1(t) + I_2(t) \rightarrow Q_2(t) + I_3(t) \rightarrow$$
$$Q_3(t) + I_4(t) \rightarrow Q_4(t) + I_5(t) \rightarrow Q(t) \tag{14.13}$$

where t is the increment of time of the excess hyetographs and hydrographs and the routing operations are indicated by arrows. $Q(t)$ is the final storm hydrograph at the outfall of the catchment.

Since its inception, the runoff-routing model has been developed further, and some simple modification have made it more readily useful in solving specific engineering problems. The division of the catchment into the sub-catchments of the major tributaries instead of travel time areas has made the model much more flexible (Mein et al., 1974). The improved model again uses non-linear catchment response and it is a distributed model since it can incorporate non-uniform rainfall and losses over the catchment. It can also account for existing impounding reservoirs or natural storages. The effects of simulated reservoirs can be determined or, at an even earlier planning stage, hydrographs can be produced for a series of possible dam sites.

An example of the improved, more realistic, application of the runoff-routing technique is shown in Fig. 14.3. The catchment area is drained by a main river having four tributaries. The subcatchments are defined by drawing in the boundaries along the watersheds from a topographical map. Nodes are marked at points of the stream where stored water is concentrated. In each subcatchment, the effective rainfall is deemed to collect at the point on the stream nearest to the centroid of the subarea. Nodes are placed at all stream junctions and at locations where flow hydrographs are required. In between the nodes, the flow is routed along the stream or through the storage. The configuration of nodes and routing is shown in diagrammatic form under the catchment map in Fig. 14.3. The rainfall excess hyetographs are inputs to the nodes of the mainstream headwater, to the four tributaries and to the nodes on the mainstream which are centroids of the riverine areas draining directly into the river. Nodes numbered 3, 6, 8 and 11 are marked at the confluence of the tributaries and the main river.

If the non-linear storage equation is given more simply by:

$$S = KQ^m \tag{14.14}$$

instead of by Equation 14.9, K is the parameter related to lag times as before. The exponent m was found to vary between 0.6 and 0.8, and from studies of natural catchments in Australia values of m between 0.71 and 0.77 have been found. The delay time parameter K is considered in two parts such that

$K = k_1 C_1$. C_1 is a catchment constant obtained by fitting the model to measured rainfall and discharge data. Values of k_1 are related to the subdivisions of the catchment and are determined from physical characteristics of the channel between the nodes. In practice, it has been found adequate to make $k_1 = L$, the length of the channel for each subarea.

Once the two parameters, K and m, have been determined, computations of sequential values of $Q(t)$ are made using Equation 14.12, and then the total catchment outflow, $Q(t)$, is obtained from continuous routing following

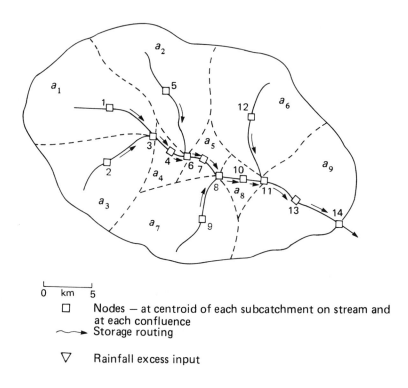

0   km   5

☐   Nodes — at centroid of each subcatchment on stream and at each confluence

⌒→   Storage routing

▽   Rainfall excess input

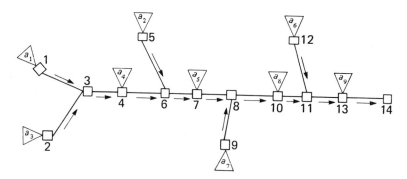

Fig. 14.3   Application of the runoff-routing model.

Equation 14.13. The runoff-routing model is now contained in an operational computer package (RORB).

An application of the runoff-routing model has been made in a catchment in Venezuela (Fig. 14.4) (Ponte, 1981). The operation of the RORB model of the upper three stretches of the river is demonstrated here.

From autographic records of a 20-h rainstorm, hourly rainfall values are obtained for the three subareas A, B and C. An initial loss of 5 mm and a runoff coefficient of 0.25 are used to assess the effective rainfall. Data values for the subareas are shown in Table 14.1. There are five storage routing sections of the river from A the top centroid to the outfall of the lowest subarea C (see Table 14.2). In this particular example, a design run of the model is made with $m = 0.60$ and $k(C) = 160$, constant for the whole catchment. In practice, m and $k(C)$ are obtained by test fit runs of the model using matching rainfall and runoff data. Results for A, B and C subareas are shown in Fig. 14.5. The bargraphs are of rainfall excess weighted according to subarea and the hydrographs relate to the outfalls of each subarea. It will be noted that the later start for the storm in subarea A results in a later peak.

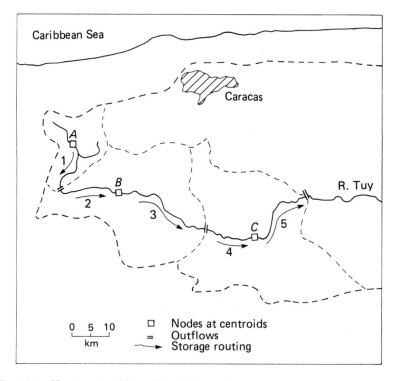

Fig. 14.4   Headwaters of River Tuy, Venezuela.

Table 14.1

	Area (km^2)	Total areal rain (mm)	Rainfall excess (mm)
A	244	29.3	3.6
B	921	78.2	15.8
C	1220	95.5	20.1

Table 14.2

	Length (km)	Relative delay time
1	13.75	0.137
2	21.25	0.212
3	27.75	0.277
4	15.00	0.150
5	22.50	0.224

The second column of values relates to the whole Tuy catchment.

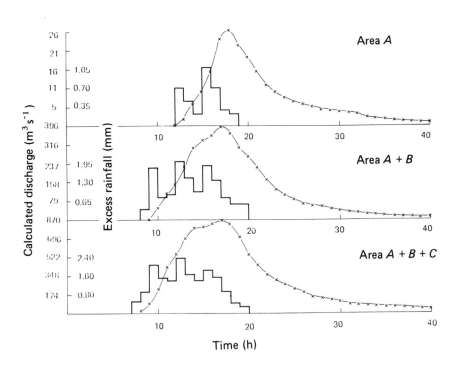

Fig. 14.5   Hydrographs from RORB model.

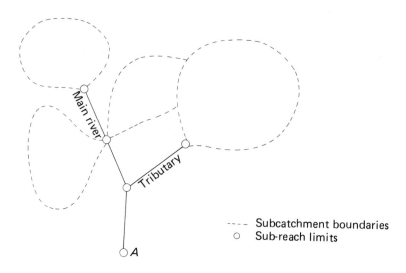

Fig. 14.6   Schematic catchment.

14.5   River Catchment Flood Model (FLOUT)

In conjunction with the Natural Environment Research Council Flood Studies, the Hydraulics Research Station at Wallingford developed a simple method to derive flood hydrographs from storm rainfall (Price, 1978). The first stage of FLOUT employs the unit hydrograph method. Based on the analysis of records from many UK catchments, the unit hydrograph is computed from recorded rainfall and runoff data for gauged catchments or from catchment characteristics for ungauged basins. To compute a flood hydrograph, particulars of a storm rainfall must be selected and assessment of losses made according to the methods recommended by the *Flood Studies Report*, Vol. 1 (NERC, 1975).

The second stage of FLOUT routes the flood hydrograph from the outfall of the gauged catchment (or contributing subcatchment) to the point on the mainstream where the flood information is required. Two flood-routing methods have been used, one based on a diffusion process and the second, now more favoured, the Muskingum – Cunge method (see Chapter 16) with variable parameters related to discharge and channel geometry.

A typical catchment upon which the FLOUT procedure would operate is shown schematically in Fig. 14.6. Several subcatchments are linked together by the subreach channels of a main river and a tributary. The separate runoff hydrographs resulting from a storm event are routed down the subreaches to give a flood hydrograph at the lowest outfall, A.

14.6   Conceptual Models

It has been emphasized that the movement of water in the land phase of the hydrological cycle is a complex process. For simplication, the hydrology of a drainage basin, from the precipitation through to a stream discharge at the lowest outfall, can be conceived as a series of interlinked processes and storages. In conceptual modelling, the catchment processes are described mathematically (e.g. by an equation for evaporation or a routing procedure for overland flow), and the storages are considered as reservoirs, for which water budgets are kept. Many conceptual catchment models have been formulated over the last two decades, each one structured by different combinations of processes, storages and interchanges, and each requiring some specific sets of input data. These models are deterministic models, in that each stage in the sequence is specifically determined over a selected time interval. Only a limited selection of models can be mentioned. To describe the structure and operation of a conceptual model, one of the simpler and earlier models, is given in greater detail (Dawdy & O'Donnell, 1965), followed by briefer accounts of other models (Fleming, 1975).

14.6.1   The O'Donnell Model

The O'Donnell model is built round four storages whose contents vary with time; a surface storage, R, in which rainfall detained on the surface or intercepted by vegetation provides the variable contents; a channel storage, S, whose contents represent the volume of water in the surface streams, rivers and lakes; a soil moisture storage, M, whose contents represent the water contained in the unsaturated soil layers; and a groundwater storage, G, in which water in the deeper saturated zone below the water table is represented. The inputs to and the processes of interaction between the storages resulting in the final total discharge are best described diagrammatically (Fig. 14.7).

The operation of such a generalized catchment model is controlled by the parameters (coefficients and exponents) in equations describing the way each component or interchange acts. The process of fitting a general model to a particular catchment makes use of known catchment records of precipitation, evaporation and runoff on that catchment and consists of finding numerical values of the controlling parameters such that the model reproduces the catchment behaviour.

The O'Donnell model has nine control parameters and operates on a predetermined time period depending on catchment size. Excess rainfall, P, over evaporation, E_R, and infiltration, F, remains in storage, R, until a threshold, R^*, is reached when the surplus, Q_I, runs into channel storage, S. The infiltrated water joins soil moisture storage, M, which can be depleted by transpiration, E_t, and, when M is greater than a threshold, M^*, by percolation, D. The soil moisture storage can also gain from the groundwater by capillary rise, C, unless G attains a value of G^*, when the two storages, M and G, are combined. In this model there is no direct feeding of channel flow from soil moisture storage. Both the surface storage, S, and the groundwater storage, G, are assumed to act as linear reservoirs, i.e. with storage directly proportional to outflow ($S = KQ$).

The nine parameters whose values have to be found in the fitting stage before the model can be applied to a problem are as follows:

R^* Surface water storage threshold.
K_s Linear reservoir constant for channel storage.
f_c Minimum rate of infiltration.
f_o Maximum rate of infiltration.
k Exponential exponent in infiltration equation.
M^* Soil moisture storage threshold.
K_G Linear reservoir constant for groundwater storage.
G^* Groundwater storage threshold.
C_{max} Maximum rate of capillary rise.

Thus, there are two parameters associated with the assumed linear reservoir discharges Q_S and Q_B, and three threshold values in the storages shown in the Fig. 14.7. The three parameters f_o, f_c and k come from the Horton infiltration equation used. The ninth parameter is a maximum value of the capillary rise, C_{max}, a constraint on the groundwater contribution to the soil moisture storage.

The methodology of fitting a conceptual catchment model is illustrated in Fig. 14.8. The large amount of data to be handled (several years of daily rainfall, evaporation and runoff data, for example) and the essentially simple arithmetic book-keeping operations for each time step, make model fitting an obvious task for the digital computer. For any trial set of model parameter values, the model is fed with sets of catchment rainfall and evaporation data and it computes corresponding runoff values, $\hat{Q}_t$. These are then compared with the corresponding recorded runoff values, Q_t. If the comparison is not acceptable, the parameter values are altered, and the process repeated. When a satisfactory comparison is achieved, another *independent* set of data

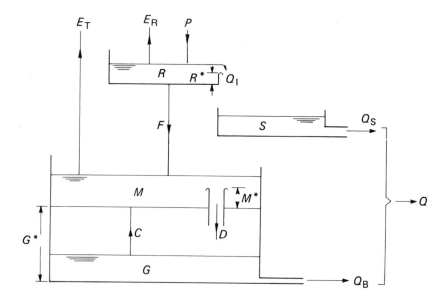

Fig. 14.7   The O'Donnell model. Discharge at gauging station $Q = Q_s + Q_B$.

from the same catchment is then used (a split-record test) to verify that the fitting achieved is indeed acceptable.

Dawdy and O'Donnell (1965) introduced the idea of using the computer itself to search for the best set of parameter values, not merely to do the model calculations. Such automatic optimization makes use of quantitative measures of goodness-of-fit (objective functions) such as:

$$F = \sum_{t=1}^{n} (Q_t - \hat{Q}_t)^2$$

for the n runoff data being optimized.

Considerable research studies on different objective functions and automatic optimization techniques have been carried out in recent years. Indeed,

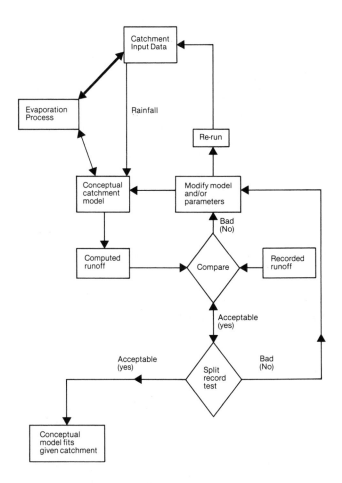

Fig. 14.8   Fitting a conceptual catchment model. (Reproduced from R.T. Clarke, A.N. Mandeville & T. O'Donnell (1975) In *Engineering Hydrology Today* (Ed. Monro) by permission of the Institution of Civil Engineers.)

the O'Donnell model was developed primarily to investigate such functions and techniques and was kept simple in hydrological terms compared with other models.

14.6.2   The Stanford Watershed Model

The first major computer study to synthesize the discharge of a river was made by Linsley and Crawford at Stanford University in the late 1950s and early 1960s. They aimed to simulate the whole of the land phase of the hydrological cycle in a catchment. The first Stanford Watershed Model was soon superseded by new versions as development and experience in application brought about improvements in performance and accuracy. Fig. 14.9 shows the flow chart for Model IV. A great variety of data is fed into the model

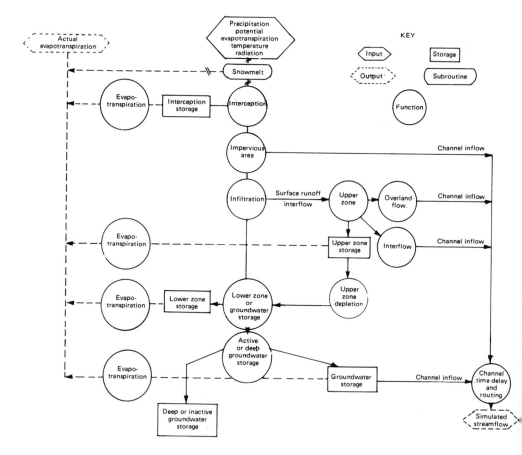

Fig. 14.9   Stanford watershed model IV flowchart. (Reproduced from N.H. Crawford & R.K. Linsley (1966) *Digital Simulation in Hydrology*, Stanford Watershed Model IV, TR 39, by permission of Stanford University.)

which is usually programmed to produce hourly river flows. Provision is made for dealing with snowmelt and, in incorporating particulars of impervious areas, the model can be applied to urban studies. Further elaboration compared to the O'Donnell model is the subdivision of the soil moisture storage into an upper zone, from which interflow feeds into channel flow, and a lower zone which feeds down to the groundwater storage. Evapotranspiration is allowed at the potential rate from the upper zone soil moisture storage but at a rate less than potential from the lower zone and groundwater storage. The total stream flow is the sum of overland flow and interflow, derived by separate procedures, and baseflow from the groundwater storage.

There are 34 parameters in the Stanford Watershed Model IV, but most of these are obtained from physical measurements either of initial conditions or of specific catchment characteristics. Four of the parameters, infiltration and interflow indices together with the capacities of the upper and lower zone soil moisture storages, must be determined by calibration of the model with recorded data. If the snowmelt routing is omitted, the number of parameters reduces to 25.

The model can be used for all sizes of catchment, and where there are data shortages, regional values of the required inputs may be used. It has been applied to catchments throughout the world and with its great flexibility has helped to provide hydrological information for problems in civil engineering design and agricultural engineering. Fleming (1975) gives full details of the Stanford Model.

14.7   Conceptual (Deterministic) Modelling in the UK

Since the advent of catchment modelling, practising engineering hydrologists and researchers have tended to devise their own models to solve a particular problem or to attempt an improvement in the representations of the different phases in the water cycle. In the beginning, there were constraints in the capacity of the existing computers, but as larger computer storages became more readily available models gradually grew more complex. More recently, there has been a trend towards simplification, since it is sometimes found that the big sophisticated models do not always give justifiable improvements in performance commensurate with the rising cost of computing. Hence there is an increasing tendency to use existing models that have cost many man hours in development and most authorities are ready to make available their computer software for application in other areas.

In the UK the Water Resources Board's model DISPRIN (Dee Investigation Simulation Program for Regulating Networks) was developed for the River Dee regulation research programme. The scheme of the basic simulation model is shown in Fig. 14.10. The catchment is envisaged in three hydrological zones, the uplands, the hillslopes and the lower valley areas, designated bottomslopes. In each of the three zones, non-linear storages are interconnected by linear routing procedures representing overland flow and interflow (quick return flow), and these feed into a common ground water

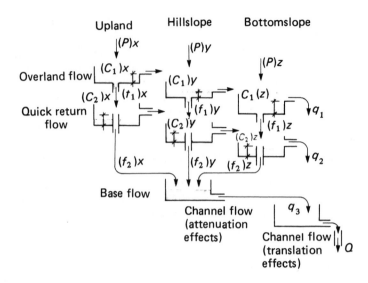

Fig. 14.10 DISPRIN subcatchment simulation model. P = precipitation. C_1 = depression storage. C_2 = available water capacity. f_1 = infiltration. f_2 = base flow recharge. q_1 = overland flow. q_2 = quick return flow. q_3 = base flow. Q = total subcatchment flow. (Reproduced from D.G. Jamieson & J.C. Wilkinson (1972) *Water Res. Res.* **8**(4), 911, by permission of the American Geophysical Union.)

storage from which there is baseflow. There are 21 parameters for the DISPRIN model, but seven of these are starting values for the seven storages. The basic form of the model is used for small or medium sized catchments, but the drainage of a large basin can be simulated in a sequence of applications. Fig. 14.11 shows how the DISPRIN model is applied to the 1040 km² of the River Dee system in North Wales. Discharges from the Upper Dee, Tryweryn and Upper Alwen subcatchments are routed through the reservoirs Llyn Celyn, Llyn Tegid and Alwen Reservoir, as indicated, and then linked up in sequence with the other tributary contributions down the main channel to the Erbistock river gauging station. The operation of the model aids the control of the reservoirs for regulating the flow of the River Dee in times of flood and drought, taking into account required abstractions from the system for water supply.

Other models developed or used in the UK are:

(a) The Institute of Hydrology model, essentially a research tool, has several different forms and can be applied over hourly or daily time periods. Although classed as a simple model, it pays particular attention to the complexities of soil moisture storage, which it represents in several layers. In addition to numerous reported studies at the Institute of Hydrology, a modified form of the model was used to investigate the effects of change in land use on East African catchments (Blackie, 1972).

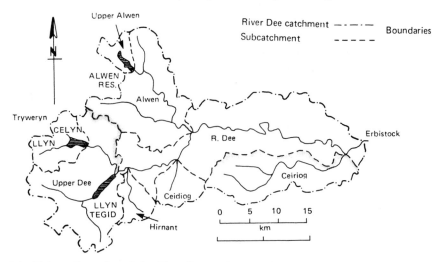

(a) Major tributaries of the River Dee and subcatchments

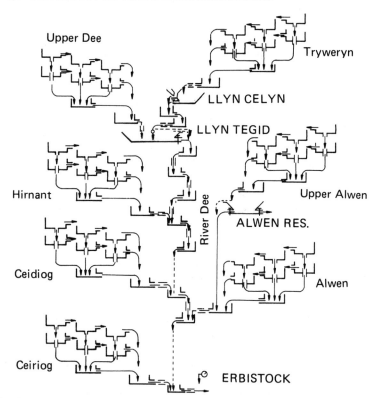

(b) Catchment simulation model applied to the River Dee system

Fig. 14.11   Application of DISPRIN to the River Dee catchment. (Reproduced from D.G. Jamieson & J.C. Wilkinson (1972) *Water Res. Res.* **8**(4), 911, by permission of the American Geophysical Union.)

(b) The Lambert model, developed in the former Dee and Clwyd River Authority essentially for small upland catchments, was a forerunner of DISPRIN for the Dee Research Programme and is proving simpler to operate in practice (Lambert, 1969).

(c) HYSIM, developed by Manley and used in the Directorate of Operations of the Severn Trent Water Authority, is one of a suite of programmes for hydrological analysis and provision of information for design and operational purposes (Manley, 1975). It operates mainly on daily values of areal rainfall and potential evaporation, and produces daily values of stream flow, but the time period can be flexible. It may be used for the extension of flow data records and data validation, real time flow forecasting and flood studies, modelling of groundwater, and has also simulated successfully daily and monthly flows on ungauged catchments.

(d) The Boughton model for small or medium sized catchments was originally developed in Australia for assessing water yield from catchments in dry regions (Boughton, 1966). Hence its immediate concern was with quick runoff. Murray modified the model to include a delayed response interflow and baseflow, and applied it to the Brenig catchment in North Wales as part of the study programme carried out by the Water Research Association in the late 1960s. The model operates on daily rainfall and evaporation to produce daily runoff (Murray, 1970).

Boughton has recently extended his work to include the concept of the varying areal contribution to runoff by designating different spatially defined surface storages (Boughton, 1987). His latest water balance model identifies three storages for a small test catchment with their capacities assessed from the hydrograph analysis of several storm events with particular attention paid to the initial state of the baseflow contribution. The operation of the model on a daily basis produced most satisfactory monthly totals of runoff over eight years of records. The value of this reported development is the demonstration of the application of hydrograph analysis to identify and quantify significant hydrological response areas of a catchment.

14.8   Other Conceptual Models

The Stanford Watershed Model and the Boughton Model have already been mentioned as originating in the USA and Australia, respectively. In addition to these pioneering studies and those of MIT, a great amount of work on conceptual or deterministic modelling has been carried out by the large Federal Government agencies of the USA with their teams of hydrologists, mathematicians and engineers working full-time with considerable resources. However, notable contributions have been made in other countries in smaller government departments, in commercial organizations and by individuals or small groups in universities. For reference purposes, a selected list is provided in Table 14.3. Further information on the models can be obtained from items in the bibliography or from the authorities concerned.

the two dimensional cross-section of the hillside (the $Y - Z$ plane in the block diagram Fig. 14.12). The unsaturated zones involved could contain a great variation of soil layers and geological strata. The flow equation is based on the equation of continuity for transient flow through a saturated-unsaturated porous medium and on Darcy's law (Chapter 7).

(c) *Groundwater flow* is discussed in Chapter 7.

(d) *Channel flow equations* are presented in Chapter 16 and an extra term for lateral inflow is added. The boundary conditions and initial state of the system must be defined. Solution of the equations is complex and several methods are outlined in the literature (e.g. Henderson, 1966).

Developing further the separate component models, researchers have coupled two together to extend the mathematical representation.

(a) *Overland flow* modelled in the form of a kinematic cascade over an infiltrating surface is coupled to a subsurface flow model by Smith and Woolhiser (1971), as presented in Fig. 14.13 in which:

$$r \qquad\qquad = \text{rainfall}$$
$$L_1, L_2, L_3 = \text{lengths of 3 cascade elements}$$
$$C_1, C_2, C_3 = \text{Chezy coefficients of 3 cascade elements}$$
$$S_1, S_2, S_3 = \text{slopes of 3 cascade elements}$$
$$f \qquad\qquad = \text{infiltration}$$
$$z \qquad\qquad = \text{soil depth}$$
$$q \qquad\qquad = \text{discharge, runoff}$$
$$q(x,t) \qquad = r(t) - f(x,t) \text{ at each time } t \text{ at each distance } x$$

(b) *Overland flow* coupled to channel flow has been investigated by several workers. Several catchment shapes have been tried, all assuming no infiltration, and the effects of differing surface roughnesses and slopes tried for variations of storm inputs. Wooding (1965), Overton and Brakensiek (1970), Woolhiser (1969) and Woolhiser et al. (1971) are

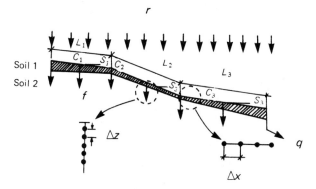

Fig. 14.13   Overland flow and infiltration. (Reproduced from R.E. Smith & D.A. Woolhiser (1971) *Water Res. Res.* **7**(4), 899–913, by permission of the American Geophysical Union.)

major contributors in this field. The team at MIT produced a catchment – stream model in which the complete open channel flow
equations are linearized in order to obtain a solution. A natural
catchment is represented by a series of catchment – stream elements
(Fig. 14.14). The catchment or overland flow is produced from
spatially uniform excess rainfall of intensity $i(t)$ by summing the
contributions from strips of unit width. These lateral inflows are
incorporated into the stream flow in a wide rectangular channel.
While the basic thinking for this component model required the
application of the two dimensional non-steady hydraulic flow
equations, the difficulties of the analytical solution of the fundamental
equations restricted the operation of the model to using the linear
system of the IUH.

(c) *Subsurface flow* coupled to a channel flow model has been used by
Freeze (1972) for studying the contributions of baseflow (q_b) to river
flow (with overland flow $q_s = 0$) and the role of subsurface flow in the
upstream reach areas of a catchment in generating surface runoff
(Fig. 14.15).

The development and application of component models to small catchments, experimental plots and laboratory catchments is advancing the
understanding of the relative importance of the different processes in the
movement of water in a drainage basin. The amounts of hydrological data

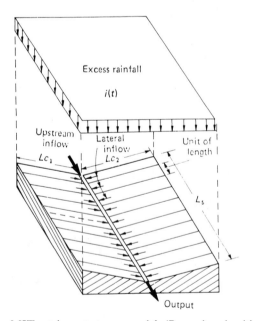

Fig. 14.14   The MIT catchment stream model. (Reproduced, with modifications,
from S.C.A. Bravo, B.M. Harley, F.E. Perkins & P.S. Eagleson (1970) *A Linear Distributed Model of Catchment Runoff*, MIT Department of Civil Engineering, Hydrodynamics Lab., Report No. 123, by permission.)

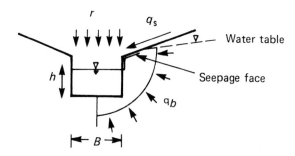

Fig. 14.15   Subsurface flow and channel flow (cross-section). (Reproduced from R.A. Freeze (1972) *Water Res. Res.* **8**(3) 609–623; **8**(5) 1272–1283, by permission of the American Geophysical Union.)

and catchment measurements required have prohibited the use of these models for practical problem solving, except under very restricted conditions. The complex iterative techniques required for solving the dynamic equations have needed to be carried out on large computers and thus limited their applicability. However technological advances in the computer industry are bringing about the means whereby the large quantities of data can be assembled and the vast number of computations be handled with ease at reasonable cost. If the data can be provided, component models such as the SHE model may become increasingly operational.

14.10   The SHE model

SHE (Système Hydrologique Européen) is a physically based component model accounting directly for spatial variations in hydrological inputs and catchment responses (Jonch-Clausen, 1979). It has been developed jointly by hydrologists in the UK, France and Denmark. Finite difference methods are used to obtain the solutions of the non-linear flow equations representing overland and channel flow, unsaturated and saturated subsurface flow. The model structure is shown in Fig. 14.16 (Abbott et al., 1986). The spatial variability over a catchment is represented horizontally by up to 2000 grid points and up to 30 in the vertical. In the unsaturated zone, the system is simplified with a one-dimensional vertical flow component used to link the two-dimensional surface water and ground water flow components. The computational sequence is as follows.

The *Precipitation rate* is the data input to the Interception Model (4 parameters); a layered snow melt model may be applied next).

The *Evapotranspiration loss model* operating from several vertical zones (2 parameters) requires 4 meteorological inputs.

The *Overland-Channel flow model* (3 parameters) requires boundary flow and initial flow depth conditions, topography of the overland flow plane and channels and particulars of any man-made discharge alterations.

The *Unsaturated flow model* has 3 parameters according to soil layer or type

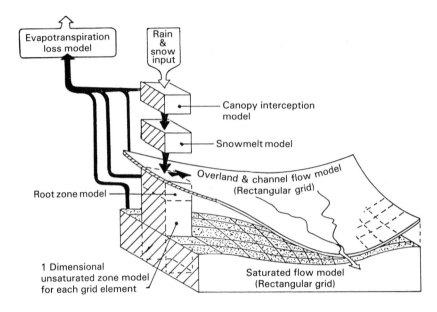

Fig. 14.16   Schematic representation of the structure of the European Hydrological System. (Reproduced from Abbott et al. *J. Hydrology*, *87*, 61–77. Elsevier.)

of vegetation and requires the initial water content profile.

The *Saturated flow model* (2 parameters in two-dimensions) requires input data on boundary conditions and topography of the base aquiclude, initial water table levels or saturated thicknesses and any man-made interference to natural conditions must be inserted.

The operation is made flexible by a Frame structure controlling the models of the individual components so that they may be applied to a variety of catchment conditions. The spatial capability can accept the variability of rainfall over a catchment and can be used to assess the effects of changes in land use. SHE has widespread application potential in the future.

References

Abbott, M.B., Bathurst, J.C., Cunge, J.A., O'Connell, P.E. and Rasmussen, J. (1986). 'An Introduction to the European Hydrological System—Système Hydrologique Européen "SHE" 2. Structure of the physically based, distributed modelling system.' *J. Hydrol.* **87**, 61–77.

Blackie, J.R. (1972). 'The application of a conceptual model to two East African catchments.' Unpublished M.Sc. dissertation, Imperial College.

Boughton, W.C. (1966). 'A mathematical model for relating runoff to rainfall with daily data.' *Trans. Inst. Eng. (Australia)*, **7**, 83.

Boughton, W.C. (1987). 'Hydrological Analysis as a Basis for Rainfall-Runoff Modelling.' *Trans. Inst. Eng. (Australia)*, CE 29, 1.

Bravo, S.C.A., Harley, B.M., Perkins, F.E. and Eagleson, P.S. (1970). *A Linear Distributed Model of Catchment Runoff.* MIT Dept. of Civil Engineering, Hydrodynamics Lab., Report No. 123.

Chow, V.T. (ed). (1964) *Handbook of Applied Hydrology.* McGraw-Hill.

Clarke, R.T. (1973). *Mathematical Models in Hydrology.* UN FAO Irrigation and Drainage Paper 19, 282 pp.

Clarke, R.T., Mandeville, A.N. and O'Donnell, T. (1975). 'Catchment modelling to estimate flows.' *Engineering Hydrology Today, ICE,* 37–44.

Crawford, N.H. and Linsley, R.K. (1966). *Digital Simulation in Hydrology.* Stanford Watershed Model IV, TR 39. Department of Civil Engineering, Stanford.

Dawdy, D.R. and O'Donnell, T. (1965). Mathematical models of catchment behaviour. *Proc. ASCE* **HY4**, **91**, 123–137.

Diskin, M.H. (1964). 'A basic study of the linearity of the rainfall runoff process in watersheds.' Ph.D. thesis, University of Illinois, Urbana, USA.

Dooge, J.C.I. (1959). 'A general theory of the unit hydrograph.' *J. Geophys. Res.,* **64**(2), 241–256.

Fleming, G. (1975). *Computer Simulation Techniques in Hydrology.* Elsevier, 333 pp.

Freeze, R.A. (1971). 'Three dimensional, transient, saturated-unsaturated flow in a groundwater basin.' *Water Res. Res.,* **7**(2), 347–366.

Freeze, R.A. (1972). Role of subsurface flow in generating surface runoff. 1. 'Baseflow contributions to channel flow.' *Water Res. Res.,* **8**(3), 609–623. 2. 'Upstream source areas.' *Water Res. Res.,* **8**(5), 1272–1283.

Henderson, F.M. (1966). *Open Channel Flow.* MacMillan, 522 pp.

Henderson, F.M. and Wooding, R.A. (1964). 'Overland flow and groundwater flow from a steady rainfall of finite duration.' *J. Geophys. Res.,* **69**, 1531–1540.

Jamieson, D.G. and Wilkinson, J.C. (1972). 'River Dee research program.' *Water Res. Res.,* **8**(4), 911.

Jonch-Clausen, T. (1979). *SHE, Système Hydrlogique Européen, A Short Description.* Danish Hydraulic Institute.

Lambert, A.O. (1969). 'A comprehensive rainfall-runoff model for an upland catchment area'. *J. Inst. Water Eng.,* **23**(4), 231–238.

Laurenson, E.M. (1964). 'A catchment storage model for runoff routing.' *J. Hydrol.,* **2**, 141–163.

Maddaus, W.O. and Eagleson, P.S. (1969). *A Distributed Linear Representation of Surface Runoff.* MIT Dept. of Civil Engineering Hydrodynamics Lab., Report No. 115.

Manley, R.E. (1975). 'A hydrological model with physically realistic parameters'. *Proc. Bratislava Symposium,* IAHS Pub. No. 115.

Mein, R.G., Laurenson, E.M. and McMahon, R.A. (1974). 'Simple non-linear model for flood estimation.' *Proc. ASCE* HY11, 1507–1518.

Murray, D.L. (1970). 'Boughton's daily rainfall-runoff model modified for the Brenig catchment.' *Proc. Wellington Symposium,* IAHS Pub. No. 144–161.

Nash, J.E. (1957). 'The form of the instantaneous unit hydrograph.' IASH Pub. No. 45, 3, 114–121.

Nash, J.E. (1960). 'A unit hydrograph study, with particular reference to British catchments.' *Proc. Inst. Civ. Eng.*, **17**, 249–282.

Natural Environment Research Council (NERC). (1975). *Flood Studies Report*, Volume I.

Overton, D.E. and Brakensiek, D.L. (1970). A kinematic model of surface runoff response. IASH Pub. No. 96, 100–112.

Ponte, Ramirez R.R. (1981). 'Storm rainfall and runoff in Venezuela.' Unpub. M. Phil. Thesis, Imperial College, University of London.

Price, R.K. (1978). *FLOUT — A River Catchment Flood Model*. Hydraulics Research Station Report No. IT 168, 2nd imp. Wallingford, England.

Smith, R.E. and Woolhiser, D.A. (1971). 'Overland flow on an infiltrating surface.' *Water Res. Res.*, **7**(4), 899–913.

Snyder, F.F. (1938). 'Synthetic unit graphs.' *Trans. AGU*, **19**, 447–454.

Wooding, R.A. (1965). 'A hydraulic model for the catchment stream problem. 1. Kinematic wave theory'. *J. Hydrol.*, **3**, 254–267.

Woolhiser, D.A. (1969). 'Overland flow on a converging surface'. *Trans. A.S. Agric. Eng.*, **12**, 460–462.

Woolhiser, D.A. and Liggett, J.A. (1967). 'Unsteady, one-dimensional flow over a plane — the rising hydrograph'. *Water Res. Res.*, **3**(3), 753–771.

Woolhiser, D.A., Holland, M.E., Smith, G.L. and Smith, R.E. (1971). 'Experimental investigation of converging overland flow'. *Trans. A.S. Agric. Eng.*, **14**, 684–687.

15

Stochastic Hydrology

The term 'stochastic' is much used in scientific studies and in operational practices within modern society today. It has the connotation of the chance of occurrence of an event within space or time, so that a stochastic process contains elements of randomness. In the hydrological cycle, the movement of water is seen to follow a now well understood deterministic path, but the magnitude and timing of the various processes are partly stochastic. The irregularity of the atmospheric circulation, still not fully explained, accounts for the randomness in the variations in the timing and magnitude of precipitation amounts which initiate the hydrological land phase. In particular, extreme rainfalls may be considered as random events since their future occurrences cannot be readily predicted. The resultant river flows reflect this component of uncertainty.

Stochastic hydrology considers chronological sequences of hydrological events with the aims of attempting to explain the irregularities of occurrence and, in particular, of forecasting the incidence of outstanding important extremes. The optimum requirements for studies in stochastic hydrology are long records of continuous homogeneous measurements of the main hydrological variables. Records of rainfall and river flow form suitable data sequences that can be studied by the methods of *time series analysis*. The tools of this specialized topic in mathematical statistics provide valuable assistance to engineers in solving problems involving the frequency of occurrence of major hydrological events. In particular, when only a relatively short data record is available, the formulation of a time series model of those data can enable long sequences of comparable data to be generated to provide the basis for better estimates of hydrological behaviour. In addition, the time series analysis of rainfall, evaporation, runoff and other sequential records of hydrological variables can assist in the evaluation of any irregularities in those records. Cross-correlation of different hydrological time series may help in the understanding of hydrological processes.

15.1 Time Series

The measurements or numerical values of any variable that changes with time constitute a time series. In many instances, the pattern of changes can be ascribed to an obvious cause and is readily understood and explained, but if there are several causes for variation in the time series values, it becomes difficult to identify the several individual effects. In Fig. 15.1, the top graph shows a series of observations changing with time along the abscissa; the ordinate axis represents the changing values of y with time, t. From visual

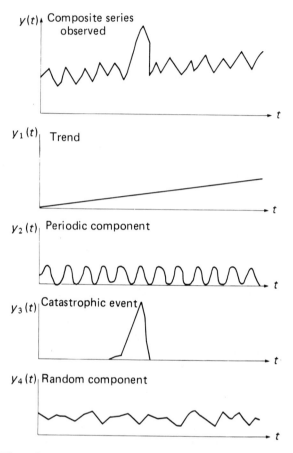

Fig. 15.1   The series components. (Reproduced from C.C. Kisiel (1969) in *Advances in Hydroscience, Vol. 5*, by permission of Academic Press.)

inspection of the series, there are three discernible features in the pattern of the observations. Firstly, there is a regular gradual overall increase in the size of values; this *trend*, plotted as a separate component $y_1(t)$, indicates a linear increase in the average size of y with time. The second obvious regular pattern in the composite series is a cyclical variation, represented separately by $y_2(t)$, the *periodic* component. The third notable feature of the series may be considered the most outstanding, the single high peak half way along the series. This typically results from a rare catastrophic event which does not form part of a recognizable pattern. The definition of the function $y_3(t)$ needs very careful consideration and may not be possible. The remaining, hidden feature of the series is the partly random stochastic component, $y_4(t)$, which represents an irregular but continuing variation within the measured values and may have some persistence. It may be due to instrumental or observational sampling errors or it may come from random unexplainable fluctuations in a natural physical process.

The complete observed series, $y(t)$, can therefore be expressed by:

$$y(t) = y_1(t) + y_2(t) + y_3(t) + y_4(t)$$

The first two terms are deterministic in form and can be identified and quantified fairly easily; the last two are stochastic with major random elements, and some minor persistence effects, less easily identified and quantified.

15.2   Hydrological Time Series

The composition of a hydrological time series depends on the nature of the variable and its mode of measurement. The variable may be continuous, like the discharge of a perennial stream, or intermittent, as with precipitation. There are differences according to climatic regime since river flow may not be continuous in semi-desert regions. For a more unusual example, the time series of water loss by evaporation and transpiration is fundamentally affected by vegetation type, and in those regions with scarce vegetation daily evaporation losses become intermittent and depend on the infrequent rains.

The sampling of a continuous variable usually results in the data being recorded at discrete times or over discrete time intervals; most time series of hydrological variables are considered in the discrete form. The length of the time interval used in assembling the data series is quite fundamental to the resultant composition of the time series. This important factor is highlighted in describing some time series of hydrological data for the UK. The examples used in Fig. 15.2 are for different locations and over unrelated time periods.

15.2.1   Rainfall

Typical patterns of rainfall data for different discrete time intervals are shown in Fig. 15.2a. Four samples each of 20 time units in length are given for time units of a year, a month, a day and an hour. The annual rainfall totals appear randomly scattered around some mean value, but no distinctive features of the series are immediately discernible. The monthly data, beginning in a January, show clear signs of a periodic component, viz, an annual cycle, with minima in the summer months and maximum monthly totals in winter. Such a cycle is a regular feature of the UK rainfall patterns, although individual monthly totals can be variable from year to year. The sequence of daily rainfall totals (Fig. 15.2a) shows a very different pattern. There are many nil values. Such a different time series requires special treatment in any analysis. The discontinuity of short-period rainfalls is emphasized in the time series over 20 h. Discrete and continuous representations have been drawn to demonstrate that even on a day with rain, the rainfall is not continuous throughout the day.

15.2.2   Evaporation

Time series of this variable sampled in Fig. 15.2b, again over a range of time units, are representative of a climatological station in the UK. The annual

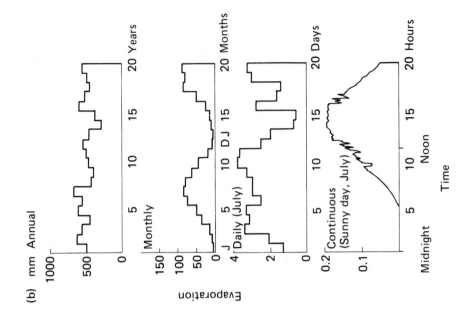

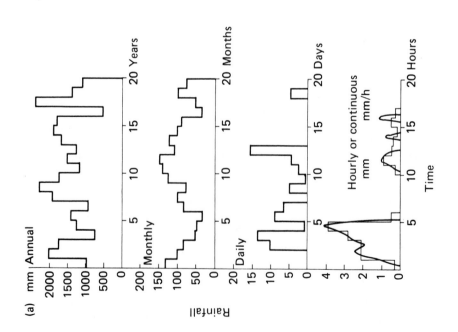

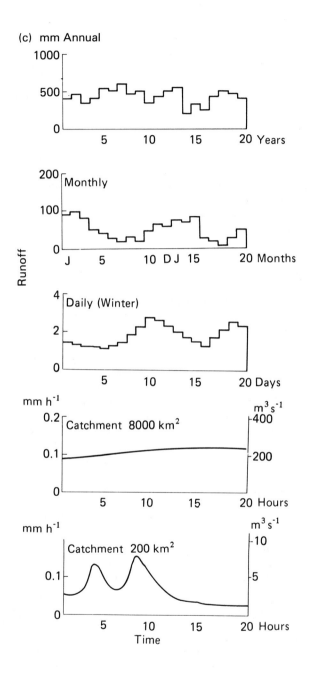

Fig. 15.2 Rainfall, evaporation and runoff in the UK.

values are less variable than those of rainfall. The monthly evaporation totals have a more regular periodic component with the maxima in the warm summer months. Winter values are very low. On the whole, evaporation values are more conservative than rainfall, i.e. there is less variability from year to year. The summer month of July is selected to give a sample of likely daily totals; the occasional days with low evaporation would correspond to spells of dull, cool weather, which do occur in the English summer, but the daily values provide a continuous discrete time series. The continuous curve of the 20-h graph represents the actual rate of evaporation (mmh⁻¹) which is likely to occur on a sunny day in July. Evaporation, negligible during the night, begins with the dawn and reaches a maximum in the warmest part of the day in the afternoon. Certain irregularities in the curve would be caused by clouds inhibiting the incoming solar radiation. Thus hourly evaporation sequences also form an interrupted time series with regular diurnal nil values.

15.2.3   Runoff

River flow records are represented here as time series of runoff expressed in millimetres of equivalent water depth over a catchment area, for ease of comparison with the other variables (Fig. 15.2c). The annual values resemble those for rainfall or evaporation with no pronounced time series component; the individual values are less variable than the rainfall. The monthly values again show a periodic component, the annual cycle following the rainfall pattern, but the minima tend to be delayed to the late summer. There is a lag effect from the low early summer rainfalls and peak evaporation months. The discrete values of the daily mean flows for a sequence of 20 days emphasize the continuing nature of river flows in the UK. There is no regular pattern in the time series over the 20 days shown; the two increases in runoff would be directly related to the primary causes, two significant rainfall events. (Cross-correlation between two daily time series of rainfall and runoff could give a measure of catchment response to rainfall.) In the last two runoff graphs, both on an hourly basis, the distinction is drawn between the continuous records from a large and a small catchment. The units of mm h⁻¹ are also shown as m³s⁻¹ for the two stated catchment sizes. At a gauging station on a river draining 8000 km², a rainfall event can be expected to have a very slow and smoothed response, but from a small catchment of 200 km² area, distinct hydrographs are quickly produced in response to distinct rainfall amounts. The presentation of the hourly runoff sequences in Fig. 15.2c emphasizes the persistence between hourly discharge values. Daily discharges also exhibit this feature, but to a lesser extent. Hourly and daily discharges are not independent of each other. Even on a monthly basis, the mean discharge is generally dependent on that of the previous months to some extent. This correlation between data in a plotted time series, although possibly less obvious to the eye than trend and periodicity, has nevertheless to be identified and quantified in any time series analysis.

15.3   Time Series Analysis

The identification and mathematical description of the several components forming the structure of a time series constitute time series analysis.

If a hydrological time series is represented by $X_1, X_2, X_3, \ldots, X_t, \ldots$, then symbolically, one can represent the structure of the X_t by:

$$X_t \rightleftarrows [T_t, P_t, E_t] \tag{15.1}$$

where T_t is the trend component, P_t is the periodic component and E_t is the stochastic component. The first two components are specific deterministic features and contain no element of randomness. The third, stochastic, component contains both the random fluctuations and the self-correlated persistence within the data series. These three components form a basic model for time series analysis. (The catastrophic component representing the effects of a very unusual occurrence is omitted in this basic discussion.)

15.3.1   Stationarity

The hydrological time series, as a sample of a data population, must be first considered in relation to the statistical population of that variable. If the statistics of the sample (mean, variance, etc) are not functions of the *timing* or the *length* of the sample, then the time series is said to be *stationary*. (The sample statistics may still exhibit the usual sampling variability.) If a definite trend is discernible in the series, then it is a non-stationary series. Similarly, periodicity in a series means that it is non-stationary. The modelling of a time series is much easier if it is stationary, so identification, quantification and removal of any non-stationary components in a data series is undertaken, leaving a stationary series to be modelled.

15.3.2   Trend Component

This may be caused by long-term climatic changes or, in river flow, by gradual changes in a catchment's response to rainfall owing to land use changes. Sometimes, the presence of a trend cannot be readily identified. Any smooth trend that is discernible may be quantified and then subtracted from the sample series.

T_t may take the form:

$$T_t = a + bt \text{ (a linear trend, as in Fig. 15.1)}$$

or

$$T_t = a + bt + ct^2 + dt^3 + \ldots \text{ (a non-linear trend)}$$

The coefficients $a, b, c, d \ldots$ are usually evaluated by least-squares fitting. The difficulty with trends is in deciding whether or not any trend so quantified will continue unchanged into the future; it may only be part of a long-cycle of change. In many sample time series, there is no obvious shift in the sample mean (m) over the period of record and the series is assumed to be trend-free.

15.3.3   Periodic Component

In most annual series of data in the UK, there is no cyclical variation in the annual observations, but in sequences of monthly data distinct periodic seasonal effects are at once apparent. In a different climate, however, significant annual periodicities have been identified (Fig. 15.8). The existence of periodic components may be investigated quantitatively by constructing a *correlogram* of the data. For a series of data, X_t, the *serial correlation coefficients* r_L between X_t and X_{t+L} are calculated and plotted against values of L, (known as the *lag*), for all pairs of data L time units apart in the series:

$$r_L = \frac{1}{n-L} \sum_{t=1}^{n-L} (X_t - \overline{X})(X_{t+L} - \overline{X}) \bigg/ \frac{1}{n} \sum_{t=1}^{n} (X_t - \overline{X})^2 \qquad (15.2)$$

where $\overline{X}$ is the mean of the sample of n values of X_t and L is usually taken for values up to $n/4$. A plot of r_L versus L forms the correlogram. Examples of correlograms are given in Fig. 15.3. In a completely random series with a large number of data, $r_L \simeq 0$ for all lags > 0, but it is more usual to obtain a correlogram with small values of r_L varying around 0, particularly if n is not large. With some short term positive correlation between the data values, the

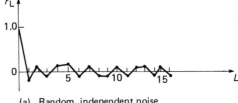

(a)   Random, independent noise

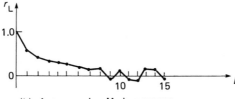

(b)   Autoregressive, Markov process

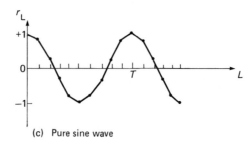

(c)   Pure sine wave

Fig. 15.3   Correlograms.

values of r_1, r_2, r_3, etc, gradually decrease from $r_0 = 1$ and finally vary closely around 0. This is representative of an *autoregressive* process. Typically, such a correlogram could be produced from a series described by the Markov model:

$$X_t = a_1 X_{t-1} + a_2 X_{t-2} + a_3 X_{t-3} \ldots + \epsilon_t \tag{15.3}$$

where a_i are related to the serial correlation coefficients r_i and ϵ_t is a random independent element.

The third example gives the correlogram for a data series taken from a pure sine wave (which could represent the periodic component in a time series) viz:

$$P_t = m + C \sin(2\pi t / T) \tag{15.4}$$

where C is the amplitude of the sine wave about a level m and of wavelength T.

The serial correlation coefficients for such a P_t are given by:

$$r_L = \cos(2\pi L / T) \tag{15.5}$$

The cosine curve repeats every T time units throughout the correlogram with $r_L = 1$ for $L = 0$, T, $2T$, $3T$, ... Thus periodicities in a time series are exposed by regular cycles in the corresponding correlograms.

Fig. 15.4a shows a sequence of correlograms. The top correlogram of a runoff time series displays flattening or dampening of pure cyclical effects; periods of several frequencies are forming part of the time series. In the following correlograms, periodicities of 12, 6, 4 and 3 months have been successively removed until the data series contains only random, independent noise. The identification of the various periodicities can be assisted by *spectral analysis* a complex procedure explained fully in standard texts (e.g. Chatfield, 1980). The variance spectra for the previous correlograms are seen in Fig. 15.4b (Roesner & Yevjevich, 1966).

The regular periodic seasonal variations in rainfall and river flow series, having been identified by correlogram and spectral analysis, need to be quantified through the course of a year. For each 'season', time period t, throughout the year (t from 1 to 12 for monthly values, 1 to 365 for daily values), the two statistics, m_t and s_t, the means and standard deviations of the values for time period t, over the n years of records are found (Fig. 15.5). The variations of m_t and s_t over the year are usually fitted by low-order harmonic series (5 or 6 terms) that smooth out any 'noisy' erratic behaviour in the m_t and s_t values (Hall & O'Connell, 1972). This quantification then allows the removal of the periodicity in the original data, one of the causes of non-stationarity.

Low-frequency cycles of 12 and 6 months have realistic physical meaning with the normal seasonal climatic fluctuation; evidence of higher order harmonics in any analysis should be treated with great reserve unless supported by recognised causes.

Once the significant periodicities, P_t, have been identified and quantified by m_t and s_t, they can be removed from the original time series along with any trend, T_t, so that a new series of data is formed given by:

(a) Correlogram

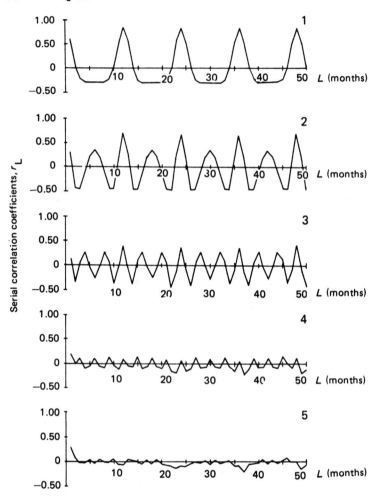

$$E_t = \frac{X_t - T_t - m_t}{s_t} \tag{15.6}$$

15.3.4   Stochastic Component

E_t represents the remaining stochastic component of the time series free from non-stationary trend and periodicity and usually taken to be sufficiently stationary for the next stage in simple time series analysis. This E_t component is analysed to explain and quantify any persistence (serial correlation) in the data and any residual independent randomness. It is first standardized by:

(b) Variance spectrum

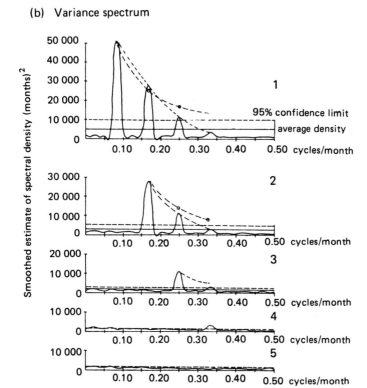

Fig. 15.4   Periodic components of 30 years runoff record, ELK River at Clark, Colorado, USA. (1) 12, 6, 4 and 3-month periods present. (2) 12-month period removed. (3) 12 and 6-month periods removed. (4) 12, 6 and 4-month periods removed. (5) All periods removed. (Reproduced from L.A. Roesner & V.M. Yevdjevich (1966) *Mathematical Models for Time Series of Monthly Precipitation and Monthly Runoff*, Hydrological Paper No. 15, Colorado State University, by permission.)

$$Z_t = \frac{E_t - \overline{E}}{s_E} \qquad\qquad (15.7)$$

where $\overline{E}$ and s_E are the mean and standard deviation of the E_t series. The series, Z_t, then has zero mean and unit standard deviation. The serial correlation coefficients of Z_t are calculated and the resultant correlogram is examined for evidence and recognition of a correlation and/or random structure.

For example, in Fig. 15.6a for the monthly flows of the River Thames at Teddington Weir, the correlogram of the Z_t stationary data series (with the periodicities removed) has distinctive features that can be recognised. Comparing it with the correlograms in Fig. 15.3, the Thames Z_t correlogram

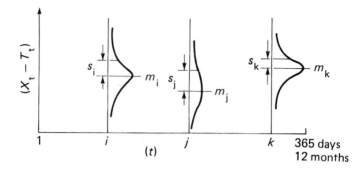

Fig. 15.5   Periodicity. m_t = average of N flows in day t/month t; s_t = standard deviation of N flows on day t/month t about mt; where t = 1 to 365 days or 1 to 12 months.

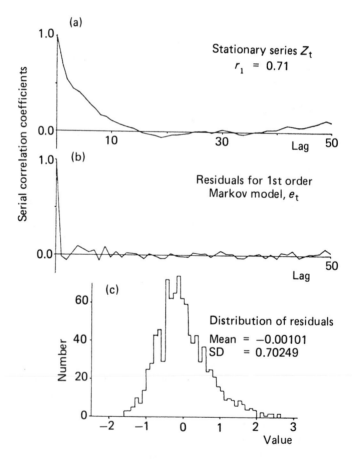

Fig. 15.6   River Thames at Teddington Weir (82 years of monthly flows).

resembles that of an autoregressive (Markov) process. For a first order Markov model

$$Z_t = r_1 Z_{t-1} + e_t \tag{15.8}$$

where r_1 is the serial correlation coefficient of lag 1 of the Z_t series and e_t is a random independent residual. A series of the residuals e_t may then be formed from the Z_t series and its known lag 1 serial correlation coefficient, r_1:

$$e_t = Z_t - r_1 Z_{t-1}$$

The correlogram of residuals is finally computed and drawn (Fig. 15.6b). For the Thames data this resembles the correlogram of 'white noise', i.e. independently distributed random values. If there are still signs of auto-regression in the e_t correlogram, a second-order Markov model is tried, and the order is increased until a random e_t correlogram is obtained. The frequency distribution diagram of the first order e_t values (Fig. 15.6c) demonstrates an approximate approach to the normal (Gaussian) distribution. The fully defined components of the time series model for the monthly flows of the Thames were found to have explained the variance of the original raw data in the following proportions: 51.6% by periodicity, 32.4% by auto-regression (first-order Markov) and 16% by the residuals.

At this stage, the final definition of the recognisable components of the time series has been accomplished including the distribution of the random residuals. As part of the analysis, the fitted models should be tested by the accepted statistical methods applied to time series. Once the models have been formulated and quantified to satisfactory confidence limits, the total mathematical representation of the time series can be used for solving hydro-logical problems by synthesizing non-historic data series having the same statistical properties as the original data series.

15.4   Time Series Synthesis

The production of a synthetic data series simply reverses the procedure of the time series analysis. For as many data items as are required, a comparable sequence of random numbers, drawn from the e_t distribution, is generated using a standard computer package. These then form the e_t values for the synthetic series from which are recursively calculated the corresponding synthetic Z_t values using Equation 15.8 (starting the series with the last value of the historic Z_t series as the Z_{t-1} value).

The E_t series then derives from Equation 15.7 in reverse:

$$E_t = Z_t s_E + \overline{E}$$

The periodic component P_t represented by m_t and s_t for time period t is then added to the E_t values to give:

$$X_t - T_t = E_t s_t + m_t \quad \text{(from Equation 15.6)}$$

The incorporation of the trend component T_t then produces a synthetic series of X_t having similar statistical properties to the historic data sample.

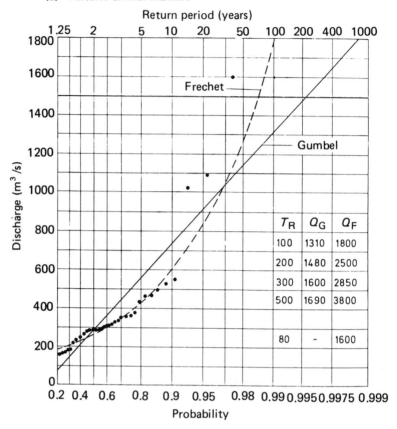

(a) Historic annual maxima

15.5   Application of Time Series Analysis and Synthesis

Generated sequences of hydrological data provide an extended data base on which to make design decisions. If historic series of rainfall or streamflow measurements are long enough to contain a range of patterns of occurrence then the generation of further long sequences of data with similar characteristics can be expected to provide better estimates of the frequency or return period of the infrequent notable events in the historic series simply by providing an increased number of such events in the longer combined record. When applied to flood flows, such improved return period estimates are of assistance in the choice of suitable protection schemes. The sequential extension of a historic flow record can also help with the assessment of future water resources and with their management by simulation of the behaviour of a water resources system.

Two contrasting applications of time series modelling are presented to demonstrate the flexibility in the use of the method.

(b) Generated annual maxima

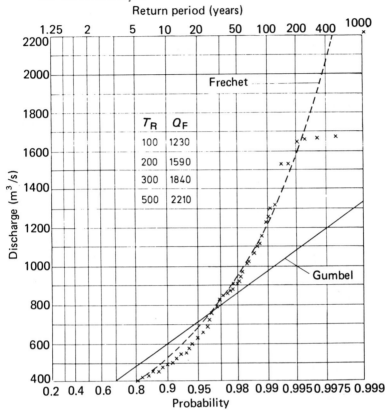

Fig. 15.7 The Vardar scheme. T_R = return period. Q_G = Gumbel values. Q_F = Frechet values. (Reproduced from T. O'Donnell, M.J. Hall & P.E. O'Connell (1972) *International Symposium Mathematical Modelling Techniques in Water Resource Systems, Vol. 1*, Environment Canada, by permission.)

15.5.1   *The Vardar Scheme*

Following a major flood and a disastrous earthquake, a reconstruction programme was planned for Skopje in the south-east of Yugoslavia. At Skopje, the River Vardar has a catchment area of 4900 km² and thereafter it flows through northern Greece into the Aegean Sea. To ensure adequate flood control works in the redevelopment of Skopje, a reliable flood magnitude – frequency relationship was required. Initial analysis of 42 years of historic annual maximum floods using extreme value distributions gave widely differing return periods for the estimated 1600 m³s⁻¹ biggest flood peak in the record. Using the Gumbel distribution, T_R, the return period for that flood was 300 years, but using the Frechet distribution T_R was only 80 years (if the variable *x* follows the Gumbel distribution, then e^x conforms to the Frechet distribution). The plotted points and the two fitted curves for the 42 historic data points are seen in Fig. 15.7a, with the Frechet curve clearly having a superior fit.

Time series analysis of the historic (42 × 365) daily mean discharges for the 42 years made fuller use of the record, including all the information in the flow data rather than just the annual peak flows. The data series was found to be free of trend by split sample testing of the means and standard deviations of the annual flows. The periodic components were quantified and removed, and the stochastic component, after several trials, was best described by a third-order Markov model. The amounts of variance in the historic daily flows explained by the various components of the model were 43.9% periodic, 51.9% autoregressive and 4.2% by the residuals.

Twenty-five data sets of daily flows, each 42 years long, were generated by the model, and the annual maximum daily flows abstracted. These were converted to peak flows by the daily-to-peak flow relationship of the historic data. The 1050 generated annual maxima were then fitted by Gumbel and Frechet distributions. These are shown on Fig. 15.7b for flows over 400 m^3s^{-1}. The Frechet distribution is seen to give a much more satisfactory fit, with the historic peak flood of 1600 m^3s^{-1} having a return period of just over 200 years. Thus a more confident estimation of the major flood return period could be made from the greater number of points provided by the data generation, while continuing to preserve the main statistical properties of the historic data (O'Donnell *et al.*, 1972).

15.5.2   The Maiduguri Rainfall

As part of a study of available water resources in the Lake Chad area of Africa, the 63 years of annual rainfalls at Maiduguri in north east Nigeria were analysed to identify the notable hydrological sequences in the Chad basin. In particular, an attempt was to be made to assess the frequency of occurrence of the severe rainfall deficiencies which has been affecting the whole Sahel region.

The correlogram of the annual rainfall series showed clearly a cycle of 30 years. A smoothed variance spectrum confirmed this result and also indicated minor periodicities of 7.5 and 3 years. However, only the 30 year and 7.5 year periodicities were found to be significant. A downward trend in the Sahel rainfall has been suspected by climatologists, so the data series was studied further after several trials by combining a third-order polynomial trend function with the periodic components. The analysis showed that only the cubic term gave any significant contribution to the model although the trend as a whole was not significant. The results of the analysis are shown in Fig. 15.8. The continuous line represents the model and the broken line the mean trend (which cannot be more specifically explained within the limits of the record). The scattered data values indicate the wide variations in the historic rainfall from year to year. On this annual basis the model is able to explain only 20% of the variance. The major component contributing 80% of the variance is the stochastic component which has no autoregressive features and is entirely random. The application of the model for evaluating future supplies has yet to be tested fully but an extension of the deterministic

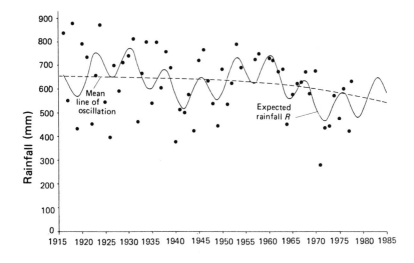

$$R = 652 - 59.1 \cos \frac{2\pi}{30} t + 27.5 \sin \frac{2\pi}{30} t$$

$$+ 62.2 \cos \frac{2\pi}{7.5} t + 23.3 \sin \frac{2\pi}{7.5} t$$

$$- 0.000325 \, t^3$$

t = Year − 1915

● = Observed data

Fig. 15.8   Maiduguri annual rainfall 1915–1977.

part of the record beyond 1977 is indicated in the figure. With the shortage of runoff data in the Lake Chad basin and the possible inaccessibility of current discharge measurements, hydrologists may be dependent on the Maiduguri rainfall for further water resources planning (Guganesharajah, 1981).

15.6   Rainfall Models

Techniques for modelling rainfall data are required, particularly for augmenting short duration rainfall records. For many problems, the hydrologist needs information on the incidence of rainfall and on high intensities, measured by recording instruments operating only for short periods.

15.6.1   Daily Series

The time series analysis techniques of the previous sections are not applicable to discontinuous series of data in which groups of non-zero values are interspersed irregularly with periods of zeros. In Fig. 15.2a, it is seen that annual and monthly rainfall series in the UK in general consist of non-zero totals of rainfall depths to which the time series techniques described could be applied, but the series of *daily* rainfall totals is discontinuous with irregularly spaced groups of rainy and dry days. The modelling of such discontinuous data requires a recognition of the radically different statistical properties of the zero and non-zero groups.

The main features of a daily rainfall record can be specified according to the following considerations:

(a) a day is WET or DRY (the definition of WET may be taken as a day with measurable rain or a day with a total greater than a selected threshold);

(b) the length of sequences of WET days and DRY days;

(c) the sequence of rainfall totals on the WET days.

The description of a daily rainfall time series thus requires the modelling of two distinct components: (i) the occurrence or non-occurrence of rainfall, namely the WET and DRY day sequences, and (ii) the rainfall amounts on the WET days.

Since the formation and incidence of rainfall varies according to weather types and is therefore largely dependent on season, there are variations in the pattern of rainfall during a year. Thus daily rainfall studies should be made on a seasonal or monthly basis over several years. The seasonal basis is defined in northwest Europe as follows:

Spring — March, April, May,
Summer — June, July, August,
Autumn — September, October, November,
Winter — December, January, February.

There have been many studies in the past decade with different rainfall series modelling methods, and only an outline of those most successfully applied in the UK type climate will be given (Cole & Sherriff, 1972; Buishand, 1977; Kottegoda & Horder, 1980).

The WET – DRY day sequences.   There are two approaches to this problem: the occurrence of a wet or dry day can be considered to be dependent on the situation of the previous day (or days) or it can be assumed independent.

(a) Markov chains can relate the probability of occurrence of an event (a wet day) to the state of the previous day (Markov order 1) or to n previous days (Markov order n). This principle was first applied to probabilities of daily rainfall occurrence in Israel (Gabriel & Neumann, 1962). First- and second-order Markov models have been used successfully when applied to the two states, wet, or dry, taken together. Expressed in the probabilistic form, a matrix of transition probabilities is built up from the historic record, e.g. for a first-order Markov chain:

		Next state, t	
		Wet (W)	Dry (D)
Initial	Wet (W)	a	$1 - a$
state, $(t - 1)$	Dry (D)	$1 - b$	b

in which if the series of daily rainfalls are X_t, then:

Prob. $(X_t = W | X_{t-1} = W) = a$,
Prob. $(X_t = D | X_{t-1} = D) = b$, etc

This approach could have the advantage of being able to model a sequence of wet days in correspondence with a weather pattern producing rainfall which may extend over 5 or 6 days, but, however, the number of parameters needed to define the higher-order model required then becomes impracticable.

(b) The second approach assumes that the lengths of the wet and dry spells are independent. Of the several probability distributions conforming to this assumption of independence, the truncated negative binomial distribution (TNBD) has been found to be the most suitable.

If $X = k$, the length of a *wet spell* then:

$$P(X = k) = \binom{k + r - 1}{k} \frac{p^r q^k}{1 - p^r}, \; k = 1, 2, 3, \ldots \text{ days}$$

with p the probability of a wet day $0 < p \leqslant 1$, $r \geqslant -1$, $r \neq 0$ and $q = (1 - p)$ the probability of a dry day (ending the wet spell).

The expression may also be used to fit *dry spells*, then p and q are probabilities of dry and wet days, respectively. When fitting the model on a monthly basis, there are then 12×2 parameters (p and r) for each type of spell (wet or dry). On a seasonal basis, there are only 8 parameters. In the application of this model to a Dutch record, it was found suitable to keep the monthly values of the parameter r the same throughout the year. For dry spells r was 0.246 and for wet spells, 0.656. This reduced the number of parameters to be determined for each month or season. The TNBD distribution provided a good fit for the lengths of both wet and dry spells (Buishand, 1978). In another study, with constant r and 4 seasonal values of p for either application, the TNBD distribution was acceptable for both wet and dry spells (Kottegoda & Horder, 1980).

The rainfall amounts on wet days.   The rainfall depths on sequential days within a wet spell may be interrelated especially when associated with a particular weather pattern. In northwest Europe there is a serial correlation coefficient of only 0.10 to 0.15 between successive daily rainfalls during winter months, whereas in summer, rainfall on wet days is random. Thus the direct use of a Markov model is not practicable. Taking all wet days together, the frequency distribution of rainfall amounts is the reverse J shape (see Chapter 10), but if the wet day is defined as a day with rainfall amounts above a small threshold then the rainfall amounts minus the threshold value may be fitted by the Γ distribution. The parameters of the Γ distribution are determined over monthly or seasonal periods.

Considerable developments in daily rainfall modelling using a Markov chain for occurrence and gamma distributions for amounts have been made in the investigation of the probabilities of rainfall quantities in rain spells and of length of dry spells for application in agricultural practices in West Africa and India (Stern, 1980).

Another method for determining the frequency of rainfall amounts considers separately the falls on isolated wet days, on wet days bounded by a wet and dry day and on wet days bounded by wet days. The frequencies of daily rainfall totals in the three categories are extracted from the historic record on a monthly or seasonal basis and probability distributions fitted. There is a marked increase in the mean daily rainfall in the three categories from isolated wet days to wet days bounded by wet days (Buishand, 1978).

The wet days can also be grouped according to rainfall amounts and different sequences of grouped amounts studied. This method has been applied satisfactorily by several workers (Pattison, 1965; Sherriff, 1970).

To generate a daily rainfall sequence, a pattern of wet and dry spells is synthesized from the most satisfactory model obtained from the historic data record. The model may be a Markov chain of probabilities of occurrence of a wet or dry day when the sequences are related or a model assuming independence of wet and dry days. The parameters of the model used will have been determined from an analysis of the historic series. Then in the generated lengths of the wet spells, the model describing the probability distributions of daily rainfalls, with the parameters derived from the historic data, is applied to synthesize daily rainfall values. These are usually selected at random from the appropriate probability distribution. Model parameters are obtained initially on a seasonal or monthly basis but in some climatic circumstances, certain parameters may be applicable throughout a year.

15.6.2   Shorter Duration Series

The analysis of individual rainfall events on, say, an hourly basis can follow the same procedure as for daily rainfall. Of the many studies that have been made in recent years only a few examples can be mentioned.

Significant features of sequential rainfall events must be identified. With the rainfall occurrences represented in discrete time intervals, as in Fig. 15.9, the main characteristics of the rainfall pattern can be defined:

(a) Rainfall duration;
(b) Rainfall depth;
(c) Time between falls.

These are important rainfall characteristics which influence streamflow and therefore their analysis and modelling can be most helpful for relating to areas of sparse data. The time intervals are usually given in hours, but they could be for shorter periods if required. Examples of the rainfall characteristics from Fig. 15.9 are enumerated in Table 15.1. Similar data were abstracted from a 5-year record of hourly data and, on a monthly basis, all

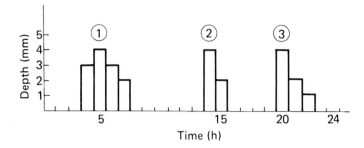

Fig. 15.9   Sequences of hourly rainfalls.

were found to be fitted by the two parameter Weibull distribution (Weldu, 1981). The probability density function for the latter is given by

$$f(x;\theta,\gamma) = \frac{\gamma}{\theta} x^{\gamma-1} \exp\left(-\frac{x}{\theta}\right)^{\gamma} \text{ for } x > 0$$

where γ and θ are the shape and scale parameters respectively. When $\gamma < 1$, the shape of the distribution is the reversed J shape which matches the frequency distributions of the rainfall variables. The fitting of this distribution was used to test the performance of data generating models.

Table 15.1   Rainfall Characteristics as Variables

'Storm' No.	Rainfall duration (h)	Rainfall depth (mm)	Time between falls (h)
1	4	12	
2	2	6	6
3	3	7	4

One very detailed analysis and modelling of rainfall took a time interval of 10 min, which is about the shortest period that can be read satisfactorily from standard autographic rainfall charts. The duration of rainfall was defined as the consecutive sequence of n periods of 10 min, each of which had a rainfall equal to or greater than 0.25 mm. The rainfall characteristics analysed were as before, the duration, total rainfall in an event and the time interval between falls, with the additional consideration of the distribution of rain within an event. Appropriate models were found from the statistics of the data series calculated on a monthly basis, and the final computer package generated durations, rainfall yields and intervals between events (Raudkivi & Lawgun, 1974).

Although the foregoing analyses of sequential rainfalls pertain to relatively minor rainfall events, the sequential falls within major storms that produce floods have been studied and modelled for catchments in North Carolina and

Illinois, USA (Ramaseshan, 1971). The storm duration was assumed to be 36 h, comparable to the largest recorded in the study area. The sample storms were matched in time by ensuring that the cross-correlation coefficients of their hourly rainfalls with one another were maximum values. This time shifting of the storms improved the statistical models. The rainfall sequences within the storms over the catchment areas conformed to a first order Markov model:

$$X_t = r_t X_{t-1} + \epsilon_t \text{ with } t = 2, 3, \ldots, 36$$

and with the random components ϵ_t lognormally distributed. It was found that the last 8 h of storm pattern contributed little to cumulative effects of the storm, and thus a duration of 28 h was deemed sufficient for modelling storms to relate to flood flows from the catchments studied.

15.7   River Flow Models

Time series analysis and synthesis can be used to model most river flow records, but there are several models based on other statistical techniques which have been applied successfully in solving hydrological problems.

15.7.1   Thomas – Fiering model

This is one of a group of regression models that works well with monthly flow data. Fig. 15.10 shows a regression analysis of q_{j+1} on q_j, pairs of successive monthly flows for the months $(j + 1)$ and j over the years of record where $j = 1, 2 \ldots 12$ (Jan, Feb, $\ldots$ Dec) and when $j = 12, j + 1 = 1 = $ Jan (there would be 12 such regressions). If the regression coefficient of month $j + 1$ on j is b_j, then the regression line value of a monthly flow, $\hat{q}_{j+1}$, can be determined from the previous months flow, q_j, by the equation:

$$\hat{q}_{j+1} = \bar{q}_{j+1} + b_j(q_j - \bar{q}_j)$$

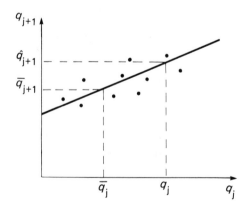

Fig. 15.10   Thomas – Fiering model.

To account for the variability in the plotted points about the regression line reflecting the variance of the measured data about the regression line, a further component is added:

$$Z.s_{j+1}\sqrt{(1-r_j^2)}$$

where s_{j+1} is the standard deviation of the flows in month $j+1$, r_j is the correlation coefficient between flows in months $j+1$ and j throughout the record, and $Z = N(0, 1)$, a normally distributed random deviate with zero mean and unit standard deviation.

To generate a continuous sequence of n years of synthesized monthly flows, q_i, where $i = 1, 2, \ldots 12\,n$ with each i value having a corresponding j value, $j = 1, 2 \ldots 12$, and starting from the last flow value of the historic data, then:

$$q_{i+1} = \bar{q}_{j+1} + b_j(q_i - \bar{q}_j) + Z_i s_{j+1}\sqrt{(1-r_j^2)}$$

with $b_j = r_j \times s_{j+1}/s_j$, there are 36 parameters for the monthly model ($\bar{q}$, r and s for each month) but these are simple to compute and the model is easy to apply.

The Thomas – Fiering (1962) model was applied to the 82 year record of monthly flows for the River Thames at Teddington Weir. From 25 generated monthly sequences, each of 82 years length, the monthly means and standard deviations were calculated and are plotted and compared with the historic values (Fig. 15.11).

The means and standard deviations are well preserved when the individual monthly flows are normally distributed. When the frequency distributions of the monthly flows are skewed, they can often be normalized by taking the logarithms of each of the values before analysis, as was done for Fig. 15.11 (Hall, 1970).

15.7.2   ARMA models

Within the last decade, a family of models has come into prominence in stochastic hydrology known as Auto-Regressive Moving Average (ARMA) models. They combine any direct serial correlation properties of a data series with the smoothing effects of an updated running mean through the series. The two components of the model for a data series X_t, e.g. annual river flows, are described by:

Auto-regression (AR(p))

$$X_t = \alpha_1 X_{t-1} + \alpha_2 X_{t-2} + \ldots \ldots \alpha_p X_{t-p} + Z_t$$

Moving average (MA(q))

$$X_t = Z_t + \beta_1 Z_{t-1} + \beta_2 Z_{t-2} + \ldots \ldots \beta_q Z_{t-q}$$

where Z_t are random numbers with zero mean and variance σ_Z^2 (Chatfield, 1980). If the first-order terms of each component are combined, the series X_t may be expressed by:

$$X_t = \alpha_1 X_{t-1} + Z_t + \beta_1 z_{t-1} \qquad (-1 < \alpha_1 < 1) \text{ and } (-1 < \beta_1 < 1)$$

Estimates of the parameters α_1 and β_1 can be obtained from the serial corre-

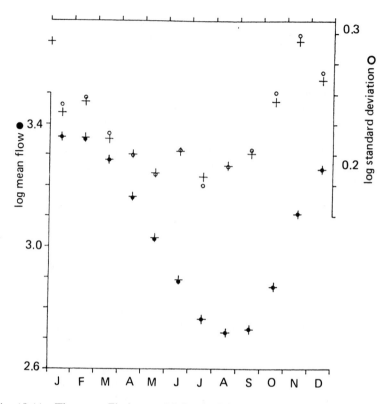

Fig. 15.11   Thomas – Fiering model for the River Thames at Teddington Weir. (Reproduced by permission of M.J. Hall.)

lation coefficients r_1 and r_2 of lags 1 and 2, respectively:

$$r_1 = \frac{(1 - \alpha_1\beta_1)(\alpha_1 - \beta_1)}{1 + \beta_1^2 - 2\alpha_1\beta_1}$$

$$r_2 = \alpha_1 r_1$$

More efficient methods of estimating ARMA parameters are to be found in advanced texts (e.g. Box & Jenkins, 1970).

It has been shown that such a model preserves features of long-term persistence in series of data, with large values of α_1 (0.80 – 0.99) giving a long 'memory' in the interrelated sequential values (O'Connell, 1974). The ARMA (p, q) models do not need to have many terms in the component sequences; with values of $p = 1$ and $q = 1$, series of annual river flows are well fitted. By determining the parameters from a limited historic record and using a random number generator, many sequences of comparable synthetic flows can be generated and required statistics abstracted from the greatly extended data sample for designing water resource systems.

A recent overview of stochastic hydrology has been given by one of the foremost exponents of the subject (Yevjevich, 1987).

References

Box, G.E.P. and Jenkins, G.M. (1970). *Time Series Analysis, Forecasting and Control*. Holden-Day.

Buishand, T.A. (1977). *Stochastic Modelling of Daily Rainfall Sequences*. Agricultural Univ. Wageningen Report 77.3, 211 pp.

Buishand, T.A. (1978). 'Some remarks on the use of daily rainfall models'. *J. Hydro.*, **36**, 295–308.

Chatfield, C. (1980). *The Analysis of Time Series; An Introduction*. Chapman & Hall, 2nd ed, 268 pp.

Cole, J.A. and Sherriff, J.D.F. (1972). 'Some single and multiple models of rainfall within discrete time increments'. *J. Hydrol.*, **17**, 97–113.

Gabriel, K.K. and Neumann, J. (1962). 'A Markov chain model for daily rainfall occurrence at Tel. Aviv.' *Quart. Jl. Roy. Met. Soc.*, **88**, 90–95.

Guganesharajah, K. (1981). 'Analysis of the hydrology of Chad Basin.' Unpublished M.Sc. dissertation, Imperial College.

Hall, M.J. (1970). 'The analysis and generation of hydrological time series.' Unpublished notes, post experience course, Imperial College.

Hall, M.J. and O'Connell, P.E. (1972). 'Time series analysis of mean daily river flows.' *Water and Water Engng.*, April, 125–133.

Kisiel, C.C (1969). 'Time series analysis of hydrologic data,' in *Advances in Hydroscience*, Vol. 5, Academic Press, pp. 1–119.

Kottegoda, N.T. and Horder, M.A. (1980). 'Daily flow model based on rainfall occurrences using pulses and a transfer function.' *J Hydrol.*, **47**, 215–234.

O'Connell, P.E. (1974). 'A simple stochastic modelling of Hurst's Law.' IAHS Pub. No. 100. Symposium, Mathematical Models in Hydrology, pp. 169–187.

O'Donnell, T., Hall, M.J. and O'Connell, P.E. (1972). 'Some applications of stochastic hydrological models.' *Int. Symp. Mathematical Modelling Techniques in Water Resource Systems*, Vol. 1, *Environment Canada*, pp. 227–239.

Pattison, A. (1965). 'Synthesis of rainfall data.' *Water Res. Res.*, **1**, 489–498.

Ramaseshan, S. (1971). 'Stochastic modelling of storm precipitation'. *Nordic Hydrol.*, **2**, 109–129.

Raudkivi, A.J. and Lawgun, N. (1974). 'Simulation of rainfall sequences.' *J. Hydrol.*, **22**, 271–294.

Roesner, L.A. and Yevjevich, V.M. (1966). *Mathematical Models for Time Series of Monthly Precipitation and Monthly Runoff*. Hydrological Paper No. 15, Colorado State Univ., 50 pp.

Sherriff, J.D.F. (1970). *Synthetic Rainfall Sequences*. Water Research Assoc. TP 72, 65 pp.

Stern, R.D. (1980). 'The calculation of probability distributions for models of daily precipitation.' *Arch. Met. Geoph. Biokl. Ser.* **B28**, 137–67.

Thomas, H.A. and Fiering, M.B. (1962). 'Mathematical synthesis of streamflow sequences for the analysis of river basins by simulation'. Chapter 12 in *Design of Water Resource Systems*, ed. Maass *et al.*, Harvard Univ. Press, 620 pp.

Weldu, F. (1981). 'The probability distribution of storm rainfall variables'. Unpublished M.Sc. dissertation, Imperial College.

Yevjevich, V. (1987). 'Stochastic Models in Hydrology.' *Stochastic Hydrol. Hydraul.* **1**, 17–36.

Part III
Engineering Applications

16

Flood Routing

One of the most common problems facing a practising civil engineer is the estimation of the hydrograph of the rise and fall of a river at any given point on the river during the course of a flood event. The problem is solved by the techniques of *flood routing*, which is the process of following the behaviour of a flood hydrograph upstream or downstream from one point to another point on the river. It is more usual (and more natural) to work downstream from an upstream point where a flood hydrograph is specified; the reverse procedure against the flow is more complex and is less often required.

A flood hydrograph is modified in two ways as the storm water flows downstream. Firstly, and obviously, the time of the peak rate of flow occurs later at downstream points. This is known as *translation*. Secondly, the magnitude of the peak rate of flow is diminished at downstream points, the shape of the hydrograph flattens out, and the volume of flood water takes longer to pass a lower section. This modification to the hydrograph is called *attenuation* (Fig. 16.1).

The derivation of downstream hydrographs like B in Fig. 16.1 from an upstream known flood pattern A is essential for river managers concerned with forecasting floods in the lower parts of a river basin. The design engineer also needs to be able to route flood hydrographs in assessing the capacity of reservoir spillways, in designing flood-protection schemes or in evaluating the span and height of bridges or other river structures. In any situation where it is planned to modify the channel of a river, it is necessary to know the likely effect on the shape of the flood hydrograph in addition to that on the peak stage, i.e. the whole hydrograph of water passing through a section, not just the peak instantaneous rate.

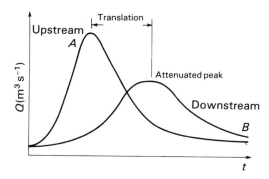

Fig. 16.1   Flood translation and attenuation.

Flood routing methods may be divided into two main categories differing in their fundamental approaches to the problem. One category of methods uses the principle of continuity and a relationship between discharge and the temporary storage of excess volumes of water during the flood period. The calculations are relatively simple and reasonably accurate and, from the hydrologist's point of view, generally give satisfactory results. The second category of methods, favoured by hydraulicians, adopts the more rigorous equations of motion for unsteady flow in open channels, but in the complex calculations, assumptions and approximations are often necessary, and some of the terms of the dynamic equation must be omitted in certain circumstances to obtain solutions.

The choice of method depends very much on the nature of the problem and the data available. Flood routing computations are more easily carried out for a single reach of river that has no tributaries joining it between the two ends of the reach. According to the length of reach and the magnitude of the flood event being considered, it may be necessary to assess contributions to the river from lateral inflow, i.e. seepage or overland flow draining from, and distributed along, the banks. Further, river networks can be very complex systems, and in routing a flood down a main channel, the calculations must be done for separate reaches with additional hydrographs being introduced for major tributaries. In order to develop an operational flood routing procedure for a major river system, detailed knowledge of the main stream and the various feeder channels is necessary. In addition, the experience of several major flood events with discharge measurements made at strategic points on the drainage network is an essential requirement.

In this chapter a selection of flood routing methods will be presented, beginning with the simplest using the minimum of information and progressing through to the more complex methods requiring the larger digital computers for their application.

16.1   Simple Non-storage Routing

It has been said that 'engineering is the solution of practical problems with insufficient data'. If there are no gauging stations on the problem river and therefore no measurements of discharge, the engineer may have to make do with *stage* measurements. In such circumstances it is usually the flood peaks that have been recorded, and indeed it is common to find the people living alongside a river have marked on a wall or bridge pier the heights reached by notable floods. Hence the derivation of a relationship between peak stages at upstream and downstream points on a single river reach may be made (Fig. 16.2) when it is known that the floods are caused by similar notable conditions.

This is a very approximate method, and there should be no major tributaries and very low lateral inflows between the points with the stage measurements. However, with enough stage records it may be possible to fit a curve to the relationship to give satisfactory forecasts of the downstream peak stage from an upstream peak stage measurement.

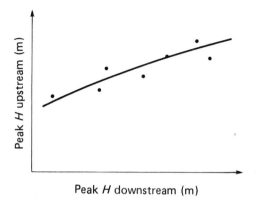

Fig. 16.2   Peak stage relationship.

For example, on the River Irrawaddy in Burma a linear relationship exists between the peak stages of an upstream gauging station at Nyaung Oo and a station at Prome, 345 km downstream. 35 comparable stages (m) for irregular flood events over 5 years (1965–69) are shown in Fig. 16.3. An equation $H_D = 1.3 \, H_U + 1.4$ relating H_D, the downstream stage to H_U, upstream stage, can then give forecast values of H_D from H_U (Maung, 1973).

The time of travel of the hydrograph crest (peak flow) also needs to be determined; curves of upstream stage plotted against time of travel to the required downstream point can be compiled from the experience of several flood events. (The time of travel of the flood peaks between Nyaung Oo and Prome on the Irrawaddy ranged from 1 to 4 days.)

A typical stage – time of travel plot in Fig. 16.4 shows the time of travel at a minimum within the stage range; this occurs when the bankfull capacity of

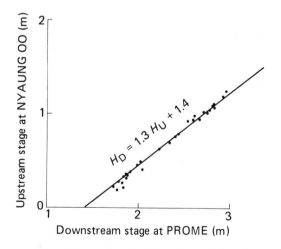

Fig. 16.3   Comparable peak stages on the River Irrawaddy.

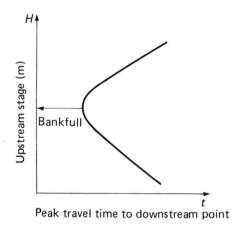

Fig. 16.4   Flood peak travel times.

the river channel is reached. After reaching a minimum at this bankfull stage the time of travel tends to increase again as the flood peak spreads over the flood plain and its downstream progress is retarded owing to storage effects on the flood plain.

The complexities of rainfall – runoff relationships are such that these simple methods allow only for average conditions. Flood events can have very many different causes that produce flood hydrographs of different shapes. Flood hydrographs at an upstream point, with peaks of the same magnitude but containing different flood volumes, in travelling downstream will produce different peaks at a downstream point. Modifications to the flow by the channel conditions will differ between steep, peaky flood hydrographs and gentle fat hydrographs with the same peak discharges.

The principal advantages of these simple methods are that they can be developed for stations with only stage measurements and no rating curve, and they are quick and easy to apply especially for warning of impending flood inundations when the required answers are immediately given in stage heights. The advantages of speed and simplicity are less important now that fast computers are available and more accurate and comprehensive real-time techniques can be used.

16.2   Storage Routing

When a storm event occurs, an increased amount of water flows down the river channel and in any one short reach of the channel there is a greater volume of water than usual contained in temporary storage. If at the beginning of the reach the flood hydrograph (above a normal flow) is given as I, the inflow (Fig. 16.5), then during the period of the flood, T_1, the channel reach has received the flood volume given by the area under the I hydrograph. Similarly, at the lower end of the reach, with an outflow hydrograph O, the flood volume is again given by the area under the curve. In a flood situation relative quantities may be such that lateral and tributary inflows can be neglected, and thus, by the principle of continuity, the volume of inflow

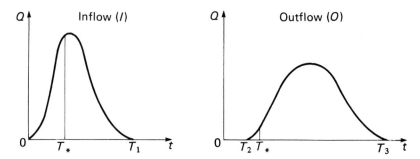

Fig. 16.5   Flood hydrographs for a river reach.

equals the volume of outflow, i.e. the flood volume $V = \int_0^{T_1} I \, dt = \int_{T_2}^{T_3} O \, dt$.

At some intermediate time T_*, in Fig. 16.5, an amount $\int_0^{T_*} I \, dt$ has entered the reach and an amount $\int_0^{T_*} O \, dt$ has left the reach. The difference must be stored within the reach, so the amount of storage, S_*, within the reach at time $t = T_*$ is given by:

$$S_* = \int_0^{T_*} (I - O) \, dt \tag{16.1}$$

where I and O are the corresponding rates of inflow and outflow.

An alternative statement of this equation is that the rate of change of storage within the reach at any instant is given by:

$$\frac{dS}{dt} = I - O \tag{16.1}$$

This, *the continuity equation*, forms the basis of all the storage routing methods. The routing problem consists of finding O as a function of time, given I as a function of time, and having information or making assumptions about S. Equation 16.1 cannot be solved directly.

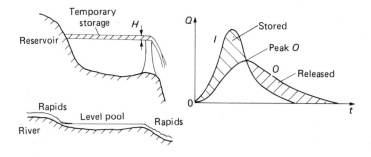

Fig. 16.6   Level pool routing.

Any procedure for routing a hydrograph generally has to adopt a finite difference technique. Choosing a suitable time interval for the routing period, ΔT, the continuity equation in its finite difference form becomes:

$$\frac{(I_1 + I_2)\,\Delta T}{2} - \frac{(O_1 + O_2)\,\Delta T}{2} = S_2 - S_1 \qquad (16.2)$$

The routing period, ΔT, has to be chosen small enough such that the assumption of a linear change of flow rates, I and O, during ΔT is acceptable (as a guide, ΔT should be less then $\frac{1}{6}$ of the time of rise of the inflow hydrograph). The subscripts 1 and 2 refer to the start and end of any ΔT time step. At the beginning of a time step, all values are known except O_2 and S_2. Thus with two unknowns, a second equation is needed to solve for O_2 at the end of a time step. A second equation is obtained by relating S to O alone, or to I and O together. The two equations are then used recursively to find sequential values of O through the necessary number of ΔT intervals until the outflow hydrograph can be fully defined. It is the nature of the second equation for the storage relationship that distinguishes two methods of storage routing.

16.2.1   Reservoir or Level Pool Routing

For a reservoir or a river reach with a determinate control upstream of which a nearly level pool is formed, the temporary storage can be evaluated from the topographical dimensions of the 'reservoir', assuming a horizontal water surface (Fig. 16.6). For a level pool, the temporary storage, S, is directly and uniquely related to the head, H, of water over the crest of the control. The discharge from the 'pool' is also directly and uniquely related to H. Hence S is indirectly but uniquely a function of O.

It is convenient to rearrange Equation 16.2 to get the unknowns S_2 and O_2 on one side of the equation and to adjust the O_1 term to produce:

$$\left(\frac{S_2}{\Delta T} + \frac{O_2}{2}\right) = \left(\frac{S_1}{\Delta T} + \frac{O_1}{2}\right) + \frac{I_1 + I_2}{2} - O_1 \qquad (16.3)$$

Since S is a function of O, $[(S/\Delta T) + (O/2)]$ is also a specific function of O (for a given ΔT), as in Fig. 16.7. Replacing $[(S/\Delta T) + (O/2)]$ by G, for simplification, Equation 16.3 can be written:

$$G_2 = G_1 + I_m - O_1 \qquad (16.3a)$$

where $I_m = (I_1 + I_2)/2$. Fig. 16.7 defines the relationship between O and $G = (S/\Delta T + O/2)$, and this curve needs to be determined by using the common variable H to fix values of S and O and then G, for a specific ΔT.

Equation 16.3a and the auxiliary curve of Fig. 16.7 now provide an elegant and rapid step-by-step solution. At the beginning of a step, G_1 and O_1 are known from the previous step (or from conditions prior to the flood for the first step). I_m is also known from the given inflow hydrograph. Thus all three known terms in Equation 16.3a immediately lead to G_2 at the end of the time step, and then to O_2 from Fig. 16.7.

The recursive calculation of values for the outflow hydrograph is best carried out in a tabular solution (Table 16.1). The calculation starts with all the inflows known (column a). The mean inflows during the time intervals

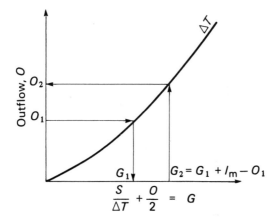

Fig. 16.7   Outflow, O, versus $(S/\Delta T + O/2)$.

ΔT given by $I_m = \frac{1}{2}(I_1 + I_2)$, are entered in column b. The initial outflow O_1 is known from the discharge before the flood (column c). Thus the first value in column e, G_1, may be determined from Fig. 16.7. The mean inflow during the first time interval less the initial outflow gives the first entry in column d and this is added to the first column e value to give the second entry in column e, G_2, (see Equation 16.3a). Then the second value of the outflow hydrograph O_2 can be read off Fig. 16.7. This now becomes the O_1 value for the next interval. The sequence of calculations is continued until the outflow hydrograph is completed or until the required outflow discharges are known.

A useful check on the validity of any level pool routing calculation is that the peak of the outflow hydrograph should occur at the intersection of the inflow and outflow hydrographs on the same plot (Fig. 16.6). At that point, $I = 0$, so $dS/dt = 0$, i.e. storage is a maximum and therefore O is a maximum. Thereafter, the temporary storage is depleted.

Table 16.1

T	a I	b I_m	c O	d $(I_m - O)$	e G
1	I_1	$(I_1 + I_2)/2$	O_1	$\{(I_1 + I_2)/2\} - O_1$	G_1
2	I_2		O_2		G_2
⋮	⋮				
i	I_i	$(I_i + I_j)/2$	O_i	$\{(I_i + I_j)/2\} - O_i$	G_i
j	I_j		O_j		G_j

Example.   Discharge from a reservoir is over a spillway with discharge characteristic $Q(\text{m}^3\text{s}^{-1}) = 110\ H^{1.5}$, where H m is the head over the spillway crest. The reservoir surface area is 7.5 km² at spillway crest level and increases linearly by 1.5 km² per metre rise of water level above crest level. The design storm inflow, assumed to start with the reservoir just full, is given by a triangular hydrograph, base length 36 h and a peak flow of 360 m³s⁻¹ occurring 12 h after start of inflow. Estimate the peak outflow over the spillway and its time of occurrence relative to the start of the inflow (from Scott-Moncrieff, 1977).

Solution.   A level water surface in the reservoir is assumed. Temporary storage above crest level is given by:

$$S = \int_0^H A\,dh = 10^6 \int_0^H (7.5 + 1.5h)\,dh\ \text{m}^3$$
$$= 10^6\ (7.5H + 0.75H^2)\ \text{m}^3$$

Outflows over the crest are given by $O = 110\ H^{1.5}$ m³s⁻¹. Take a value of $\Delta T = 2$ h $= 7200$ s. Then:

$$G = \left(\frac{S}{\Delta T} + \frac{O}{2}\right) = \frac{10^6}{7200}\ (7.5H + 0.75H^2) + 55H^{1.5}\ \text{m}^3\text{s}^{-1}$$

i.e.:

$$G = 104H\ (10 + 0.53\ \sqrt{H} + H)\ \text{m}^3\text{s}^{-1}$$

Table 16.2   Derivation of O and G for the auxiliary curve of O versus G

H	O	$104H$	$0.52\sqrt{H}$	$(10 + 0.53\sqrt{H} + H)$	G
0.2	10	20.8	0.24	10.26	213
0.4	28	41.6	0.34	10.38	432
0.6	51	62.4	0.41	10.47	653
0.8	79	83.2	0.47	10.55	878
1.0	110	104.0	0.53	11.53	1199
1.2	144	124.8	0.58	11.78	1470
1.4	182	145.6	0.63	12.03	1752
1.6	223	166.4	0.67	12.27	2042
1.8	265	187.0	0.71	12.51	2339
2.0	312	208.0	0.75	12.75	2652

The derivation of O and G values are shown in Table 16.2. O is plotted against G and a curve drawn through the points (Fig. 16.8). (Note that the relationship is for $\Delta T = 2$ h.)

The inflows into the reservoir are evaluated for each 2-h period from the beginning of the storm inflow and the outflows are computed recursively using the storage equation (Equation 16.3a and the auxiliary curve). The results are given in Table 16.3 and plotted on Fig. 16.9. The peak outflow over the spillway is 180 m³s⁻¹ and it occurs 24 h after the commencement of the inflow. The approach to the peak outflow becomes obvious during the

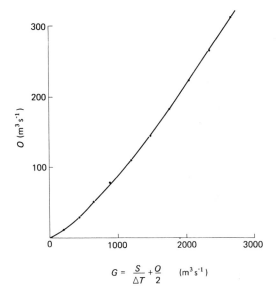

$$G = \frac{S}{\Delta T} + \frac{Q}{2} \quad (\text{m}^3\,\text{s}^{-1})$$

Fig. 16.8   Auxiliary curve. Outflow v, G (for $\Delta T = 2$h).

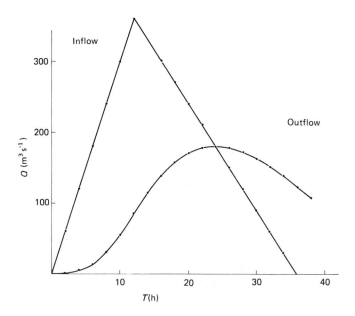

Fig. 16.9   Outflow over reservoir spillway.

Table 16.3

T	I	I_m	O	$I_m - O$	G
0	0		0		0
		30		30	
2	60		1		30
		90		89	
4	120		6		119
		150		144	
6	180		14		163
		210		196	
8	240		31		459
		270		239	
10	300		55		698
		330		275	
12	360		85		973
		345		260	
14	330		114		1233
		315		201	
16	300		138		1434
		285		147	
18	270		158		1581
		255		97	
20	240		171		1678
		225		54	
22	210		178		1732
		195		17	
24	180		180		1749
		165		− 15	
26	150		178		1734
		135		− 43	
28	120		173		1691
		105		− 68	
30	90		163		1623
		75		− 88	
32	60		152		1535
		45		− 107	
34	30		139		1428
		15		− 124	
36	0		123		1304
		0		− 123	
38			108		1181

computations as the net inflow rate $(I_m - O)$ declines, and after the maximum O is reached, $(I_m - O)$ becomes negative.

16.2.2   River Routing

For a river channel reach where the water surface cannot be assumed horizontal, the stored volume becomes a function of the stages at both ends of the reach, and not at the downstream (outflow) end only.

In a typical reach, the different components of storage may be defined for a given instant in time as in Fig. 16.10.

Again, the continuity equation holds at any given time:

$$\frac{dS}{dt} = I - O \tag{16.1}$$

where the total storage, S, is the sum of prism storage and wedge storage. The prism storage is taken to be a direct function of the stage at the downstream end of the reach; the simple assumption ignores the effects of the slope of the water surface and takes the downstream stage and the outflow to be uniquely related, and thus the prism storage to be a function of the outflow, O. The wedge storage exists because the inflow, I, differs from O and so may be assumed to be a function of the difference between inflow and outflow, $(I - O)$.

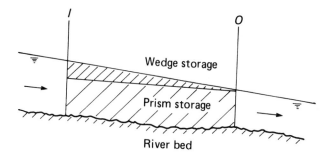

Fig. 16.10   River reach storages.

Three possible conditions for wedge storage are shown in Fig. 16.11: during the rising stage of a flood in the reach, $I > O$, and the wedge storage must be added to the prism storage; during the falling stage, $I < O$, and the wedge storage is negative to be subtracted from the prism storage to obtain the total storage. The total storage, S, may then be represented by:

$$S = f_1(O) + f_2(I - O) \tag{16.4}$$

with due regard being paid to the sign of the f_2 term.

In the level pool method, S was a function only of O. Here it is a more complex function involving I as well as O at the ends of a river reach. However, there are again two equations, Equation 16.1 and Equation 16.4 and again, it should be possible using a finite-difference method to solve for the unknown, S_2 (Equation 16.2), at the end of a routing interval, ΔT. One such method of solution will be described.

The Muskingum Method.   McCarthy (1938) made the bold assumption that in Equation 16.4, $f_1(O)$ and $f_2(I - O)$ could be both straight-line functions, i.e. $f_1(O) = K.O$ and $f_2(I - O) = b(I - O)$. Thus:

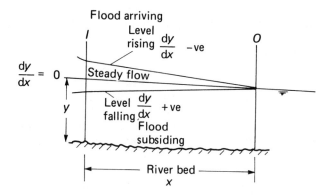

Fig. 16.11   Storage and non-steady flow.

$$S = bI + (K - b)O$$
$$= K\left[\frac{b}{K}I + \left(1 - \frac{b}{K}\right)O\right] \tag{16.5}$$

and writing $x = \dfrac{b}{K}$, Equation 16.5 becomes:

$$S = K[xI + (1 - x)O] \tag{16.6}$$

Thus x is a dimensionless weighting factor indicating the relative importance of I and O in determining the storage in the reach. The value of x has limits of zero and 0.5, with typical values in the range 0.2 to 0.4. K has the dimension of time.

Substituting for S_2 and S_1 in the finite difference form of the continuity equation (Equation 16.2):

$$\tfrac{1}{2}(I_1 + I_2)\,\Delta T - \tfrac{1}{2}(O_1 + O_2)\,\Delta T = K[xI_2 + (1 - x)O_2] - K[xI_1 + (1 - x)O_1]$$

Collecting terms in O_2, the unknown outflow, on to the LHS:

$$O_2(-0.5\Delta T - K + Kx) = I_1(-Kx - 0.5\Delta T) + I_2(Kx - 0.5\Delta T) + O_1(-K + Kx + 0.5\Delta T)$$

Then:

$$O_2 = c_1 I_1 + c_2 I_2 + c_3 O_1 \tag{16.7}$$

where:

$$\left.\begin{aligned}
c_1 &= \frac{\Delta T + 2Kx}{\Delta T + 2K - 2Kx}\\[2mm]
c_2 &= \frac{\Delta T - 2Kx}{\Delta T + 2K - 2Kx}\\[2mm]
c_3 &= \frac{-\Delta T + 2K - 2Kx}{\Delta T + 2K - 2Kx}
\end{aligned}\right\} \tag{16.8}$$

$\Sigma c = 1$ and thus when c_1 and c_2 have been found $c_3 = 1 - c_1 - c_2$. Thus the outflow at the end of a time step is the weighted sum of the starting inflow and outflow and the ending inflow, as per Equation 16.7.

Application of the Muskingum Method

(a) In order to use Equation 16.7 for O_2, given inflows, I_1 and I_2, and the value of outflow, O_1, at the ends of a time interval, ΔT, it is necessary to know K and x in order to calculate the c coefficients.

(b) Using recorded hydrographs of a flood at the beginning and end of the river reach, trial values of x are taken, and for each trial the weighted flows in the reach, $[xI + (1 - x)O]$, are plotted against actual storages determined from the inflow and outflow hydrographs as indicated in Fig. 16.12. Trial plots are shown in Fig. 16.13 a, b and c. The correct value for x will be the one giving the best approximation to a straight-line plot as implied by Equation 16.6. When the looping plots of the weighted discharges against storages have been narrowed down so

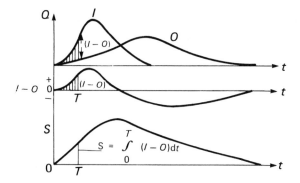

Fig. 16.12   Determining storage in a river reach.

that the values for the rising stage and the falling stage for a particular value of x merge together to form the best approximation to a straight line, then that x value is used, and the slope of the straight line gives the required value of K (Fig. 16.3c). For natural channels, the best plot is often curved, making a straight line slope difficult to estimate.

(c) The Muskingum coefficients c_1, c_2 and c_3 are next evaluated. Then beginning with the initial inflow and outflow, sequential values of O are computed from Equation 16.7.

Example.   The data in Table 16.4 represent observed inflow and outflow flood hydrographs for a channel reach. Derive the Muskingum constants K

Table 16.4

T(h)	0	12	24	36	48	60	72	84	96	108	120	132
$I(m^3 s^{-1})$	22	35	103	109	86	59	39	28	22	20	19	18
$O(m^3 s^{-1})$	22	21	34	55	75	85	80	64	44	30	22	20

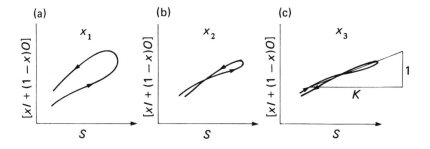

Fig. 16.13   Trial plots for Muskingum x values.

and x for the reach. Using the derived constants in the Muskingum equations, route the observed inflow hydrograph through the reach and plot the calculated outflow hydrograph obtained, comparing it with the observed hydrograph.

Solution. The application of the Muskingum method is best carried out by setting out the computations in a table (Table 16.5). ΔT in this example is taken as 12 h and trials of $x = 0.3$ and $x = 0.4$ are made.

Table 16.5

T (h)	I (m³s⁻¹)	O (m³s⁻¹)	$I - O$ (m³s⁻¹)	S_i (m³s⁻¹h)	$(0.3I + 0.70)$ (m³s⁻¹)	$(0.4I + 0.60)$ (m³s⁻¹)
0	22	22	0	0	22	22
12	35	21	14	168	25	27
24	103	34	69	996	55	62
36	109	55	54	1644	71	77
48	86	75	11	1776	78	75
60	59	85	− 26	1464	77	75
72	39	80	− 41	972	68	64
84	28	64	− 36	540	54	50
96	22	44	− 22	276	37	35
108	20	30	− 10	156	27	26
120	19	22	− 3	120	21	21
132	18	20	− 2	96	19	19

From an inspection of the plots of S_i against $[xI + (1 - x)O]$ in Fig. 16.14, the graph of the relationship with $x = 0.4$ is a reasonable approach to a straight line and such a line is drawn through the points. The slope of the line $K = 34$ h. The coefficients are now calculated:

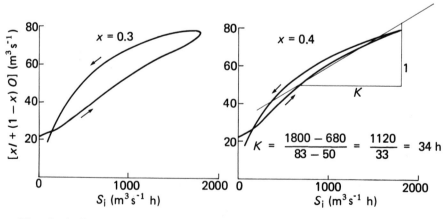

Fig. 16.14   Determining x (Muskingum example).

$$c_1 = \frac{\Delta T + 2Kx}{\Delta T + 2K - 2Kx} = \frac{12 + 2 \times 34 \times 0.4}{12 + 2 \times 34 - 2 \times 34 \times 0.4} = \frac{39.2}{52.8} = 0.74$$

$$c_2 = \frac{\Delta T - 2Kx}{\Delta T + 2K - 2Kx} = \frac{12 - 2 \times 34 \times 0.4}{12 + 2 \times 34 - 2 \times 34 \times 0.4} = \frac{-15.2}{52.8} = -0.29$$

$$c_3 = 1 - c_1 - c_2 = 1 - 0.74 + 0.29 = 0.55$$

$$\left(\text{Check } c_3 = \frac{-\Delta T + 2K - 2Kx}{\Delta T + 2K - 2Kx} = \frac{28.8}{52.8} = 0.55 \right)$$

Hence using $O_2 = 0.74I_1 - 0.29I_2 + 0.55O_1$ sequential values of outflow can be calculated (Table 16.6). The hydrographs are plotted in Fig. 16.15.

Table 16.6

t(h)	0	12	24	36	48	60	72	84	96	108	120	132
I(m³s⁻¹)	22	35	103	109	86	59	39	28	22	20	19	18
Routed O	22	18	6	48	82	92	83	66	51	38	30	25

It will be noted that the routed outflow hydrograph is not a particularly good reconstruction of the observed outflow. As often happens, the computed outflow appears to decrease initially. However, the positive error in the forecast peak is acceptable to the engineer and the timing of the peak is correct. The peak outflow does not lie on the inflow recession curve due to the effect of the wedge storage. At the intersection point where $I = O$ and therefore $dS/dt = 0$, at that moment S is a maximum, but since S is a function of I and O, the maximum value of S does not imply a maximum for O at that point.

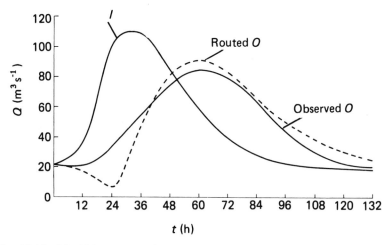

Fig. 16.15   Muskingum example.

16.3   Hydraulic Routing

The hydrological methods of flood routing that have been considered so far have been based on the principle of continuity plus a second equation linking storage to flow rates, but it can be appreciated that few rivers are truly sequences of level pools and that the Muskingum method in its conventional trial-and-error form is approximate as well as very tedious. It is thus reasonable to expect that civil engineers would attempt to solve the problems of flood routing in rivers by considering the hydraulics of open channel flow. The hydraulic methods of flood routing are based on the solution of the two basic differential equations governing gradually varying non-steady flow in open channels (the Saint Venant equations). Again, one of these is an expression of the continuity principle, but the second is no longer an empirical approximate equation.

16.3.1   The Continuity Equation

For a small length of channel, dx, (Fig. 16.16) and considering a small time interval, dt, from continuity:

$$\frac{\partial S}{\partial t} = Q - \left(Q + \frac{\partial Q}{\partial x}\, dx \right) \tag{16.9}$$

where Q is the inflow and $\left(Q + \dfrac{\partial Q}{\partial x}\, dx \right)$ is the outflow.

If A is the average cross-sectional area in the reach, then $S = A\, dx$, whence

$$\frac{\partial S}{\partial t} = \frac{\partial A}{\partial t}\, dx = -\frac{\partial Q}{\partial x}\, dx$$

(from Equation 16.9) and so

$$\frac{\partial Q}{\partial x} + \frac{\partial A}{\partial t} = 0 \tag{16.10}$$

Now $Q = A.v$, i.e. cross-sectional area times mean velocity, so differentiat-

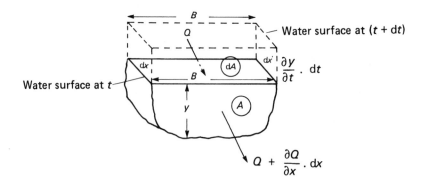

Fig. 16.16   Dimensional changes with time in a short river reach.

ing with respect to x and substituting in Equation 16.10 gives

$$A\frac{\partial v}{\partial x} + v\frac{\partial A}{\partial x} + \frac{\partial A}{\partial t} = 0 \tag{16.11}$$

If $\partial A = B\partial y$, where B is the width of the water surface, then

$$A\frac{\partial v}{\partial x} + Bv\frac{\partial y}{\partial x} + B\frac{\partial y}{\partial t} = 0 \tag{16.12}$$

16.3.2   The Dynamic Equation

The second fundamental equation can be derived from considering either the energy or momentum equation for the short length of channel, dx. In Fig. 16.17, v is a mean velocity and g is the gravity acceleration. The loss in head over the length of the reach, dx, has two main components:

$h_f = S_f dx$, the head loss due to friction, and:

$h_a = S_a dx = \dfrac{1}{g}\dfrac{\partial v}{\partial t}\, dx$, the head loss due to acceleration.

If it is assumed that the channel bed slope is small and the vertical component of the acceleration force is negligible, then the combined loss of head is $(h_f + h_a)$. Using the Bernoulli expression for total head, H:

$$H = z + y + \frac{v^2}{2g} \tag{16.13}$$

then the change in H over dx is $-dH = h_f + h_a$. Thus:

$$\frac{dH}{dx} = -S_f - S_a = \frac{d}{dx}\left[z + y + \frac{v^2}{2g} \right] \tag{16.13}$$

whence

$$S_f = \frac{-\partial z}{\partial x} - \frac{\partial y}{\partial x} - \frac{v\partial v}{g\partial x} - \frac{1}{g}\frac{\partial v}{\partial t}$$

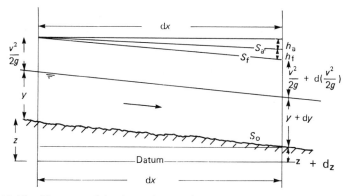

Fig. 16.17   Elements of the dynamic equation.

i.e.

$$S_f = S_0 - \frac{\partial y}{\partial x} - \frac{v \partial v}{g \partial x} - \frac{1}{g}\frac{\partial v}{\partial t} \tag{16.14}$$

The friction slope, S_f, can be determined from the Chezy formula $v = C\sqrt{(RS_f)}$, where R is the hydraulic mean radius given by cross sectional area divided by wetted perimeter ($R = A/P$) and C is a coefficient depending on the nature of the channel and the flow. It is assumed to be valid for non-steady flow. Then:

$$S_f = \frac{v^2}{C^2 R} \tag{16.15}$$

can be used in Equation 16.14. Manning's equation can also be used, when $S_f = v^2 n^2 / R^{\frac{4}{3}}$, with n the Manning roughness coefficient. For more rigorous derivations of the equations describing unsteady open channel flow, the reader is referred to the hydraulic texts listed at the end of the chapter (e.g. Chow, 1959; Henderson, 1966).

The application of the hydraulic equations (Equations 16.12 and 16.14, also known as the Saint Venant equations) in flood routing by their integration down the length of a channel is difficult and complex. Many methods for obtaining numerical finite difference solutions of the equations abound in the hydraulics literature. Two of the simpler and more usual techniques are the method of finite increments and the method of characteristics (Amein, 1966). For simplification of the dimensional terms, the channel geometry is generally reduced by assuming a rectangular cross-section with a uniform bed profile. Numerical solutions using finite difference techniques are being used increasingly with the help of computers and several flood routing packages based on these techniques are now readily available for application to real problems. One such package is the FLOUT procedure mentioned in Chapter 14.

16.4   Developments in Flood Routing Practice

The expansion of the practice of engineering hydrology in the UK in the last 15 years has stimulated research in the application of flood routing techniques (Price, 1974). An excellent review of methods is given in Volume III of the *Flood Studies Report* (NERC, 1975) together with examples of their use on British rivers.

The necessary discharge data for a series of points down a river necessary to calibrate any given flood routing method are often lacking, but since the establishment of more gauging stations after the Water Resources Act, 1963, it has generally been possible to apply the standard methods and their refinements to good effect in the UK. In particular, the routing of flood flows down a river system (essential in any river regulation scheme) has been demonstrated successfully on the Rivers Dee, Severn and Wye in Wales.

16.4.1   *Muskingum – Cunge*

One of the modified methods that has worked well is the Muskingum – Cunge method (Cunge, 1969). As shown previously, the Muskingum method is based on the storage equation with the coefficients K and x derived by trial and error. Cunge showed that K and x could be determined by considering the hydraulics of the flow.

From $S = K[xI + (1 - x)O]$ and $dS/dt = I - O$ the differential with respect to t gives

$$K \frac{d}{dt} [xI + (1 - x)O] = I - O$$

If this is expressed in finite-difference form with subscripts 1 and 2 representing the beginning and end of a time increment, ΔT, then

$$\frac{K}{\Delta T} x (I_2 - I_1) + \frac{K}{\Delta T} (1 - x) (O_2 - O_1) = \frac{(I_2 + I_1)}{2} - \frac{(O_2 + O_1)}{2}$$

(16.16)

K may be shown to be approximately equal to the time of travel of a flood wave through the reach and this assumption was used by Cunge (1969), i.e. $K = \Delta L/c$, where c is the average speed of the flood peak and ΔL is the length of the river reach. Substituting for K in Equation 16.16 gives:

$$\frac{\Delta Lx}{c\Delta T} (I_2 - I_1) + \frac{\Delta L}{c\Delta T} (1 - x) (O_2 - O_1) + \tfrac{1}{2}(O_2 - I_2 + O_1 - I_1) = 0$$

Multiplying by $\dfrac{c}{\Delta L}$ and rearranging gives:

$$\frac{x(I_2 - I_1)}{\Delta T} + \frac{(1 - x) (O_2 - O_1)}{\Delta T} + \frac{c}{2\Delta L} [(O_2 - I_2) + (O_1 - I_1)] = 0$$

With $K = \Delta L/c$ as an acceptable approximation, a means is required of obtaining x, the coefficient governing the discharge weighting. Cunge derived the following expression for x from channel properties:

$$x = \tfrac{1}{2} - \frac{\overline{Q}_p}{2s\overline{B}c\Delta L}$$

where $\overline{Q}_p$ is the mean peak discharge, s average bed slope, $\overline{B}$ mean channel width and with the other variables as defined previously. In practice, the Muskingum coefficients are evaluated according to each reach forming the subdivisions of the total length of the river reach being considered. Then the routing of the inflow hydrograph can proceed as before to obtain the outflow hydrograph by recurrent application of the Muskingum equation:

$$O_2 = c_1 I_1 + c_2 I_2 + c_3 O_1$$

with the coefficients c_1, c_2 and c_3 being evaluated from K and x for each reach as before.

The success and accuracy of the Muskingum – Cunge flood routing

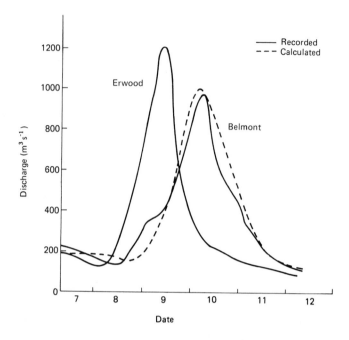

Fig. 16.18   Flood flow at Belmont calculated by the Muskingum – Cunge method. River Wye, December 1960. (Reproduced from National Environment Research Council (1975) *Flood Studies Report, Vol. III*, by permission of the Institute of Hydrology.)

method depend on the choice of ΔL and ΔT. For example, on the 69.8 km stretch of the River Wye from Erwood to Belmont, with average bed slope of 0.88×10^{-3}, there are no tributaries and a very small lateral inflow to the river. Thus with good records available at each station, flood routing techniques can be demonstrated. The application of the Muskingum – Cunge method to this river reach is abstracted from the *Flood Studies Report* results. The flood hydrograph calculated for Belmont, the downstream station, for the flood of December, 1960, is shown in Fig. 16.18. For that severe storm $\bar{Q}_p = 1210$ m^{3}s^{-1} and $c = 0.98$ ms^{-1}. ΔL was taken as 6975 m (i.e. there were 10 reaches) and ΔT as 7200 s. The calculated outflow shows a good fit to the recorded discharge curve. The lack of fit with the rising and falling limbs of the measured outflow hydrograph is probably due to the out-of-bank flood plain inundations. The error in the predicted peak discharge was only 3.7 % with a positive 1.95 % error in the predicted speed, ensuring a forecast of the time of the peak on the safe side, before the actual peak arrival.

16.4.2   3-parameter Muskingum

Both the original Muskingum method of Section 16.2.2 and the modified Muskingum-Cunge method (as demonstrated for the River Wye in Fig. 16.18) apply to situations where there is no lateral inflow to the river

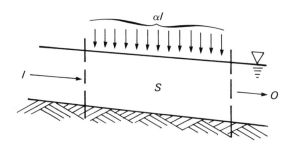

Fig. 16.19   Inflow, storage, outflow and lateral inflow for a reach of river. (Reproduced from O'Donnell, T. 1985.)

reach between the upstream and downstream gauging stations. In most rivers, this constrains the routing reaches to be rather short, generally terminating at tributaries, and requires gauged or estimated tributary inflows to be added to the main channel inflow term. In turn, this means using many reaches in the total routing procedure. A second modified Muskingum method has been developed by O'Donnell (1985) that incorporates a simple lateral inflow model. This modified method has two further advantages: (a) it replaces the tedious and subjective graphical trial-and-error estimation of the K and x parameter values described in Section 16.2.2 by a numerical and direct best-fit solution technique; and (b) by treating the whole river as one reach, it avoids the need for multiple routings (and multiple parameter determinations) over many sub-reaches.

The lateral inflow model used is shown in Fig. 16.19 (O'Donnell, 1985) and assumes that the total rate of lateral inflow over the whole reach is directly proportional to the upstream inflow rate. The proportionality constant, α, is taken to be fixed for any one event but takes different values for different events. The original 2-parameter Muskingum model (K, x) is thus extended to a 3-parameter model (K, x, α).

The three coefficients, c_i, of the routing equation 16.7 can be related to K, x and α (O'Donnell, 1985) and *vice versa*. A direct least squares solution by a matrix inversion technique yields a set of best-fit values for the c_i coefficients from the set of equations formed *via* equation 16.7 applied to all the ΔT intervals in an observed event. The three coefficients no longer sum to unity as in the original 2-parameter Muskingum method.

O'Donnell *et al* (1987) have applied this extended Muskingum procedure in a split-sample mode to a number of flood events over a single 50 km reach on the Grey River, New Zealand. Half the events were used for calibration, i.e. to establish average values for K and x for the reach. (The latter parameters are postulated to be fixed properties of the reach whereas each event has its own α value, a property of the causative storm). The average K and x values were then applied to reconstruct the outflow hydrographs for the events not used in the calibration from their individual inflow hydrographs and α values. Fig. 16.20 shows such a reconstruction for an event in which the value of α was 6.92. The volume of outflow at Dobson for this event was nearly eight times the volume of inflow at Waipuna due to the very substantial lateral inflow between the two stations.

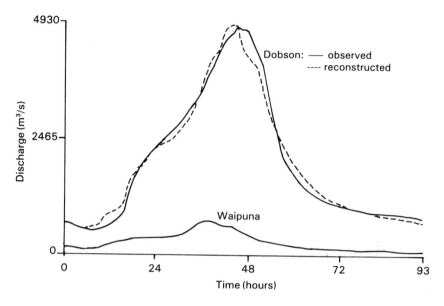

Fig. 16.20   Grey River flood event hydrographs at Waipuna (upstream) and Dobson (downstream). K = 5.06 h, x = 0.158, α = 6.92. (Reproduced from O'Donnell, T., Pearson, C.P. and Woods, R.A., 1987.)

16.5   Conclusion

This chapter has presented the outlines of the basic methods adopted in flood routing. As indicated previously, the analytical description of the physical processes from first principles necessitates the simplification of the channel conditions. Much of the research in flood routing is concentrated upon including considerations of the irregularities in natural rivers. Inflow from the river banks and seepage from the ground to the channel cannot be measured directly, and if the quantities are significant in flood flow situations, then they must be estimated or calculated by some means. The contributions of well defined tributaries are less difficult to determine, especially if they are gauged, but the relative timing of their contributory hydrographs always adds to the difficulties in a design problem.

The application of flood routing methods depends very much on the nature of the river, its channel and flood plain, and the existence of reliable discharge measurements. For work in the UK, the experience of the engineers who produced the *Flood Studies Report* (NERC, 1975) is invaluable, but notable contributions to the subject have been made world wide. The latest developments, to be found in the scientific journals, usually have computer packages available with comprehensive user manuals explaining the theory embodied in the program.

References

Amein, M. (1966). 'Stream flow routing on computer by characteristics.' *Water Res. Res.*, **2**, (1), 123–130.

Chow, V.T. (1959). *Open Channel Hydraulics*. McGraw-Hill, p. 540.

Cunge, J.A. (1969). 'On the subject of a flood propagation method.' *J. Hydraulics Res. IAHR*, **7**, 205–230.

Henderson, F.M. (1966). *Open Channel Flow*. Macmillan, 522 pp.

Maung, O. (1973). 'Flood level forecasting method.' Unpublished, DIC, Imperial College, London.

McCarthy, G.T. (1938). 'The unit hydrograph and flood routing.' Unpublished m/s conference of US Army Corps of Engineers.

Natural Environment Research Council (NERC).(1975). *Flood Studies Report*, Vol. III. Flood Routing Studies.

O'Donnell, T. (1985). 'A direct three-parameter Muskingum procedure incorporating lateral inflow.' *Hydrol. Sci.* J30, **4**, 479–96.

O'Donnell, T., Pearson, C.P. and Woods, R.A. (1987). An Improved Three-Parameter Muskingum routing procedure. *J. Hydraulic Eng. ASCE*, **114**, 5.

Price, R.K. (1974). 'Comparison of four numerical methods for flood routing.' *Proc. ASCE. Hydraulics Div.*, **100**, (HY7), 879–899.

Scott-Moncrieff, A. (1977). Unpublished lecture notes, Imperial College.

17

Design Floods

A major objective in water management is to see that excess water from extreme flood events is controlled so as to minimize distress and hardship to the population and damage to the environment. Both the river engineer and the water resources engineer need the skills of the hydrologist to evaluate flood flows. The information requirements of the former vary according to the nature of the river channel and the adjoining flood plain; data on floods for the water resources engineer, apart from managing reservoir storages, are primarily needed for the design of reservoir spillways.

17.1 Land Drainage

Some of the duties of the river engineer come under the heading of *land drainage*, an activity that has occupied man throughout history. In the UK, the draining of areas such as Romney Marsh and the Fenlands to rid the land of extraneous water began in the Middle Ages, and the maintenance and extension of the drainage channels has been continued through to the present day. The Land Drainage Act of 1930 established Catchment Boards throughout the country to look after major river catchments or groups of smaller river basins. The duties of the Catchment Boards included responsibility for the draining of their areas and for the control of specified water courses designated 'main' rivers. These responsibilities passed to River Boards in 1948, to River Authorities in 1963, and since 1973, they lie with Regional Water Authorities. Land drainage activities in the country since 1930 have been encouraged and financially supported in central government by the Ministry of Agriculture.

The principal objectives of land drainage works are as follows:

(a) *Drainage*, to reduce the water content of agricultural land to prevent water-logging.
(b) *Flood protection*, to minimize the overflowing of rivers on to agricultural crops and on to built-up areas.
(c) *Conservation*, to contain the rivers in their channels for the benefit of riparian users and navigation.
(d) *Disposal of surface water*, to provide outfalls and proper drainage channels for surface water flowing from urban areas.
(e) *Sea defences*, to prevent sea water penetrating on to land areas.

In this chapter on design floods, the first three of the above aims of the land drainage engineer will be considered. Surface water drainage from urban areas is dealt with more fully in Chapter 18, but treatment of sea defence works is beyond the scope of this book.

17.1.1   Drainage

Drainage of land to improve its usefulness for agricultural purposes must take into account the incidence of high intensity rainstorms as well as river water flooding of the land. Thus an engineer may be required to design a system of field drains in addition to the drainage channels necessary to carry off the surplus water. In low-lying areas such as the Fenlands, which are often below sea level, the water in the channels must be pumped up into the main rivers. Estimates of the quantities of water to be involved are needed before the pumped drainage scheme can be designed. The 18th and 19th century drainage engineers adopted a figure of about 2 in (50 mm) of rainfall per week for calculating channel storages and pumping capacities. More recent runoff calculations take 73.5 to 110.2 $ls^{-1}km^{-2}$, equivalent to 6.35 to 9.5 mm of rain in 24 h, respectively (Johnson; 1966). An up-to-date design manual for field drainage systems has been produced by the Agricultural Development and Advisory Service (ADAS) (MAFF, 1983).

Standards for field drainage and flood protection of agricultural land depend on the quality of the soil and possible land use. Table 17.1 defines classes of land potential and the degree of protection recommended. The period of the year from March to November is critical for many crops, especially those of high value such as winter and spring sown cereals and potatoes, but short duration flooding of the fields during winter months would not be cause for great concern. However, the standard of protection should not fall below the minimum recommended on the basis of the whole year which includes the winter flood events. For horticultural crops, particularly those with high crop value such as fruits, bulbs and winter vegetables, the land should be protected to give absolute minimum risk of flooding at any time of the year. Flood plains are often left in rough pasture for grazing, and these can be expected to flood on average once a year. Provided flood water is not allowed to stand for long on such land, little damage is done to the grass or soil structure. Forewarning of possible flooding is appreciated by the farmers, so that they may remove grazing stock before an inundation.

Table 17.1   Agricultural Land Drainage Standards

(Reproduced from Ministry of Agriculture, Fisheries and Food (1974) *Arterial Drainage and Agriculture*, by permission. © Crown copyright.)

		Design flood frequency — not more than	
Land potential	Crops	Mar – Nov	Whole year
Very high	All agricultural and horticultural crops	'No flood allowed' — say 1 in 100 years	
High	Root crops, cereals, grass	1 in 25 yrs	1 in 10 yrs
Medium	Cereals, grass	1 in 10 yrs	1 in 5 yrs
Low	Grass	1 in 5 yrs	1 in 2 yrs
Very low	Grazing land	1 in 3 yrs	1 in 1 yr

In order to provide the degree of protection recommended, the field drainage system must be adequate to remove intense rainfall and prevent surface ponding. The character of the soil, its infiltration rate and hydraulic conductivity need to be studied. The drainage channels must be designed to accommodate high rates of runoff and therefore discharges for the required return periods must be evaluated. Design floods related to drainage areas form the basic data supplied to the land drainage engineer by the hydrologist.

17.1.2   Flood Protection

Flood protection works are most often designed to prevent or at least mitigate flooding of centres of population where lives would be at risk and where damage to property would be serious. A typical flood problem is shown in Fig. 17.1. A settlement has grown up at the confluence of a small tributary with a main river, a very common situation. A major flood down the main river breaks out of the normal channel and floods the sections of the town built on the flood plain. In the days of wide urban expansion in the Industrial Revolution, the level land near rivers provided ideal sites for new factories, and extensive uncontrolled development of flood plains took place.

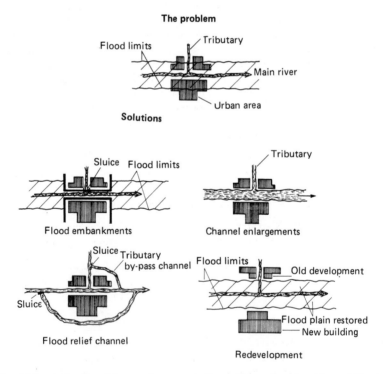

Fig. 17.1   A flood problem and some solutions. (Reproduced, with modifications from M. Nixon (1966) in R.B. Thorn (Ed.) *River Engineering and Water Conservation Works*, by permission of Butterworths.)

To protect the settlement from future inundations, several schemes can be adopted. It is at this stage that the ingenuity of the river engineer, with his skill and experience, is challenged. In the illustration, four solutions to the problem are suggested. The straightforward building of protective flood embankments with a controlling sluice for the tributary may solve the problem satisfactorily, but the size of the embankments necessary to exclude the design flood may be too unsightly and unacceptable on environmental grounds. The second solution simply excavates the existing channels to the dimensions that would contain the design flood. Thus a wide artificial channel which would certainly need constructed side walls and perhaps a man-made channel bed, is driven through the settlement. In normal flow conditions, most of the channel would be bare, unused and again unlikely to be approved by the community. Flood relief channels constructed outside the town, to take the surplus flood water via controlling sluice gates when required, is a solution that results in the least modification to the channels through the built-up area. The final sketch labelled 'redevelopment' demonstrates the most drastic solution to the problem. The whole width of the natural flood plain is cleared of obstructions and the settlement is extended beyond its original limits away from the river. The normal channel takes the regular flows, but the major floods spread over the flood plain with no direct harmful effects.

Many other methods may be devised; often a combination of several methods is adopted. One of the most complex of problems has been the flood protection scheme for Kidderminster on the River Stour. A design discharge of 90 m^3s^{-1} with a return period of 100 years was agreed, but the provisions for passing this flow have stimulated much discussion. The existing river channel systems between X and Z (see Fig. 17.2) has a maximum conveyance of 28 m^3s^{-1} before flooding occurs so that accommodation has to be found for 62 m^3s^{-1}. A detailed study of alternative flood alleviation schemes including detention reservoirs, diversion works or river improvement has been carried out. Fig. 17.2 illustrates three possible schemes as solutions to the problem:

(a) Improvement of the existing river system by introducing a pressure culvert at Y and deepening the west and east channels.
(b) The construction of single or double tunnels to pass 62 m^3s^{-1} on routes A or B.
(c) The construction of a twin reinforced concrete box culvert beneath the existing canal to pass 62 m^3s^{-1}.

Each problem area has its own local site difficulties and novel solutions are always acclaimed by the civil engineering profession.

In choosing the best solution to a flood problem, the first requirement is to establish the magnitude of the design flood. Several different flood discharges of different return periods are usually estimated. Feasibility studies of schemes to accommodate each flood magnitude are made and estimated costs produced. The responsible authority must then decide on the scheme most acceptable to the community, balancing the benefits with the estimated costs. A great many aspects of the problem need consideration before the optimum solution is found, but the initial responsibility for the evaluation of the flood magnitude – frequency relationship lies with the hydrologist.

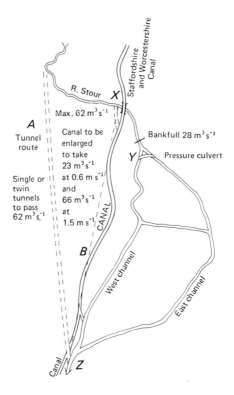

Fig. 17.2   Kidderminster, Design discharge 90 m³s⁻¹. T_p = 100 years.

(By permission of C.H. Dobbie and Partners)

17.1.3   Conservation

The maintaining of rivers in their channels does not attract quite the same attention as the big flood protection scheme. However, along certain critical stretches of a river it may be eminently desirable to contain discharges up to a manageable limit within the normal channel, e.g. the River Avon through the historic centre of Bath. Estimates of the existing magnitude – frequency relationship of bankfull and greater floods are required, and an optimum limit to suit the location becomes the design flood which, after conservation works, is not to be allowed to extend over the flood plain. Straightforward engineering works related to conservation include dredging the river bed and strengthening the river banks. More extensive and expensive schemes may necessitate widening and realigning the river channel to cut off over-developed and sluggish meanders. The conservation of navigable rivers may entail more costly works in the form of controlling sluices for which design flood flows must be assessed. However, in tidal reaches, knowledge of the fresh water floods is supplementary to the more important extremes of tidal surges, as in the design of the Thames Barrier.

17.2   Assessing a Design Flood

The frequency of occurrence of floods of different magnitude can be estimated by a variety of methods depending on the availability of hydrometric data. A logical procedure for obtaining flood flows for selected return periods is given by the *Flood Studies Report* (NERC, 1975) which is now recommended for practice in the UK (see Section 17.3). However, before the publication of this report, experienced design engineers adopted a variety of other techniques, some hydraulic and some hydrological in derivation. Several of the major flood alleviation schemes in the UK were designed from analyses of insufficient basic hydrological measurements and decisions on the frequency (return period) of a particular flood magnitude have often been made from very scanty information.

The actual choice of a design flood magnitude with its assessed return period depends both on the expected life of the scheme and on the degree of protection required. In the limit, flood defences should never fail, but the cost of providing complete protection could be prohibitive. The consideration of the chance of failure is fundamental in the design of reservoir spillways, since impounding dams should never be overtopped or the structure be allowed to fail. Any future urban effects on a design flood should also be considered in preparing a scheme for flood protection. The final choice of a scheme based on flood magnitude and return period and the associated costs and benefits, is made by the client, the authority commissioning the scheme. However, the engineer must assess the benefits of the protection and weigh these against the estimated costs for each of a range of schemes. A method for evaluating the cost – benefit relationship is introduced in Section 17.4.

Protection of valuable inner city properties and dense housing areas near a river may merit a 1 in 200 year scheme with the authorities being obliged to meet the cost. For land drainage works incorporating flood banks, sluices and weirs, an estimated life of 75 years is considered reasonable, whereas channel regrading should last 50 years (Nixon, 1966a). In making the decisions, local authorities are much influenced by the severity of any recent major flood that has just occurred. Thus the floods of 1947 due to snow melt in Eastern England, estimated to be a 75-year event, governed the design of such major undertakings as the Great Ouse Flood Protection Scheme; in the south west, the floods of 1960 were the yardstick against which many schemes were measured, to be replaced more recently by 1979 inundations. In some locations, such as the example in Fig. 17.1, a 30-year flood might be recommended for the design, but less vulnerable sites may only be worth a 1 in 5-year protection. At this lower end of the design flood range, the degree of protection for settlements and developed land joins the recommended standards for agricultural land (Table 17.1).

17.3   The Flood Studies Report

At the instigation of the Institution of Civil Engineers and with the financial support of the Natural Environment Research Council (NERC) of the government Department of Education and Science, a special team of hydrologists was set up at the Institute of Hydrology to update the technical bases of the 1933 report on *Floods in Relation to Reservoir Practice* (Reprinted ICE, 1960). Working in collaboration with colleagues in the Meteorological Office, the Hydraulics Research Station and the Irish Department of Public Works, they assembled all available precipitation and river flow data in the British Isles and abstracted the relevant information on maximum rainfalls and river peak discharges. From statistical analyses of the data and using well proven hydrological techniques and their developments, they produced a methodology for estimating maximum possible floods and of the flood discharges for selected return periods for all sizes of rivers in the British Isles.

The report of this thorough and rewarding study was published in five volumes by NERC in 1975:

Vol. I   *Hydrological Studies* (570 pp.)
Vol. II   *Meteorological Studies* (91 pp.)
Vol. III   *Flood Routing Studies* (85 pp.)
Vol. IV   *Hydrological Data* (549 pp.)
Vol. V   *Maps* (24)

The first three volumes are scholarly texts that have been acclaimed internationally and they provide a valued schematic example to hydrologists in the developing countries. Experience in the application of the methods of the *Flood Studies Report* by hydrologists and engineers in the UK has led to further studies and some improvements in detail. The five volumes form a basic manual for estimating a design flood and they warrant a place in all design offices concerned with schemes in water engineering.

The recommended procedure for evaluating the frequency of a given flood magnitude for a particular river depends on the availability of data and the amount of discharge detail required. A good long record of over 25 years of annual maximum flows would provide, by extreme-value statistical analysis, satisfactory estimates of floods of return periods up to say 500 years. A requirement for the detailed shape of a flood hydrograph from only two or three years data, or even none at all, would need estimates to be found by a combination of statistical and deterministic approaches. The main deterministic technique used in the *Flood Studies Report* is the unit hydrograph method of relating rainfall to stream flow.

An outline of the various procedures that may be followed is given in Fig. 17.3. For advising on a design flood flow for a scheme, it is necessary to make estimates of flood magnitudes for a selection of return periods by more than one method and then to compare results. Especially when there are only limited data, gross margins of error can be avoided by comparing design floods evaluated by different techniques. A summary of the methods developed by the Flood Studies team for application in the British Isles follows.

17.3.1   Statistical Methods

(a) River flow records over 25 years.   The peak flows are abstracted for each year of record, and this annual maximum series is fitted by the extreme value

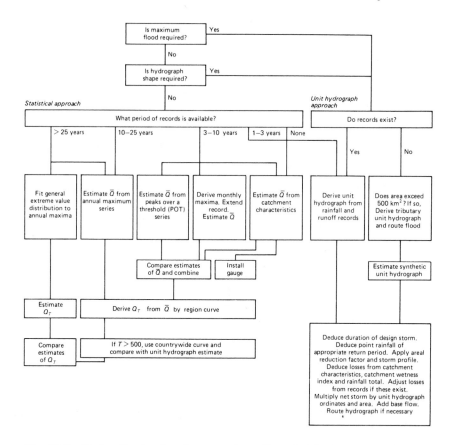

Fig. 17.3   Design flood estimation procedure. R = rainfall. Q = river discharge. $\overline{Q}$ = mean annual flood. Q_T = flood of T-year return period. EV1 = extreme value type 1 distribution. UH = unit hydrograph. D = duration. (Reproduced from J.V. Sutcliffe (1978) *Methods of Flood Estimation: A Guide to the Flood Studies Report*, by permission of the Institute of Hydrology.)

EVI (Gumbel) distribution. The method has already been described in detail in Chapter 12. Estimated values of flood discharges, Q_T, for any return period T, can then be obtained. In practice, for river flood protection works, return periods of over 500 years are rarely required.

(b) Records over 25 years and 10 – 25 years.   The mean annual flood, $\overline{Q}$, is estimated from the annual maximum series. Then the required Q_T is taken from a $Q/\overline{Q}$ plot against return period for the appropriate region (Figs. 17.4 and 17.5). These curves were derived from all the assembled records, the regions being defined by comparable records which have been combined together to give common $Q/\overline{Q}_T$ relationships. For the long (> 25 years) period records this estimate of Q_T can be compared to the direct estimate from the fitted EVI distribution in (a).

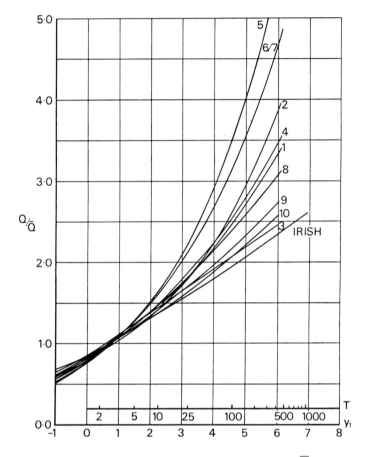

Fig. 17.4   Region curves showing average distribution of $Q/\overline{Q}$ in each region (Reproduced from National Environment Research Council (1975) *Flood Studies Report*, by permission of the Institute of Hydrology.)

(c) Records of 3 – 10 years.   Estimates of $\overline{Q}$ should be obtained by three methods.

(i) Use of the peak-over-threshold (POT) series. This method has been defined in Chapter 12, but now needs a more detailed specification of the selected threshold and of independence of peaks as given in Fig. 17.6. The number of exceedances per year is assumed to follow a Poisson distribution whose parameter, λ, is estimated by:

$$\hat{\lambda} = M/N \qquad \text{(See Appendix)}$$

where M is the number of exceedances in N years of record.

The peak magnitudes q_i are treated as an exponential distribution to give an estimate of the parameter, β, by:

$$\hat{\beta} = \overline{q} - q_0 = \sum_{i=1}^{M} (q_i - q_0)/M$$

Fig. 17.5   Homogeneous flood regions. (Reproduced from National Environment Research Council (1975) *Flood Studies Report*, by permission of the Institute of Hydrology.)

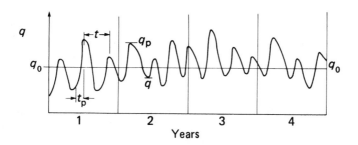

Fig. 17.6   Definition of peak over threshold series (POT). Threshold q_0 to give average of 3 – 5 peaks a year. Peaks to be independent, defined by $t > 3\,t_p$, and $q < \frac{2}{3}$ q_p. Exceedances of $q_0 = M = 13$. Years of record, $N = 4$.

(See Appendix) where $\bar{q}$ is the average peak and q_0 is the threshold. Then $\bar{Q}$ may be estimated from:

$$\bar{Q} = q_0 + \hat{\beta}\ell n\hat{\lambda} + 0.5772\,\hat{\beta}$$

(ii) The short period record is extended by taking the monthly maximum flow peaks, deriving regression equations with comparable data from neighbouring long-term stations, then using the regression relationships to estimate more peaks at the short-period station from the long-term records. An estimated value of $\bar{Q}$ is then calculated. This method needs care and competence in statistical methods.

(iii) The value of the mean annual flood, $\bar{Q}$, can also be calculated from a multiplicative equation relating $\bar{Q}$ to several catchment characteristics. The parameters of the equation were obtained by multiple regression between the assembled peak flows and the appropriate characteristics of each catchment with satisfactory records. Thus for the short-period station, the catchment area particulars are obtained from relevant maps, and the mean annual flood is calculated form:

$$\bar{Q} = 0.0201\ \text{AREA}^{0.94}\ \text{STMFRQ}^{0.27}\ \text{S1085}^{0.16}\ \text{SOIL}^{1.23}\ \text{RSMD}^{1.03}$$
$$(1 + \text{LAKE})^{-0.85}$$

where AREA is in km^2, STMFRQ is stream frequency in junctions per km^2, S1085 is the stream slope between 10 and 85% of length in m km^{-1}, SOIL is an index determined from 5 soil types, RSMD is the net 1 day 5 year rainfall in mm and LAKE is an index of lake area as proportion of total area.

For detailed explanations and determining the mapped variables, the reader is referred to Vol. I of the *Flood Studies Report* (NERC, 1975). The coefficient 0.0201 given for countrywide application of the method in the British Isles may also be modified for catchments in the different regions.

For the Thames, Lee and Essex region, the equation is simplified to:

$$\bar{Q} = 0.373\ \text{AREA}^{0.70}\ \text{STMFRQ}^{0.52}\ (1 + \text{URBAN})^{2.5}$$

where URBAN is the proportion of built up area in the catchment (Sutcliffe, 1978).

When a value of $\bar{Q}$ has been determined by averaging the estimates by the three methods, then required values of Q_T are obtained from the appropriate regional curve of $Q/\bar{Q}$.

(d) No records, ungauged catchments.   The above regression equation method (c,iii) can be used to derive an estimate of $\bar{Q}$ using the appropriate regional equation for $\bar{Q}$ and the catchment characteristics measured on topographical maps and read from the specially compiled maps of soil and rainfall indices in the *Flood Studies Report*. Then the Q_T values are taken from the corresponding regional curve of $Q/\bar{Q}$. However, before deciding on a design storm for a flood protection scheme, a river gauging station should be established at a suitable site so that some records may be obtained both to add confidence to this crude $\bar{Q}$ estimate and to enable alternative Q_T values to be determined by the unit hydrograph method.

17.3.2   The Unit Hydrograph Method

This method for deriving values of Q_T for various return periods is the best when only 1 to 3 years of records are available. During such a short period, there is usually a sufficient number of significant storms from which the rainfall – runoff relationship for a catchment can be found. The unit hydrograph method is also the recommended procedure to follow when estimates of the probable maximum flood and the shape of the flood hydrograph are required.

Application of the unit hydrograph to obtain a flood flow hydrograph assumes that the storm rainfall is uniformly distributed over the catchment. The extent of areally uniform high intensity rainfall is limited and thus the analysis of records to obtain means of deriving synthetic unit hydrographs has been restricted to catchments with areas less than 500 km². However, the unit hydrograph method can be used for catchments up to 1000 km² provided that the rain storms are spatially uniform over the area. To obtain design floods for large rivers, flood flows must be determined separately on the major tributaries and routed downstream to the required location.

Unit hydrograph derivation.   (i) Several large events ($>$ 5), with the storm rainfall well spread over the catchment area, should be selected from the short-period record. The recorded hydrographs and corresponding rainfalls are abstracted and an average unit hydrograph calculated by the method detailed in Chapter 13.

(ii) For ungauged catchments $<$ 500 km², a synthetic triangular unit hydrograph (Fig. 17.7) may be constructed from catchment characteristics using relationships derived from analysis of countrywide data. The triangular unit hydrograph in Fig. 17.7 is defined by three parameters: time to peak, T_p, in h, the peak flow, Q_p, in m³s⁻¹/10 mm and the time base, TB, in h. The Flood Studies data was used to find (by multiple regression) the following equation:

$$T_p = 46.6 \ \text{MSL}^{0.14} \ \text{S1085}^{-0.38} \ (1 + \text{URBAN})^{-1.99} \ \text{RSMD}^{-0.4}$$

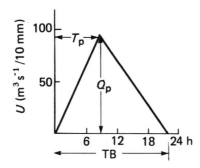

Fig. 17.7   A synthetic unit hydrograph (10 mm effective rain).

where MSL is the main stream length (km) and the other variables are as defined previously.

If there are some records for the catchment, it is more reliable to use:

$$T_p = 0.9 \text{ LAG}$$

where LAG is defined here as the time (h) from the centroid of effective rainfall to the peak runoff:

$$Q_p = 220/T_p \text{ (m}^3\text{s}^{-1}/100 \text{ km}^2/10 \text{ mm)}$$

and

$$TB = 2.52 \ T_p$$

The data interval for the ordinates of the unit hydrograph may usefully be taken from $t = T_p/5$ rounded to a convenient unit. For moderately sized catchments of 200 – 500 km², an interval t of 1 h may be expected.

The design storm.   The return period of a possible design flood is chosen at this stage and a corresponding design storm assessed. As a result of varying storm rainfall profiles combined with differing initial catchment wetness, design storms of a given return period do not produce floods having the same frequency of occurrence. It requires a storm of lesser frequency (longer return period) to produce a flood of a given return period. For the UK, the relationship between return periods of storm rainfalls and flood peak flows is shown in Fig. 17.8.

Storm duration.   A design storm duration, D, can be obtained by rounding to the nearest odd integer multiple of the unit hydrograph data interval, t, the estimate for D from:

$$D = (1.0 + \text{SAAR}/1000) \ T_p$$

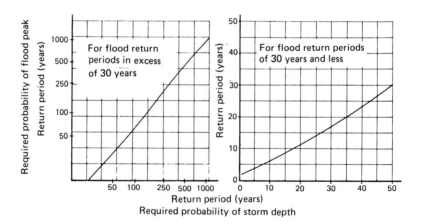

Fig. 17.8   Storm depth versus flood peak return periods.

Table 17.2   Ratios of *M5-D* h/*M5-2* day Rainfalls Related to *r*, *M5-1* h/*M5-2* day Rainfalls (Percentages)

r (%)	D h					
	2	4	6	12	24	48
12	18	26	33	49	72	106
15	21	30	37	53	75	106
18	25	34	41	56	77	106
21	28	38	45	60	80	106
24	31	41	48	63	81	106
27	35	44	51	65	83	106
30	38	48	55	68	85	106
33	41	51	57	71	87	106
36	44	54	60	73	88	106
39	47	57	63	75	89	106
42	50	60	66	77	91	106
45	53	63	68	79	92	106

where SAAR is the standard average annual rainfall (1916–50). This ensures that the storm peak is at a central data point. The design storm duration will normally be less than 48 h.

Storm depth.   The evaluation of the design storm requires data from maps and diagrams given in Vols. II and V of the *Flood Studies Report*. The Meteorological Office devised a series of relationships between storm rainfall depth, duration and frequency from a wide variety of data from 6000 rain gauges in the UK. The design task requires a $MT-D$ h rainfall i.e. the total D-h storm rainfall with a T-year return period.

The following key values and ratios are mapped:

SAAR: the standard average annual rainfall (1916–50)
$M5$-2 day rain: the depth falling in 2 days once in 5 years
r: the ratio (percentage) ($M5$-1h/$M5$-2day) × 100.

Taking the value of r for the catchment location, the ratio of $M5-D$ h/$M5$-2 day is first found from Table 17.2 as a percentage. This is applied to the $M5$-2 day rainfall to give the $M5-D$ h rain, i.e. the D h storm rainfall expected once in 5 years. Table 17.3 of growth factors, $MT/M5$, relating the desired T-year return period fall to the 5 year fall, are then applied ($MT/M5 \times M5-D$ h), to give the required design storm rainfall, $MT-D$ h.

It must be noted that this is only a point rainfall estimate, and to give the rainfall over the catchment, an areal reduction factor (ARF) must to applied (Fig. 10.8, Chapter 10). Thus the catchment storm rainfall for return period T years and duration D h is given by:

$$P(\text{mm}) = \text{ARF} \times MT\text{-}D \text{ h}$$

Table 17.3   Growth Factors, *MT/M5* as a Percentage of M5 Rainfall

M5 rainfall (mm)	0.5	1	2	*T*, return period, years 10	20	50	100	1000	10 000
England and Wales									
0.5	52	67	82	116	130	151	170	252	376
2	49	65	81	117	133	153	174	260	394
5	45	62	79	119	136	156	179	275	428
10	43	61	79	122	141	165	191	309	501
15	46	62	80	124	144	170	199	332	554
20	50	64	81	124	145	173	203	343	580
25	52	66	82	124	144	172	201	337	567
30	54	68	83	122	142	170	197	327	541
40	56	70	84	119	138	164	189	303	486
50	58	72	85	117	134	158	181	281	436
75	63	76	87	114	128	147	164	237	343
100	64	78	88	113	125	140	154	212	292
150	64	78	88	112	121	133	145	190	250
200	64	78	88	111	119	130	140	179	230
500	65	79	89	109	115	120	127	152	—
1000	66	80	90	107	112	118	123	142	—
Scotland and Northern Ireland									
0.5	55	68	83	115	131	151	171	254	378
2	55	68	83	116	132	154	175	265	401
5	54	67	82	117	135	162	186	294	466
10	55	68	82	119	139	169	197	325	536
15	55	69	83	120	139	170	198	328	544
20	56	70	84	119	138	166	193	314	512
25	57	71	84	118	137	164	189	303	485
30	58	72	85	118	136	161	185	292	460
40	59	74	86	117	134	156	177	272	416
50	60	75	87	116	130	152	172	257	385
75	62	77	88	114	127	146	162	231	330
100	63	78	88	113	124	140	154	212	292
150	64	79	89	111	120	133	145	190	250
200	65	80	89	110	118	131	140	179	230
500	66	80	89	108	114	120	127	152	—
1000	66	80	89	107	112	118	123	142	—

Antecedent catchment condition and percentage runoff.   Although the wetness of a catchment will change during a period of rainfall, it is assumed that a catchment wetness index (CWI) derived from the average annual rainfall (Fig. 17.9) will be applicable throughout the storm duration. A standard percentage runoff factor (SPR) is calculated from the SOIL and URBAN indices for the catchment, defined previously, via:

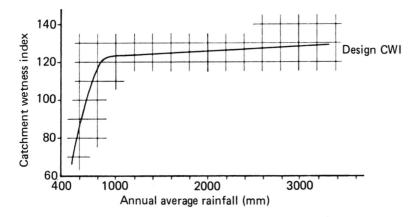

Fig. 17.9   Recommended design values for catchment wetness index. (Based on data from National Environment Research Council (1975) *Flood Studies Report, Vol. 1*, by permission of the Institute of Hydrology.)

SPR = 95.5 SOIL + 12 URBAN

and the percentage runoff PR for the design storm P mm is then given by:

PR = SPR + 0.22 (CWI − 125) + 0.1(P − 10)

This estimate of percentage runoff for a storm of P mm should be checked from runoff values in neighbouring catchments.

Storm rainfall profile.   In order to apply the P mm of rainfall in D h to the triangular unit hydrograph, a distribution of the rain depth during the rainfall period, a storm profile, is required. A family of storm profiles for different seasons in the year has been compiled from the autographic records from many stations. These all assume a symmetrical shape with the peak intensities occurring in the middle of the storm (Fig. 17.10a). The peakiness of the storm tends to be greater in the summer than in the winter owing to the greater likelihood of short intense thunderstorms. The distribution of the rainfall within the storm duration is represented in a series of cumulative percentage graphs related to the storm centre (Fig. 17.10b). The derivation of a design storm profile can best be demonstrated by an example. For a 75% winter design storm, P = 120 mm in a duration D of 9 h (odd multiple of data interval t = 1). That data interval is represented by 100/9 = 11.1% of the total duration. Corresponding percentages of rainfall and duration are taken from Fig. 17.10b for the 75% winter curve and the computations are carried out as in Table 17.4. As the profile is symmetrical, the rainfall percentages and amounts are taken in blocks two time intervals long (apart from the single central peak block) and then divided by two.

Design hydrograph.   By applying the percentage runoff, PR, uniformly to the rainfall amounts throughout the storm, the values of effective rain to be applied to the unit hydrograph are obtained. The convolution of the effective

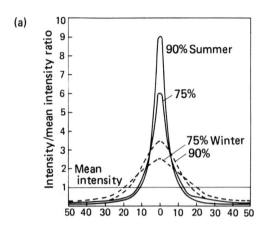

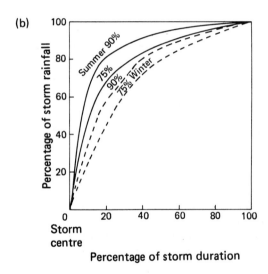

Fig. 17.10   Design profiles for storms with 90 and 75 percentile points of profile peakedness (summer and winter). (Reproduced, with modifications, from National Environment Research Council (1975) *Flood Studies Report, Vol. 1*, by permission of the Institute of Hydrology.)

rainfall with the unit hydrograph to give the surface runoff hydrograph is described in Chapter 13. The resultant hydrograph is an estimate of the design flood hydrograph, from which can be obtained the peak flow and the total volume of surface runoff. If relevant, an estimated baseflow component can be added; this may be negligible compared with the flood discharges.

Table 17.4   Derivation of Design Storm Profile (Given in Last Column)

	Percentage of duration					Intervals (h)	Rain (mm)
	100	77.7	55.5	33.3	11.1	0 1	4.2
						2	6.0
						3	11.4
Rain %	100 93	83	64	26	0 4	22.8
Increment %	7	10	19	38	26 5	31.2 Peak
Amount, mm (% of *P* mm)	8.4	12.0	22.8	45.6	31.2 6	22.8
Intervals	1 + 9	2 + 8	3 + 7	4 + 6	5 7	11.4
					Centre 8	6.0
					9	4.2
					Total	120.0

17.3.3   Flood Studies Worldwide

Since the publication of the *Flood Studies Report* (FSR), the hydrologists of the Institute of Hydrology have applied the FSR techniques to engineering problems in many countries. In evaluating the flood discharges for required return periods, the computational method used is usually determined by the availability of data. Where hydrometric measurements are limited, estimates are obtained by more than one method and engineering judgement is required in deciding on design values to recommend to clients.

Currently, research is being undertaken on the statistical analysis of long series of annual maximum floods from all over the world and wherever there are sufficient records regional growth curves are being developed as shown for the UK in Fig. 17.4. At present records from 70 countries have been analysed and it is hoped to expand the study when further data series have been assembled (Farquharson *et al*, 1987).

The series of annual maximum flood peaks for each station were converted to a non-dimensional form (q) by dividing each peak (Q) by the mean annual flood $(\overline{Q})$ so that comparisons between stations within a region could be made. The analyses of the q values in a region were combined, and regional flood frequency curves obtained by fitting the data to the general extreme value (GEV) distribution chosen for its flexibility in accommodating the conditions of different climates and terrain.

The probability of an annual maximum $q_i \leqslant q$ is given by

$$F(q) = \exp\left(-\left(1 - \frac{k(q - u)}{\alpha} \right)^{1/k} \right)$$

with three parameters u, α and k.

A selection of the results of the study are given in Table 17.5 for regions where the quality of the regional prediction growth curves was considered 'good' or 'very good'. This was assessed by the amount of data available in each region. The predicted values of q are provided for return periods (T) of

50, 100 and 500 years from which Q_T can be determined:

$$Q_T = \overline{Q} \times q_T$$

given that a mean annual flood value is available.

The shapes of the growth curves are influenced by climate and by catchment area and within each region local differences would be expected to produce divergent results. The present results provide only a guideline for planning and pre-feasibility studies at ungauged sites or in areas deficient in good quality data. More confident results are expected from further studies.

Table 17.5   Selected Results from Worldwide Flood Studies

	GEV parameters			$q(T)$ values		
	u	a	k	50	100	500
EUROPE						
Czechoslovakia	0.717	0.356	− 0.1821	2.74	3.28	4.82
Denmark	0.814	0.302	− 0.0368	2.08	2.33	2.92
Hungary and adjacent parts of Yugoslavia	0.793	0.281	− 0.1389	2.25	2.61	3.57
AMERICAS						
Brazil (Rio Grande do Sul)	0.830	0.348	+ 0.0959	1.96	2.12	2.46
AFRICA						
Kenya	0.651	0.459	− 0.1577	3.13	3.75	5.50
Malawi	0.655	0.422	− 0.1968	3.13	3.81	5.80
South Africa and adjacent parts of Botswana	0.485	0.422	− 0.3990	4.45	6.06	12.06
Togo and Benin	0.818	0.413	+ 0.1579	2.02	2.17	2.45
ASIA						
India (Kerala)	0.747	0.370	− 0.0991	2.51	2.90	3.92
Indonesia (Java & Sumatra)	0.837	0.266	− 0.0340	1.95	2.16	2.68
Jordan	0.525	0.474	− 0.3039	4.07	5.28	9.28
Korea	0.775	0.373	− 0.0256	2.31	2.60	3.29
Papua New Guinea	0.818	0.280	− 0.0682	2.07	2.33	2.98
Sri Lanka	0.656	0.328	− 0.3275	3.25	4.17	7.32
Thailand	0.762	0.343	− 0.1049	2.42	2.79	3.77

17.4   Cost – Benefit Analysis

In parallel with the Flood Studies at the Institute of Hydrology, the Natural Environment Research Council also supported an investigation previously initiated at Middlesex Polytechnic into flood damage assessment techniques. From this study, a manual has been produced for engineers (Penning-Rowsell & Chatterton, 1977), to help in the assessment of benefits from land drainage and flood alleviation schemes. By comparing the costs of schemes prepared for a range of design floods with the benefits expected from the different schemes, a more balanced judgement can be made of the merits of the various solutions to a flooding problem. Provided with cost – benefit relationships, the client or responsible authority can take full account of this particular element in selecting an optimum scheme.

The main difficulty in cost – benefit analysis is in quantifying the benefits of a flood protection scheme, i.e. benefits that are derived from assessing the damage that would be done without the protection. The flood discharge and damage relationships are demonstrated in Fig. 17.11. In the first quadrant, (a), the stage – discharge curves (H versus Q) are given for a river section before an alleviation scheme (solid line) and after a flood alleviation scheme

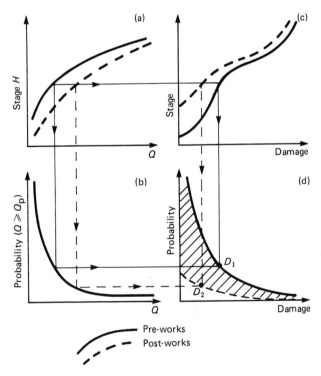

Fig. 17.11   Rivers stage – damage relationships. (Reproduced from E.C. Penning-Rowsell & J.B. Chatterton (1977) *The Benefits of Flood Alleviation: A Manual of Assessment Techniques*, by permission of Gower Publishing Co. Ltd.)

has been completed (dashed line). Quadrant (b) shows the annual exceed-ence probabilities of discharges (Q) derived from an extreme value analysis of the river flow records. With the same Q scales in (a) and (b) it is seen that for a given stage, H, the river channel can pass a higher discharge after the alleviation scheme and this has a lower probability of occurrence (i.e. a higher return period) than the flood discharge at the same stage before the scheme. For a given flood level greater than the level of bankfull discharge, a corresponding assessment of damage can be made to the buildings and land beyond the normal river channel. A stage – damage relationship is drawn in quadrant (c) for conditions without the protection and for the improved state after completion of a protection scheme. With the stage scale corresponding to that in quadrant (a), the damage for a given stage height H is seen to be less than the damage resulting from the same flood level before protection. By relating the discharge – probability curve with the stage – damage relation-ships, a plot of the probability of occurrence of a range of damages is obtained (quadrant d). The improvement in the damage prospects after the alleviation scheme is seen by comparing points D_1 and D_2 representing before and after conditions on the probability – damage curves. The shaded area represents the annual savings in damages resulting from the alleviation scheme.

The manual of assessment techniques (Penning-Rowsell and Chatterton, 1977) describes in detail a procedure for the assemblage of information to provide the data for stage – damage curves. From several sample surveys, the damage to residential property, retail shops and offices and industrial premises has been quantified according to the level of inundation. Samples of depth damage relationships are shown in Fig. 17.12 for the damage com-ponents in a single type of property in each of three categories: residential, retail shops and offices. Such detailed information on damages in the residential sector is available for 5 types of property in 4 categories according to date of building, and 4 subcategories of occupation by social class. The retail shops contain 6 categories with 6 subcategories of food shops while

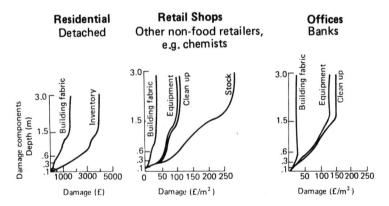

Fig. 17.12   Depth-damage relationships (1977 values). (Reproduced from E.C. Penning-Rowsell & J.B. Chatterton (1977) *The Benefits of Flood Alleviation: A Manual of Assessment Techniques*, by permission of Gower Publishing Co. Ltd.)

offices are subdivided into 6 types with 7 categories of commercial offices. Industrial properties are assessed individually. Guidance is also given for the assessment of other benefits or savings from other losses afforded by flood alleviation. Floods cause disruption of communications and services; roads, railways and telephone cables may be affected and electricity, gas and water supplies cut off. The cost of the emergency services must also be included in damage costs.

In addition to flood effects in urban areas, the assessment of agricultural benefits from protection is included in the manual. Damage to crops according to different land use categories and to agricultural buildings and machinery is related to depth and duration of inundation.

The values considered so far have constituted direct benefits, prevention of damage to property, and indirect benefits such as disruption of traffic and cost of emergency services. There are also intangible benefits difficult to quantify. These are the reduction of anxiety and inconvenience that otherwise could be caused by a flood situation. For example, in Fig. 17.13, at level 1, the river is out of its normal channel, but only the bottom of the garden is flooded; there is no inconvenience and no undue anxiety about the flood event. At level 2 above the sewer level, the flood water becomes polluted and there is considerable anxiety at level 3 when the water penetrates the house foundations. Direct damage begins at this level, since the outhouses are flooded. The ground floor is reached at level 4. With the car engine drowned at level 5, to the increased tangible damages to the property is added the extra inconvenience of lack of transport. In certain places, occupants of such properties are able to put a price on the anxiety and inconvenience caused by the different levels of flooding. These assessments, although complex, may be

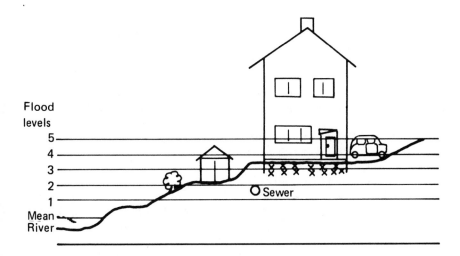

Fig. 17.13 Defining intangible damages. (Reproduced from E.C. Penning-Rowsell & J.B. Chatterton (1977) *The Benefits of Flood Alleviation: A Manual of Assessment Techniques*, by permission of Gower Publishing Co. Ltd.)

made by experienced interviewers, but estimating the costs of ill health, loss of employment and the ultimate loss of life, remains a very controversial problem.

Before recommending a flood alleviation scheme to a client, the benefits of protection at different levels corresponding to discharges of various return periods should be enumerated. From surveys of the affected areas, (the land use and buildings), the required answers can be obtained from the detailed computer programs associated with the published manual. For proper cost effectiveness, such considerations are recommended for major schemes in the UK.

17.5   Reservoirs and Floods

The dimensions of a barrage or dam across a natural river valley are governed by the amount of storage capacity required in the reservoir. When a reservoir is operated for water supply, it is expedient to keep it full for most of the time; but provision must be made for absorbing and passing flood flows from the catchment area when they cannot be absorbed by the reservoir. Therefore impounding structures must incorporate a flood spillway in their design and where the area downstream of the dam merits total protection, the capacity of the spillway plus any other outlets should be such as to take the maximum flood safely, without risk of overtopping the dam and possible structural failure.

There are strict regulations in the UK concerning the safety of dams. The Reservoirs (Safety Provisions) Act 1930 makes it a statutory requisite for all reservoirs over 5 million gallons capacity to be inspected by a qualified and experienced engineer. A later provision in the Reservoirs Act, 1975, increases the minimum capacity to 25 Ml (25 000 m^3). To guide engineers in the operation of the 1930 Act, The Institution of Civil Engineers published an interim report of a committee with recommendations as to design criteria for reservoir construction (Republished with additions, ICE, 1960). The *Flood Studies Report* (NERC, 1975) has produced much more information from the analysis of a wider range of hydrological measurements, and is a great improvement in the help available to reservoir design engineers. The Institution of Civil Engineers has subsequently published an engineering guide (ICE, 1978) recommending standards to be adopted for various categories of reservoir, and this seems to have been accepted by the engineering profession (Law, 1981).

The main factors that need to be considered when assessing the effect of extreme floods on the capabilities of a dam are the initial reservoir level, the shape of the flood inflow hydrograph and the wind speeds occurring during the flood producing storm. The latter may cause a build-up of waves on the reservoir surface and if these are aligned along the valley, they could add considerably to surcharged levels at the dam. Table 17.6 gives the recommended reservoir flood standards for the three main categories of reservoir. Category A has the most stringent standards, since for such a reservoir a breach in the dam would endanger the lives of people congregated in a town or village

Table 17.6   Reservoir Flood Standards

Category of reservoir	Initial condition	Dam design flood inflow		Wind speed Min. wave surcharge
		General	Min. standard, rare overtopping	
A. Breach endangers lives in community	Spilling long term av. daily inflow	PMF	Larger of 0.5 PMF or 10 000 year flood	Winter: max. hourly wind 1 in 10 years Summer: av. annual max. hourly wind
B. Breach may endanger lives not in a community, extensive damage	Full	Larger of 0.5 PMF or 10 000 year flood	Larger of 0.3 PMF or 1000 year flood	Wave surcharge allowance > 0.6 m.
C. Breach with negligible risk to life and causing limited damage	Full	Larger of 0.3 PMF or 1000 year flood	Larger of 0.2 PMF or 150 year flood	Av. annual max. hourly wind Wave surcharge allowance > 0.4 m.

For proportions of PMF, the ordinates of the computed PMF hydrograph are multiplied by the proportion indicated.

downstream. The worst conditions are envisaged, the spillway already taking the average daily inflow when the probable maximum flood arrives and the wave surcharge allowance must be for heights greater than 0.6 m. Categories B and C have decreasing standard requirements.

The derivation of the probable maximum flood is the factor that most concerns the hydrologist. The *Flood Studies Report* recommends that the unit hydrograph method is used since the shape of the whole flood hydrograph is required for routing the flood through the reservoir. However, extreme floods are usually caused by a combination of hydrological factors, and a certain amount of judgement is required in making the estimate of the probable maximum flood. In addition to extreme rainstorms, critical conditions of the catchment may also contribute to very high flows; there may be deep snow covering the catchment providing potential melt water or the catchment may be frozen, thus encouraging a 100% runoff. Although heavy rainfalls are most likely during the summer months in the UK, it is possible that a thunderstorm could fall on a frozen catchment with deep snow lying. With an additional rise in air temperature, these conditions could make for a very serious flood situation.

For the estimation of the probable maximum flood, the following

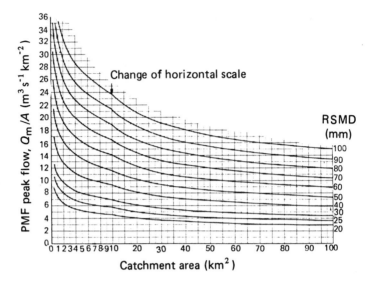

Fig. 17.14   (Reproduced from ICE (1978) *Floods and Reservoir Safety*, by permission of the Institution of Civil Engineers.)

modifications to the unit hydrograph procedure (described previously) are recommended:

(a) For the calculated storm duration and the estimated maximum rainfall, a storm profile of the greatest peakiness should be adopted, and the preceding precipitation should be used in the calculation of the antecedent wetness (CWI).

(b) Snowmelt at a uniform rate should be added to precipitation.

(c) The time-to-peak of the unit hydrograph should be reduced by one third thus raising its peak flow rate by 50%, since the base time, TB is related to T_p via TB = 2.52 T_p and the unit hydrograph must maintain unit area. Any significant baseflow should be added to the final flood hydrograph.

A rapid calculation of a flood peak inflow into a reservoir may be made from the graph of flood peak intensities (Fig. 17.14) taken from ICE, 1978. The catchment area and RSMD (net 1 day 5 year rainfall in mm) are the only data needed. Values of RSMD have been mapped for the British Isles (Fig. 17.15 is a sample for England and Wales). Thus for RSMD of 60 mm (an average value for the Pennines and Central Wales), a catchment area of 55 km² could give a flood peak of 550 m³s⁻¹. This is an approximation for an undulating, impermeable catchment. Such values should only be used as a rough guide.

Once the flood inflow has been estimated, then the other reservoir factors must be considered. The inflow must be routed through the reservoir storage and the water levels at the dam adjusted for the wind fetch effects.

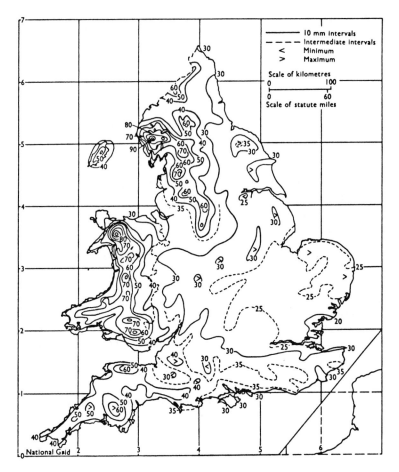

Fig. 17.15   RSMD (mm): England and Wales. (Reproduced from ICE (1978) *Flood and Reservoir Safety*, by permission of the Institution of Civil Engineers.)

There are still many areas in the study of design floods that require further investigation. A particularly challenging problem is that of storm precipitation and the varying temporal pattern of intensities throughout a storm's duration; the occurrence of the peak intensity is a variable to be considered. The initial state of the catchment is also of great significance in the generation of an extreme flood. A thorough knowledge of a catchment and of its varying responses during adverse conditions is essential for an engineering hydrologist to evaluate reliable design floods.

References

Farquharson, F.A.K., Green, C.S., Meigh, J.R. and Sutcliffe, J.V. (1987). 'Comparison of flood frequency curves for many different regions of the world.' in Singh, V.P. (ed.) *Regional Flood Frequency Analysis*, 223–56. D. Reidel Publishing Company.

Institution of Civil Engineers (ICE).(1960). *Floods in Relation to Reservoir Practice*.

Institution of Civil Engineers (ICE).(1978). *Floods and Reservoir Safety: An Engineering Guide*, 58 pp.

Johnson, E.A.G. (1966). 'Land drainage in England and Wales,' Chapter 3 in *River Engineering and Water Conservation Works*, ed. Thorn, R.B. Butterworths, 520 pp.

Law, F.M. (1981). *Flood Studies Report — Five Years On*. Thomas Telford Ltd, Chapter 7.

Ministry of Agriculture, Fisheries & Food (MAFF).(1974). *Arterial Drainage and Agriculture*, Ministry of Agriculture, Fisheries & Food, Drainage Leaflet No. 16, 22 pp.

Ministry of Agriculture, Fisheries and Food (MAFF) (1983). *The Design of Field Drainage Pipe Systems*. ADAS Reference Book 345, HMSO.

Natural Environment Research Council (NERC).(1975). *Flood Studies Report*. In five volumes.

Nixon, M. (1966a). 'Economic evaluation of land drainage works.' Chapter 4 in *River Engineering and Water Conservation Works*, ed. Thorn, R.B. Butterworths, 520 pp.

Nixon, M. (1966b). 'Flood regulation and river training.' Chapter 18 in *River Engineering and Water Conservation Works*, ed. Thorn, R.B. Butterworths, 520 pp.

Penning-Rowsell, E.C. and Chatterton, J.B. (1977). *The Benefits of Flood Alleviation: A Manual of Assessment Techniques*. Saxon House, 297 pp.

Sutcliffe, J.V. (1978). *Methods of Flood Estimation: A Guide to the Flood Studies Report*. Institute of Hydrology Report No. 49, 50 pp.

18

Urban Hydrology

Throughout history, there have been periods when populations have tended to congregate together in towns and cities. The reasons for town development are numerous; in the early empires of the Middle East, water supply was a prime factor and in the city states of the Mediterranean lands, safety from raiders and pirates dictated collective defence. Sources of fresh water also played a part in the establishment of nucleated settlements in the lowlands of Great Britain, and the advantages of safety in numbers led to the growth of the medieval walled towns. With increases in populations in more peaceful times, people were encouraged to spread out to make their homes where they could grow more food, and, often spurred on by religious persecution, many travelled across the world to establish themselves in new lands. The Industrial Revolution and the growth of manufacturing industries brought people together again. The establishment of factories meant that livelihoods became dependent on employment rather than on subsistence farming through self endeavour. This process of urbanization, the congregation of people together to live in towns, has escalated in the present century. In the older developed countries, like the UK, the well established towns and cities have continued to expand; all the benefits of a high standard of living are much more economically provided in a centralized community. In the developing countries, the cities are an attraction to expanding rural populations seeking factory employment; throughout the world, urban centres are growing and in some countries, the expansions are beyond control.

Thus with much of the world's population living in urban environments, the effect of such developments on elements of the hydrological cycle assumes a significant importance.

A high proportion of the activities of a civil engineer is directed towards urban construction, and for the specialist water engineer, this includes facilities for the provision of potable water for the community and for the removal and treatment of surplus and waste water. In England and Wales, 99% of the population receives a piped water supply from a public authority or a private water company and 95% of households are connected to a public sewerage system. In addition to these major modifications to the natural water balance of an area, there are marked effects of extensive built-up areas on the separate phases of the hydrological cycle.

18.1 Climate Modifications

The most obvious change in climate caused by urbanization is an increase in

Table 18.1   Average Annual Temperatures, London 1941–70

(Reproduced from Meteorological Office (1976) *Averages of Temperature for the United Kingdom 1941–70*, by permission of the Controller, Her Majesty's Stationery Office. © Crown copyright.)

	Mean height (m)	Maximum (°C)	Minimum (°C)	Mean (°C)
Outside urban area*	77	13.7	5.5	9.6
Suburbs**	64	14.0	6.4	10.2
Central London***	23	14.5	7.3	10.9

*   Mean of Rothamsted, St. Albans, Wisley, Shinfield.
**   Mean of Bromley, Southgate, Waddon (Croydon).
***   Mean of Kensington Palace, Regent's Park, St. James Park.

temperature. In temperate climates, there is a beneficial warming effect in winter months, although in the summer months, the town environment can be uncomfortably hotter than nearby rural areas. These contrasts are readily observed by London city workers who commute from homes in the country each day. Table 18.1 shows that these temperature differences may be quantified by comparing long-term annual averages; the warming effect is more marked by the greater range of mean minimum temperatures (1.8 °C) absolute minimum temperatures give a much larger range of $5\frac{1}{2}$ °C. A distinctive benefit of higher winter temperatures is that, for the comparable heating of similar properties, there is a saving of about 20% of fuel costs between Central London and surrounding rural areas (Chandler, 1965). This broad assessment was calculated from the additional climatic variables of wind speed and sunshine which are also modified in urban areas. On an international scale, the average changes in the elements of climate due to urbanization have been summarized by Landsberg (1970) in Table 18.2. However, it is advisable to consider the particular characteristics and location of a major city when assessing its urban climate, and the generalizations of Table 18.2 should be taken only as guidelines. The regional climate and local topography are particularly important and may override any urban effects. As an example, increases in precipitation in towns and cities have been observed in the interior of North America (continental climate), but studies in the UK (maritime climate) have failed to relate annual and seasonal rainfall differences to urban development (Tabony, 1980). In the London area, rainfall differences are influenced mainly by altitudinal differences. On a short time scale, the proportionally greater incidence of severe thunderstorms in built-up areas compared with rural areas is well noted and can be ascribed to greater concentrations of condensation nuclei in the air, increased turbulence and urban overheating. Information on such extreme events is naturally more readily available (and perhaps more likely to be recorded) in centres of population, and the hydrological consequences are

Table 18.2   Average Changes in Global Climatic Variables due to Urbanization

(Reproduced from H.E. Landsberg (1970) *Climates and Urban Planning in Urban Climates*, by permission of the World Meteorological Office.)

Variable	Compared with rural area
Temperature	
Annual mean	$\frac{1}{2}$ – 1 °C more
Winter minima (average)	1 – 2 °C more
Wind speed	
Annual mean	20 – 30% less
Extreme gusts	10 – 20% less
Calms	5 – 20% more
Contaminants	
Condensation nuclei & particles	10 times more
Gaseous admixtures	5 – 25 times more
Radiation	
Global	15 – 20% less
Short-wave, winter	30% less
Short-wave, summer	5% less
Sunshine duration	5 – 15% less
Cloudiness	
Cloud cover	5 – 10% more
Fog, winter	100% more
Fog, summer	30% more
Precipitation	
Totals	5 – 10% more
Days with < 5 mm	10% more

usually of immediate concern. Storm magnitudes and their frequency of occurrence are of greater importance than annual rainfall totals in urban hydrology. The degree of influence of the urban environment on such events is not fully understood and is under investigation.

18.2   Catchment Response Modifications

The changes made to a rural area by the construction of a concentration of buildings have a direct effect on its surface hydrology. The covering of the land surface by a large proportion of impervious materials means that a much larger proportion of any rainfall forms immediate runoff. In addition to extensive ground coverage by the buildings in a city, the paved streets and car parks contribute large areas to the impervious surfaces. Any slope of the land also greatly enhances the runoff response of a paved area. In a defined catchment area, the effect on the stream discharge is dependent on the extent of the impervious area. Contributions to groundwater are limited to rainfall on the remaining pervious surfaces, where normal infiltration into the soil and percolation into the underlying strata can take place. Thus, after major urban developments in a catchment, the following differences in the river flow from that of an equivalent rural catchment can be identified:

(a) there is a higher proportion of rainfall appearing as surface runoff, and so the total volume of discharge is increased;

(b) for a specific rainfall event, the response of the catchment is accelerated, with a steeper rising limb of the flow hydrograph; the lag time (see Chapter 13) and time to peak are reduced;

(c) flood peak magnitudes are increased, but for the very extreme events (when the rural runoff coefficient > 50%) these increases in urban areas are diminished;

(d) in times of low flows, discharges are decreased since there is a reduced contribution from the groundwater storage that has received less replenishment; and

(e) water quality in streams and rivers draining urban areas is degraded by effluent discharges, increased water temperature and danger from other forms of pollution.

Many of these modifications are promoted by structural changes made to drainage channels. It is essential to remove rain water quickly from developed areas, and surface water drainage systems are included in modern town extensions. In many old established settlements, storm water runs into the domestic waste water sewers, but in some countries, e.g. Australia, the cities have separate storm water and sewerage systems. When an area is newly developed, it is sometimes expedient to modify the natural stream channels; realignment of the water courses, lining and regrading of the channels are improvements made to facilitate drainage.

The interaction of the artificial nature of urban catchments and the need to accommodate the changed hydrological characteristics is complex. The solving of one drainage problem may easily exacerbate another feature of the catchment runoff, e.g. rain events on the planned surface drainage of a new housing estate could produce higher peaks downstream than formerly, and these might cause flooding at previously safe points along the channel.

The quantifying of urbanization effects on the rainfall – runoff relationship has been studied widely. Much of the work has concentrated on the modifications made to the volume and time distribution of surface water runoff hydrographs from single rainfall events. The various hydrograph parameters such as peak discharge, Q_p, time to peak, t_p and lag time (various definitions) are usually related to catchment characteristics including area of impervious surfaces or proportion of area urbanized, in order to obtain quantitative rainfall – runoff relationships. A thoroughly reasoned description of many of the studies is given by Packman (1980). The various formulae that have resulted from the individual studies are only applicable to the areas where they have been derived, and it is not advisable to use them for areas with different climates and topography. The use of empirical formulae in urban hydrology is being replaced by the application of mathematical simulation models of the rainfall – runoff process. Standard computer models can be adjusted to fit the problem area and the relevant hydrological and catchment data can be fed in to give required discharges according to stipulated design criteria. Examples of the application of such models to drainage problems in many countries have been assembled by Helliwell (1978).

18.3   Urban Development Planning

In the development of new urban centres, hydrological knowledge of the areas is required at two stages. The first is the planning stage when the general lay-out of the new town is being decided. Estimates of the discharge hydrographs (and corresponding stage hydrographs) for chosen return periods are wanted at a few selected points on the natural water courses, perhaps where bridges are to be constructed, and certainly at vulnerable confluences of tributary streams. Knowing the proposed extent of the new urban area, initial flood and stage estimates may indicate that the existing river channel would not contain the expected enhanced flood flows. Then it may be thought expedient to improve the channel and/or to provide for flood water storage in a retaining pond (and the necessary open space for this must be included in the overall town plan). The second stage of hydrological involvement occurs at the detailing stage, the designing of the storm water drainage channels and pipes to carry the surface water into the rivers. (This is considered fully in Section 18.4.1.)

The principal objective at the planning stage is the determination of the size of flood, with its related return period, that the developing authority is prepared to accommodate. The design of the drainage system is dependent on a satisfactory assessment of the flood magnitude – return period relationship and the subsequent choice of a design flood.

In designing the storm water drainage system for a proposed new Australian city, a Regional Stormwater Drainage Model (RSWM) has been developed (Aitken, 1975) for the simulation of the hydrological processes over a large development area (over 5 km²). A diagram of the model is shown in Fig. 18.1. The RSWM has four modules:

(a) library module — regulates the model operation. Any arrangement of catchment areas, channels and storage ponds can be processed;

(b) hydrograph module — the core of the model which uses the Laurenson non-linear runoff routing model (Chapter 14) to obtain a hydrograph from rainfall on each subcatchment. The non-linear storage delay time (K) for a subarea is given by:

$$K = BQ^{-0.285}$$

where Q is the instantaneous discharge $(m^3 s^{-1})$ into the river-reach and B is the subarea coefficient given by:

$$B = 0.285 \, A^{0.520} \, (1.0 + U)^{-1.972} \, S^{-0.499}$$

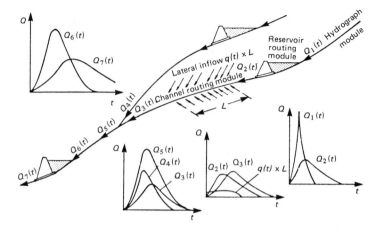

Fig. 18.1   Diagrammatic representation of the RSWM. (Reproduced from A.P. Aitken (1975) in T.G. Chapman & F.X. Dunin (Eds.) *Prediction in Catchment Hydrology*, by permission of the Australian Academy of Science.)

where A is the catchment area (km^2), S is the main channel slope ($\%$), and U is the urbanization factor. For natural catchments, $U = 0$; for fully urbanized catchments, $U = 1$;

(c) channel routing module — uses the Muskingum – Cunge procedure (Chapter 16) and includes lateral inflow; and

(d) reservoir routing module — uses the level pool flood routing procedure (Chapter 16).

The RSWM can produce, from design storms, details of hydrographs, water levels and storage volumes at required locations and graphical plots of the computed hydrographs.

Another model that may be used in this urban development planning stage is the comprehensive Stanford Watershed Model, which is available in the Hydrocomp Simulation Program (HSP) (Chapter 14).

Associated with the plans for surplus storm water drainage there is also the need for planning the disposal of waste water from the urban area. Hence the application of suitable water quality models can help in the siting and design of treatment plants and the setting of appropriate standards for effluents, both domestic and industrial.

The planning stage of hydrological studies for urban development in the UK has taken the form of feasibility studies of the required drainage systems for the new towns such as Harlow, Crawley, the extension of Horley, and Milton Keynes. Many of the recent studies of urbanization effects on the catchment response to storm rainfall have been made in the various new town development areas. One of the difficulties encountered in such investigations is the shortage of autographic rainfall and stream flow data. This is always a major problem when attempting to derive storm hydrographs of a long return period for design purposes. For the large developments scheduled to take several years to complete, the installation of special stream gauging stations is deemed worthwhile and their continued operation after development should produce valuable evidence of urbanisation effects in the future. The techniques published in the *Flood Studies Report* (NERC, 1975) are now being applied to present investigations.

The siting of retention ponds and special channels to take extreme floods has also been a feature in the planning of the English new towns. In residential areas, the land set aside to store temporarily excess water of the peak flows may be used as recreational areas, thereby serving a dual purpose. Experience of storage ponds for flood control in new urban areas has resulted in a recent guide to their design (Hall & Hockin, 1980). This also uses the *Flood Studies Report* methods for determining design floods.

18.4   Drainage Design

Once the broad outlines of the hydrological consequences of an urban development of an area have been determined at the planning stage and major remedial works considered, then the detailed design of the drainage systems is required. The engineering hydrologist is fully concerned with evaluating the runoff from the subareas to be drained in order to design the necessary storm water sewers. The peak runoff from the selected design storm determines the size of sewer pipe which is dependent on the extent of each subarea to be drained. At the head of a catchment subarea, the required pipe size may be quite small, but downstream, as the sewer receives water from a growing area through a series of junctions, the pipe size gradually needs to be increased.

The problem of estimating the runoff from the storm rainfall is very much dependent on the character of the catchment surface. The degree of urbanization (extent of impervious area) greatly affects the volume of runoff obtained from a given rainfall. Retention of rainfall by the initial wetting of surfaces and absorption by vegetation and pervious areas reduces the amount of storm runoff. These surface conditions also affect the time distribution of the runoff. Thus the computational method used to obtain the runoff from the rainfall should allow for the characteristics of the surface area to be drained.

18.4.1   Impervious Areas

These comprise the roof areas and large expanses of paved surfaces of congested city centres and industrial sites, in which there is very little or even no part of the ground surface into which rainfall could infiltrate. The calculation of the runoff from these relatively small catchments is the most straightforward, since the areas can be easily defined and measured. Over such limited areas, the storm rainfall can be assumed to be uniformly distributed with 100% runoff occurring. The response of the impervious surfaces is rapid, resulting in a short time of concentration of the flow in the drainage system. The Rational Formula (Chapter 13) can thus provide the peak drainage:

$$Q(\text{m}^3\text{s}^{-1}) = \frac{A(\text{ha}) \times i(\text{mm h}^{-1})}{360}$$

or

$$Q(\ell s^{-1}) = \frac{A(m^2) \times i(mm\ h^{-1})}{3600} \quad or \quad \frac{A(ha) \times i(mm\ h^{-1})}{0.360} \qquad (18.1)$$

A recommended method for the design of a piped sewer drainage system using the Rational Method is given in the Transport and Road Research Laboratory (TRRL) Road Note 35 (TRRL, 1976). The procedure may be explained by considering the simple pipe design in Fig. 18.2. The sequence of pipes must be numbered according to the convention shown. The first pipe of a branch is always labelled 1.0, 2.0, etc. and the following pipes in a line are labelled sequentially, 1.1, 1.2, etc. Here there is a line of three pipes leading to an outfall and a tributary area (pipe 2.0) drains into the junction at the end of the second pipe in line 1, pipe 1.1. The computations to determine the required pipe sizes are shown in Table 18.3. The first four columns give the surveyed particulars of level differences along each pipeline, the required length and the calculated gradient.

At the outset of the design procedure, the selected return period for a design storm will have been decided. Storm water sewers are usually designed for 1 in 1, 1 in 2 or 1 in 5 year storm return periods (NWC, 1976); in the example, the expected annual storm intensities are used (1-year return period). The type of pipe will also have been chosen; the internal roughness governs the flow characteristics, and a roughness coefficient, k_s, must be selected from published tables (Ackers, 1969) in order to use the Colebrook – White equation to determine the flow velocity in the pipe. Velocities and discharges for standard sized pipes computed from this complex formula are published in tabular form for different pipe sizes, assuming full bore conditions, a hydraulic gradient equal to the pipe gradient and appropriate rough-

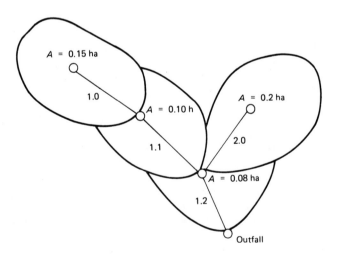

Fig. 18.2   Simple pipe design.

Table 18.3 Rational Method Drainage Design

Pipe No.	Level diff. (m)	Pipe length (m)	Grad. (1 in)	Trial pipe da. (mm)	Pipe V (ms⁻¹)	Q (ℓs⁻¹)	Time of flow (min)	Time of conc. (min)	Rate of rain (mm h⁻¹)	Imp. A cum. (ha)	Storm Q (ℓs⁻¹)	Comment
1.0	1.00	65	65	150	1.26	23.0	0.86	2.86	67.5	0.15	28.1	Surcharge
				225	1.64	67.5	0.66	2.66	69.2		28.8	Partial flow
1.1	0.90	70	78	225	1.50	61.7	0.78	3.44	63.2	0.25	43.9	Partial flow
2.0	1.50	60	40	150	1.61	29.4	0.62	2.62	69.5	0.20	38.6	Surcharge
				225	2.10	86.0	0.48	2.48	70.7		39.3	Partial flow
1.2	0.90	50	56	225	1.77	72.8	0.47	3.91	60.2	0.53	88.6	Surcharge
				300	2.13	156.0	0.39	3.83	60.7		89.4	Partial flow

ness (Ackers, 1969). Design charts for the velocities and discharges are also available and provide for easier interpolation (HRS, 1978). Flows larger than those derived from the tables or charts would require hydraulic gradients greater than the pipe gradient stipulated, and these could only occur by ponding (or surcharging) of water in the manholes at the pipe junctions. The design objective is to avoid such surcharging.

The design procedure begins with the choice of a trial pipe size for pipe 1.0, viz 150 mm (the smallest used in practice) (Table 18.3). From the published tables and for $k_s = 0.6$ for a normal concrete pipe, the velocity and discharge for a gradient of 1 in 65 are noted, 1.26 m s^{-1} and 23.0 ℓ s^{-1}, respectively. A flow greater than 23.0 ℓ s^{-1} would result in surcharging.

The time of flow along the pipe is next calculated from the velocity and length of pipe and comes to 0.86 min. The time of concentration at the end of the first pipe is then 0.86 min plus an assumed allowance of 2 min, for the time of entry, which is assumed to cover the lag time between the onset of the storm rainfall and the entry of the overland flow into the leading manhole. With the time of concentration of the drainage to the end of the first pipe known, the design return period rainfall intensity (i) over this duration to give the peak flow can be obtained from intensity-duration-frequency data published by the Meteorological Office. In this example, the rates of rainfall are taken from Table 18.4 for a location in Southern England (TRRL, 1976). The storm peak discharge for this subarea is then calculated from the rate of rainfall (i) and its cumulated impervious area using Equation 18.1 for comparison with the unsurcharged full bore pipe flow. The first trial pipe of 150 mm diameter would clearly be surcharged, so the calculations are repeated with the next size pipe, diameter 225 mm. The calculated storm discharge, 28.8 ℓ s^{-1}, would be easily contained by the larger pipe.

The calculations proceed for each pipe in turn, with the previous time of concentration being added to the new time of flow to give the combined times of concentration at the end of sequential pipes. The drainage areas are also accumulated. It will be noted that the 2.0 min time of entry is also added to the flow time of pipe 2.0 since it is at the start of a branch pipeline. The time of concentration for the last pipe 1.2, is the sum of the time of concentration of pipe 1.1 and the flow time of pipe 1.2. The extra contribution from the greatly increased area drained by the tributary pipe results in a much larger

Table 18.4   Rainfall Intensities (mm h^{-1}) for Specified Durations and Return Periods (Point Location in Southern England)

Duration (min)	Return period (years)		
	1	2	5
2.0	75.6	93.4	120.5
2.5	70.5	87.5	113.4
3.0	66.3	82.3	107.2
3.5	62.8	77.8	101.7
4.0	59.6	73.8	96.8

discharge requiring the next size larger pipe, 300 mm diameter. (The pipe diameters are given as rounded metric equivalents to the old 6, 9 and 12 in diameter pipes.)

Thus in the simple pipe design for the system in Fig. 18.2, pipes 1.0, 1.1. and 2.0 need to be 225 mm in diameter and the last pipe 1.2 must be of 300 mm diameter. These requirements conform to the normal concrete pipes specified ($k_s = 0.6$) and 1-year return period design storm intensities, with an assumed 2 min time of entry.

It will be appreciated that the computations become complicated as more branch pipe lines are incorporated into the system. This method is most satisfactory for small impervious areas, but if more pervious fractions are included within the catchments, results from the Rational Method become less acceptable.

18.4.2 Motorways and Airports

The rapid drainage of surface water from roadways and aircraft runways is very important for the safety of rapidly moving traffic. However, the areas feeding to a drainage system usually incorporate a significant proportion of pervious surfaces, e.g. in the central reservations of motorways and grassed slopes of adjoining cuttings. Flows for such areas may be calculated by the Rational formula by making an estimate of the proportion of rainfall (effective rainfall) that will form the runoff. The choice of such a runoff coefficient is very much dependent on the character of the catchment and has to be guided by the experience of the engineer.

Investigations into the rainfall and runoff of small sections of motorway with up to 50% of the drainage areas pervious resulted in two improved methods of obtaining storm hydrographs for design purposes (Swinnerton *et al.*, 1972, 1973). With the time of concentration less easily estimated on a variable catchment surface, the derivation of the storm hydrograph provides the temporal distribution of the runoff in addition to the peak discharge.

In the first method, a dimensionless hydrograph technique was derived for designing motorway storm water drainage in the UK. From a total of 40 storms on seven experimental areas ranging from 0.28 ha up to 2.63 ha, the dimensionless hydrograph (Fig. 18.3) was derived. To obtain a storm hydrograph, values of Q_p, the peak flow (m^3s^{-1}), and T_R and T_F, durations of rise and recession of the hydrograph, must be found.

The peak rate of runoff, Q_p, may be determined from a regression equation using the following measures:

maximum rainfall intensity of 15 min duration;
impervious area and total area of the catchment; and
length and average longitudinal slope of the site.

The time of rise, T_R, is obtained from the duration of rainfall of intensity greater than 5 mm h⁻¹, and the duration of recession, T_F, is calculated from Q_p, T_R and the total volume of runoff. Details of the definitions and regression equations are to be found in Swinnerton *et al.*, 1972. Alternatively, new

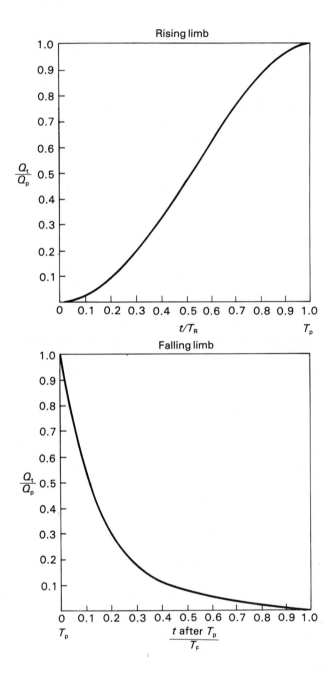

Fig. 18.3   Dimensionless hydrograph.

expressions for Q_p, T_R and T_F could be found from more recent and longer records.

In the second method, a conceptual model was tested on the same catchment data with the additional advantage of analysing more complex rainfall events with periods of no rain occurring within the total storm duration. After try-

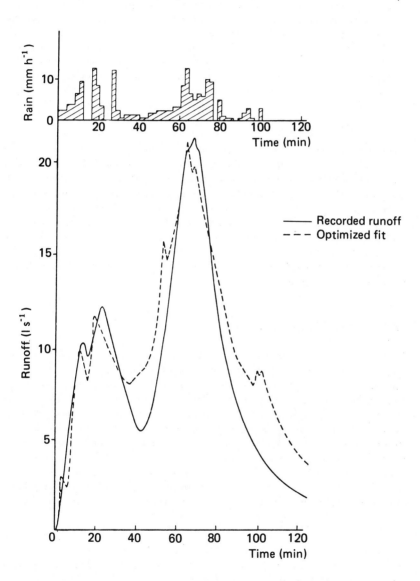

Fig. 18.4   Conceptual model application motorway catchment 2.63 ha, 51% paved. (Reproduced from C.J. Swinnerton, M.J. Hall & T. O'Donnell (1973) *Civ. Eng. Public Works Rev.,* **68**, 123–132, by permission of Morgan-Grampian.)

ing several types of models, the following linear reservoir model with two alternative storage constants, k_1 and k_2, was found to fit satisfactorily for periods of rainfall intensity i in time period T, and for periods of no rain:

For rain $q_T = i(1 - \exp(-T/k_1)) + q_0 \exp(-T/k_1)$
For no rain $q_T = q_0 \exp(-T/k_2)$

where q_0 and q_T are the discharges at the beginning and end of the time period T, respectively. From the motorway catchment data, expressions for k_1 and k_2 were found in terms of impermeable area and length of catchment, total rainfall and the maximum rainfall intensity of 15 min duration. Fig. 18.4 demonstrates how well the model simulates the peak discharges from a very irregular rainfall sequence on a motorway catchment 2.63 ha in area of which 51% of the surface is paved motorway.

18.4.3 Small Urban Catchments

In designing storm water sewerage systems for towns, city suburbs and new developments of around 200 – 400 ha with varied surface characteristics, a method is required which also takes into account differences in storm rainfall over the catchment area. Developed from the time – area concept of catchment response (Chapter 13), the Transport and Road Research Laboratory (TRRL) Hydrograph Method (Watkins, 1962) has been applied widely in the UK. In a survey of Water Authorities, consulting engineers and others concerned with sewer design (NWC, 1976), 90% reported using the Rational Method and 62% the TRRL Method. (Many authorities use both, depending on catchment size.)

18.5 TRRL Hydrograph Method

In the time – area method, the total catchment area is deemed to be contributing to the flow after the time of concentration, T_c, the time it takes for the rain on the furthest part of the catchment to reach the outfall. Thus in Fig. 18.5, for two drains receiving uniform rainfall from areas A_1 and A_2 with drain 2 joining the main channel, drain 1, a relationship of contributing area, A, versus time, T, is constructed. From the beginning of the flow in drain 1 at $T = 0$ there is a steady increase in area contributing until $T = T_1$ which is the value of T_c for area A_1. Drain 2 begins to contribute to the outfall flow at $T = T_3$ before $T = T_1$. After a further period, T_2, area 2 reaches its own T_c at time $T = (T_2 + T_3)$. Between times T_3 and T_1 both drains have been flowing and the joint contributing area (at C) at $T = T_1$ is given by:

$$A_1 + \frac{(T_1 - T_3)}{T_2} \cdot A_2$$

From $T = (T_2 + T_3)$, both areas are contributing fully. The time – area curve for the combined drains is the composite line $OBCD$.

The principle of the TRRL Hydrograph Method is outlined in Fig. 18.6. In Fig. 18.6a, a catchment area, divided into four subareas, is drained by a

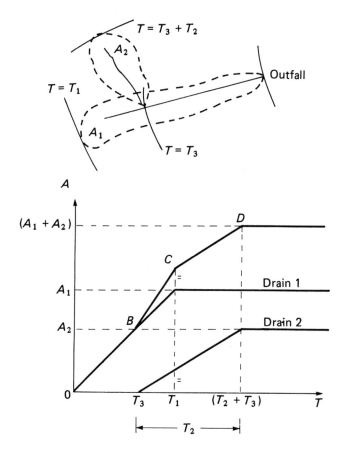

Fig. 18.5   A time – area diagram.

single channel to the outfall where the hydrograph is required. Subarea 1 begins contributing to the flow first, to be followed sequentially by the other three subareas. The individual time – area curves are shown in Fig. 18.6b and the composite curve for the whole catchment is drawn by summing the subarea contributions at regular time intervals. The choice of time unit is dependent on the surface characteristics of the catchment and may range from 1 min for highly impervious areas to about 30 min for nearly natural catchments. The incremental contributing areas after each time interval are then read from the composite curve, a_1, a_2, a_3, etc. In the diagram, the time of concentration for the whole area is $9\frac{1}{2}$ time units.

The next stage in the method involves the storm rainfall. The values of the areal rainfall are calculated from the rain gauge measurements by one of the standard methods (Chapter 10) for each of the chosen time unit intervals throughout the duration of the storm. Since some of the rainfall will infiltrate the pervious areas, not all the storm rainfall will contribute to the direct run-off from the catchment area. An effective rainfall rate must be assessed for

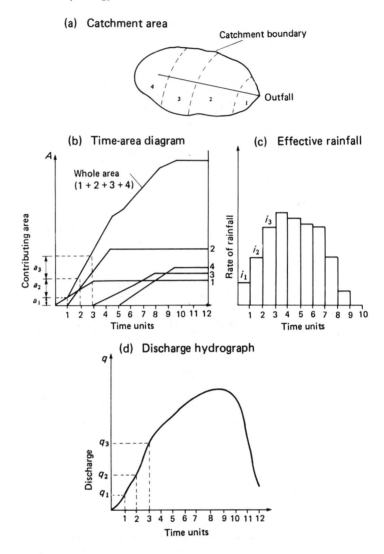

(a) Catchment area

Catchment boundary

Outfall

(b) Time-area diagram

Whole area
(1 + 2 + 3 + 4)

Contributing area

Time units

(c) Effective rainfall

Rate of rainfall

Time units

(d) Discharge hydrograph

Discharge

Time units

Fig. 18.6   TRRL hydrograph method.

each time unit. The effective rainfalls may be obtained by assuming a runoff coefficient and applying this to each time unit rainfall in turn, or a constant loss rate can be assumed and subtracted from each time unit rainfall rate. Fig. 18.6c shows the effective rates of rainfall for each time unit, i_1, i_2, i_3, etc, for the storm duration (9 time units).

The discharge rates after each time unit interval are given by:

$$q_1 = i_1 a_1$$
$$q_2 = i_2 a_1 + i_1 a_2$$

$$q_3 = i_3 a_1 + i_2 a_2 + i_1 a_3$$
etc.

Fig. 18.6d shows the sequence of discharges forming the runoff hydrograph at the outfall.

A worked example is shown in Table 18.5. There are four increments of area (ha) resulting in a time of concentration for the catchment equivalent to four time units. The storm duration extends over 10 time units. A runoff coefficient of 0.64 has been assumed and thus the total areal rainfalls in column 2 have been multiplied by 0.64 to give the corresponding effective rainfalls (i mm h^{-1}). The values of q for each area increment (a) and effective rainfall rate (i) are calculated from:

$$q(\ell \ s^{-1}) = \frac{i(mm \ h^{-1}) \cdot a(ha)}{0.36}$$

where 0.36 is the units conversion factor. The summation of the rows across a_1 to a_4 gives the discharge values after each time increment, and thus the required hydrograph. It will be noted that the peak flow occurs after the fourth time interval, the time of concentration of the catchment. This does not always happen, e.g. with late peaking rainfalls.

Two further considerations are necessary. A time of entry from the onset of the storm rainfall to the time of flow into the pipe is usually taken to be 2 min and must be allowed for in the computations. Secondly, experience has shown that there is a certain amount of retention of water in the pipe channel, and amendments to the hydrograph must be made to account for pipe storage (Watkins, 1962).

The procedure of the TRRL method, as presented, applies to one

Table 18.5   The TRRL Hydrograph Method (Runoff Coefficient 0.64)

Time unit	Areal rate of rain		Increment of area (ha)				Discharge $q(\ell \ s^{-1})$
	Total	Effective i(mm h^{-1})	a_1 0.25	a_2 0.82	a_3 0.92	a_4 0.34	
1	13.7	8.8	6.1				6.1
2	90.0	57.6	40.0	20.0			60.0
3	59.4	38.0	26.4	131.2	22.5		180.1
4	18.3	11.7	8.1	86.6	147.2	8.3	250.2
5	16.8	10.8	7.5	26.7	97.1	54.4	185.7
6	13.7	8.8	6.1	24.6	29.9	35.9	96.5
7	5.3	3.4	2.4	20.0	27.6	11.1	61.1
8	5.1	3.3	2.3	7.7	22.5	10.2	42.7
9	6.1	3.9	2.7	7.5	8.7	8.3	27.2
10	4.6	2.9	2.0	8.9	8.4	3.2	22.5
11				6.6	10.0	3.1	19.7
12					7.4	3.7	11.1
13						2.7	2.7

drainage unit, i.e. one pipe, in a system. The calculations have to be carried out for each pipe in a sewerage network as demonstrated earlier in the Rational Method of design. In practice, the application of the TRRL hydrograph method to even a simple configuration of drainage pipes becomes too complex for manual computation and a computer program was produced at an early stage in the method's development. This incorporated certain simplications, such as the assumption that only the impervious areas contributed to the runoff (with a runoff coefficient of 100%) and there would be no runoff from the pervious areas. Thus only the impermeable areas feature in the area data required. The restriction of the method to impermeable areas using 100% runoff in the calculations reduced the flexibility in the program's application to catchments with variable surface characteristics and has thereby attracted much criticism. However, the program is readily available and is the form in which the method is most usually applied. A flow chart is shown in Fig. 18.7. It will be noted that the pipe hydraulics and testing of pipe sizes are included. The TRRL program may be used for calculating flow rates in the design of new sewerage systems or in the checking of existing systems (TRRL, 1976).

18.6   Storm Water Management Model (SWMM)

This comprehensive model was developed by a consortium of American engineers for the US Environmental Protection Agency (EPA). SWMM takes the rainfall and catchment characteristics, determines quantity and quality of runoff, routes the runoff through a combined or separate sewer system and identifies the effluent impact on receiving waters. Thus it is a mathematical model capable of representing urban storm runoff including sewage storage and treatment and combined sewer overflow phenomena (Torno, 1975).

The computer package contains five blocks of subroutines.

(1) EXECUTIVE block contains service routines controlling the computational blocks and has a subroutine with the ability to combine data sets and separate outputs into a single routing, thereby enabling the modelling of large areas.

(2) RUNOFF block computes the storm water runoff and pollution loadings for a given event for each subcatchment and the results form the input into the relevant inlet to the main sewer system. Infiltration on previous areas (a loss to the urban system) is calculated by Horton's equation. Overland flow following the filling of depression storages and gutter flows are simulated using Manning's equation and the continuity equation. The runoff computations require hyetograph data, numerate characteristics for each subcatchment, Horton coefficients and pipe dimensions and roughness. Land use and other surface features and street cleaning frequency are needed for quantifying pollution.

(3) TRANSPORT block contains the centre core of the model with routines for routing flows through the main sewer system. It receives

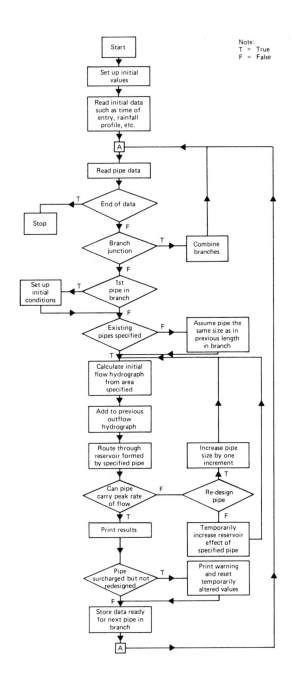

Fig. 18.7   Outline flow diagram of computer program of the TRRL hydrograph method. (Reproduced from TRRL (1976) *A Guide for Engineers to the Design of Storm Sewer Systems*, Road Note 35, by permission of the Transport & Road Research Laboratory.)

the calculated inputs from RUNOFF, the dry weather flow assessed from population statistics etc and any infiltration into the sewers.

The routing procedure uses a finite-difference solution of the St. Venant equations. When a pipe is surcharged, the excess is stored at the upstream manhole. Flows may be routed to a maximum of five outlet points. The particulars of sewer sizes, slopes, types of manholes and of any internal control structures must be entered into the block.

(4) STORAGE block contains modules to simulate the treatment of the sewage and receives the hydrographs and pollution sequences at one of the outlet points of the TRANSPORT block. This block can also calculate costs of selected treatment processes.

(5) RECEIVING WATER block receives the output from TRANSPORT or STORAGE and computes the effect of the discharges on the quality of the receiving waters which are simulated by a network of nodes connected by channels of specified constant surface and cross sectional areas. Control weirs or tidal conditions can also be modelled.

SWMM is written in FORTRAN and requires core storage of at least 512K bytes. When developed, the model was run on IBM 360 machines, but with present computer capabilities it should be possible to accommodate it now on the latest microcomputers.

The SWMM model is the national program in the United States for the design of stormwater sewerage systems and has been incorporated in regional water quality management planning. It has been applied in Europe for long-term planning of combined sewer pollution abatement. Enhancements of the model include improvements in the routing procedures of the TRANSPORT block and capabilities to treat a wider range of pollutants.

18.7   The Wallingford Procedure

Many factors have stimulated a recent appraisal of current methods used in the design of storm water drainage for urban areas. Of particular urgency is the repair or replacement of the many urban sewers in the UK built in the urban and industrial expansions of the mid 19th century and now coming to the end of their design life. In the major industrial cities of the UK, there are regular reports of sewer collapses or pipe failures after severe storms; the cost of their renewal is a crippling load on the finances of local authorities. Indeed, the most expensive item on the budget of a Water Authority is, in 1982, the disposal of waste water and part of the necessary provisions must include the design, building and maintenance of new storm water drainage systems.

The expansion of suburban development, where the proportion of paved surfaces is much less than in city centres, has meant that the methods for calculating the runoff flows that consider only impervious areas tend to result in extravagent drainage systems. The need for more precise criteria on which to base drainage design is being emphasized widely. The ready access to large computers means that simplifications of former methods are now no longer necessary and many more relevant factors can be introduced into the design considerations.

In furtherance of the assessment of design practice in the UK, the Department of the Environment and the National Water Council set up a Standing Committee on Sewers and Water Mains, and a special working party on the Hydraulic Design of Storm Sewers began work in 1974 (NWC, 1976). Combining old and new methods and adopting the most recent systems approach to the problems of design for all urban catchment conditions, a definitive work has been published (NWC, 1981). The Wallingford Procedure consists of five volumes:

Volume I   *Principles, Methods and Practice*
Volume II   *Program Users' Guide*
Volume III   *Maps*
Volume IV   *Modified Rational Method*
Volume V   *Programmer's Manual* (Supplied only with the computer programs)

The production of the Wallingford Procedure for the design and analysis of urban storm drainage has benefitted greatly from the work of the Flood Studies Report, the five volumes of which were published in 1975 (NERC, 1975). Much of the research work has been carried out and coordinated by the Hydraulics Research Station, Wallingford, with important contributions from the Institute of Hydrology and the Meteorological Office. The resultant design programs have been tested by several organizations such as Water Authorities, Scottish Regional Councils, development corporations and consultant engineers involved in storm sewerage design. Basic data required for the design techniques are included for the whole of Great Britain and Northern Ireland so that the procedure may be applied nationwide.

The Wallingford Procedure describes the hydraulic design and analysis of pipe networks for both new schemes and existing systems. It can be accommodate both independent storm water sewers and combined sewers, but the waste water flows must be given as inflows at the appropriate junction in a combined sewer. The hydraulic analysis of a range of structures controlling the flow in a pipe system can be made and certain economic factors are also incorporated into the procedure. The whole package provides a range of methods from which a series of calculation techniques can be selected to suit the conditions of any particular design scheme.

The Wallingford Procedure comprises:

(a) the Modified Rational Method incorporating the results from recent studies but still giving peak flows only. A routing coefficient in addition to the volumetric runoff coefficient is included in the formula. It is recommended for initial designs and for use in homogeneous catchments up to 150 ha in area;

(b) the Hydrograph Method models surface runoff and pipe flow and provides a pattern of discharge in time. It is useful for most situations;

(c) the Optimizing Method, which optimizes the design of pipe depth and gradient as well as diameter. This method is particularly geared to assessing minimum costs; and

Table 18.6 Methods and Models in the Wallingford Procedure

(Reproduced from National Water Council (1981) *Design and Analysis of the Urban Storm Drainage: The Wallingford Procedure*, by permission.)

Method	Model					
	Rainfall models	Overland flow models	Pipe flow models	Sewer ancillaries models	Construction cost model	Flood alleviation benefit model
Modified Rational Method	Intensity – duration – frequency relationship	Percentage runoff model + time of entry	Pipe full velocity	Storm overflow	⎫	—
Hydrograph Method	Rainfall profiles	Complete surface runoff model / Sewered sub-area model may be used for selected sub-areas	Muskingum – Cunge	Storm overflow / Storage tank / Pumping station	⎬ TRRL resource cost model	—
Optimizing Method	As for Modified Rational Method	As for Modified Rational Method	Pipe full velocity	—	⎭	—
Simulation Method	As for Hydrograph Method	Complete surface runoff model / Sewered sub-area model (without surcharging) may be used for selected sub-areas	Muskingum – Cunge and surcharged flow	As for Hydrograph Method plus Tailwater level		Middlesex Polytechnic Flood Hazard Research Project Model (not included in programs)

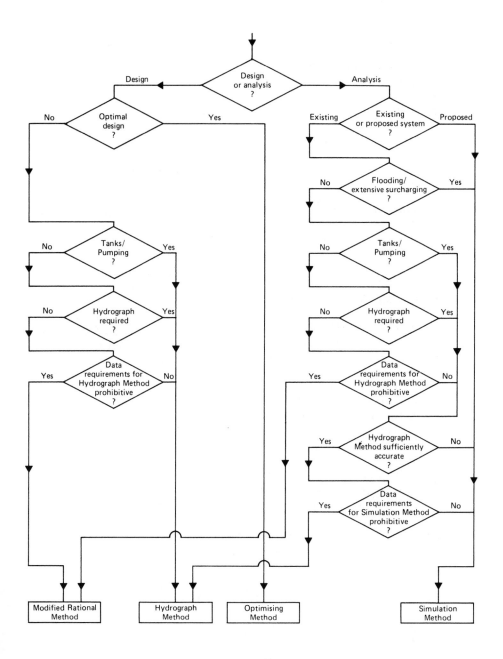

Fig. 18.8 Selection of a method. (Reproduced from National Water Council (1981) *Design and Analysis of Urban Storm Drainage: The Wallingford Procedure*, by permission.)

(d) the Simulation Method, which analyses the performance of existing systems or proposed designs operating under surcharge conditions.

A general outline of the four methods is given in Table 18.6 which itemizes component parts of each method. All the methods can be linked to the construction cost model developed at TRRL (Farrar, 1977), and, for the Simulation Method concerned with surcharging and flooding, it is recommended to apply the Flood Alleviation Benefit Model (Penning-Rowsell & Chatterton, 1977; see Chapter 17). In modelling the different components such as rainfall and overland flow, the methods incorporate many of the formulae derived for the *Flood Studies Report*. Rainfall depth – duration – frequency data and storm profiles, calculations of net rainfall and the use of catchment wetness indices have all been applied in the detailing of the Wallingford Procedure. Fig. 18.8 shows a flow chart constructed to assist in the choice of methods for particular problems in the design of a new drainage scheme or the analysis of an existing system.

The Simulation Method is proving to be most useful to practising engineers since the vexed problem of pipe surcharging is included in the analysis. As indicated in Table 18.6, the major difference from the Hydrograph Method lies in the modelling of the pipe flow (Bettess *et al*, 1978). The same flow equations are used namely the Colebrook-White equation for velocity (Ackers, 1969, and HRS, 1978) and the Muskingum-Cunge routing procedure (Chapter 16), but instantaneous discharges are calculated throughout the sewer system at a given time increment instead of complete hydrographs being routed sequentially from one pipe to another. Thus the interactions between surcharged pipes can be modelled.

The Simulation Method allows for the storage of surcharged water within manholes and on the ground surface flooded to a uniform depth over an assumed area contributing to the pipe length. The extent of the temporary surface flooding related to calculated flood volumes must be assessed on the ground.

Surcharging occurs when incoming flow is greater than full-bore pipe capacity or when a raised tailwater level causes a backwater effect. It is assumed that the manhole losses are proportional to velocity head in the pipe, then head loss, Δh, in the surcharged pipe is composed of pipe friction loss plus losses at both manholes over a time increment dt.

$$\Delta h = \left(\frac{L_\lambda}{d} + k_m \right) \frac{V^2}{2g} + \frac{1}{g} \frac{dV}{dt}$$

where L and d are pipe length and diameter (m)

V is flow velocity (ms^{-1})
g is gravity acceleration (ms^{-2})
λ is the friction coefficient ($8gRi/V^2$) ($s-1$)
   (with R hydraulic radius and i hydraulic gradient)
m is the head loss coefficient for manholes
   (with values of 0.15 for a straight manhole, 0.50 for 30° bend and 0.90 for 60° bend manhole)

From the storage equation:

$$\frac{dS}{dt} = I - O$$

where I is the total flow into the upstream manhole from the upstream pipes and direct surface runoff from its subcatchment. With h, the difference in levels in upstream and downstream manholes and the storage equation, the flow over the chosen time increment can be simulated given the manhole and pipe geometries. The transition phases to and from free surface flow and surcharged (pressurized) flow are demonstrated in Fig. 18.9.

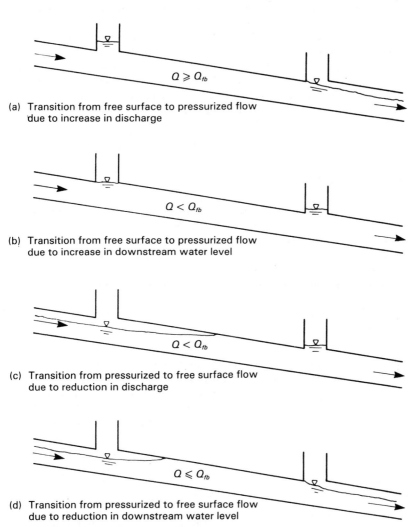

(a) Transition from free surface to pressurized flow
     due to increase in discharge

(b) Transition from free surface to pressurized flow
     due to increase in downstream water level

(c) Transition from pressurized to free surface flow
     due to reduction in discharge

(d) Transition from pressurized to free surface flow
     due to reduction in downstream water level

Fig. 18.9   Transitions between free surface and pressurized flow in a pipe. (Reproduced from National Water Council (1981) Design and Analysis of Urban Storm Drainage: The Wallingford Procedure, by permission)

The state of the sewer system with the volumes stored at each surcharged manhole is determined by repeating the calculations for each time increment until the performance of the system has been established over the whole period of the storm event.

The Wallingford Procedure package (WASSP) includes many practical refinements to the basic method. Recommendations on the roughness values of sewers and on maximum design velocities are given and particulars of ancillary structures within the sewer system such as detention tanks and pumping stations can be incorporated into the model.

The application of the Wallingford Procedure in storm water drainage design in the UK greatly assists local authority engineers faced with the complex renewal problem of old systems with severely curtailed financial resources. It also provides a comprehensive package for the developer and consulting engineer designing new urban areas when more precise evaluations of pipe systems are required to keep costs to a minimum. Although the procedure has already been tested widely, further improvements in detail are envisaged in the future. The suite of programs has been adapted for application outside the UK.

18.8   The MOUSE System

The rapid advances in computer technology have stimulated the development of a suite of programs for urban drainage analysis and design to run on a microcomputer. The MOUSE System is a microcomputer software package produced by a consortium of Danish research engineers, software specialists and practising consulting engineers. It is written in PASCAL computer language and has a series of menus providing a choice of operations selected interactively with user friendly guidelines.

The MOUSE capabilities are the simulation of surface runoff, pipe flow in urban sewer systems and pollution loads.

There are three main groups of menus:

(1) INPUT. A detailed data base is assembled in files providing a choice of information in the several menus.

Rainfall	– standard depth-duration-frequency matrix
	– intensity formula with a range of parameters
	– individual rain events
	– historic storm events.
Wastewater	– volumes and composition of wastewater
Hydrological data	– initial losses, time/area function
	– infiltration, surface roughness
Pipe and catchment data	– critical levels, weir and pump functions
Hydraulic/boundary data.	

(2) COMPUTATION. This menu contains three computational modules:
  (a) Runoff module which contains two surface runoff methods
      Level A using a single loss and the time/area function
      Level B using Horton infiltration, initial losses of surface wetting and detention and the kinematic wave description for hydraulic modelling.
  (b) Pipe flow module which provides for kinematic/diffusive/dynamic wave computations.
  (c) Pollution model – SAMBA.
(3) OUTPUT. These presentation menus provide for plots or print-outs of simulation results requiring a graphic package.

The running of MOUSE requires a minimum of 512K bytes of Random Access Memory, a hard disk with a minimum of 10 Mbytes storage and at least an 80-character width printer.

References

Ackers, P. (1969). *Tables for the Hydraulic Design of Pipes*. Hydraulics Research Station, 2nd ed. HMSO. (3rd ed. 1977.)

Aitken, A.P. (1975). 'Catchment models for urban areas', in *Prediction in Catchment Hydrology*, ed. Chapman, T.G. and Dunin, F.X. Australian Academy of Science, pp. 257–275.

Bettess, R., Pitfield, R.A. and Price, R.K. (1978). 'A surcharging model for storm sewer systems' in *Urban Storm Drainage*, ed. Helliwell, P.R., Pentech Press, 306–16.

Chandler, T.J. (1965). *The Climate of London*. Hutchinson, 292 pp.

Farrar, D.M. (1977). *A Procedure for Calculating the Cost of Laying Rigid Sewer Pipes*. TRRL, Supp. Report 333.

Hall, M.J. and Hockin, D.L. (1980). *Guide to the Design of Storage Ponds for Flood Control in Partly Urbanised Catchment Areas*. CIRIA Tech, Note 100, 103 pp.

Helliwell, P.R. (ed). (1978). *Urban Storm Drainage*, Pentech Press, 728 pp.

Hydraulics Research Station (HRS). (1978). *Charts for the Hydraulic Design of Channels and Pipes*. 4th ed, HMSO.

Landsberg, H.E. (1970). *Climates and Urban Planning in Urban Climates*. WMO Tech. Note 108.

Meteorological Office (1976). *Averages of Temperature for the United Kingdom 1941–70*, HMSO/Met. Office 883.

National Water Council (NWC). (1976). *Working Party on the Hydraulic Design of Storm Sewers: A review of progress, March 1974–June 1975*, 26 pp.

National Water Council (NWC). (1981). *Design and Analysis of Urban Storm Drainage: The Wallingford Procedure*. 5 Volumes.

Natural Environment Research Council (NERC). (1975). *Flood Studies Report*. 5 Volumes.

Packman, J.C. (1980). *The Effect of Urbanisation on Flood Magnitude and Frequency*. Institute of Hydrology, Report No. 63, 117 pp.

Penning-Rowsell, E.C. and Chatterton, J.B. (1977). *The Benefits of Flood Alleviation: A Manual of Assessment Techniques*. Saxon House, 297 pp.

Swinnerton, C.J., Hall, M.J. and O'Donnell, T. (1972). 'A dimensionless hydrograph design method for motorway stormwater drainage systems.' *J. Inst. Highway Eng.*, Nov, 2–10.

Swinnerton, C.J., Hall, M.J. and O'Donnell, T. (1973). 'Conceptual model design for motorway stormwater drainage.' *Civ. Eng. & Public Works Rev.*, **68**, 123–132.

Tabony, R.C. (1980). 'Urban effects on trends of annual and seasonal rainfall in the London area.' *Meteor. Mag.*, **109**, 189–202.

Torno, H.C. (1975). 'A model for assessing impact of stormwater runoff and combined sewer overflows and evaluating pollution abatement alternatives.' *Water Research*, **9**, 813–15.

Transport & Road Research Laboratory (TRRL). (1976). *A Guide for Engineers to the Design of Storm Sewer Systems*. TRRL Road Note 35, 30 pp.

Watkins, L.H. (1962). *The Design of Urban Sewer Systems*. DSIR Road Research Tech. Paper No. 55. HMSO, 96 pp.

19

Water Resources

The subject of water resources has provided the greatest encouragement to the growth in hydrological studies in this century. Certainly, it has been seen in Chapters 17 and 18 that the need for the appraisal of floods and for the determining of urban runoff from rainfall has stimulated developments in stochastic and deterministic hydrology. However, it is the expansion of hydrological measurements needed for the evaluation of water resources that has initiated widespread and comprehensive study of the hydrological environment. In the UK, this increased activity was given legislative support by the Water Resources Act, 1963. (Alas, the statutory hydrometric requirements were repealed in the Water Act 1973 and there is now active encouragement for their reinstatement.) The establishment of hydrological networks of measurement stations resulted from the legal requirement laid on River Authorities to evaluate the water resources of their areas. Before the new gauging stations had time enough to produce the long-term information required by the planners, hydrologists were encouraged to devise techniques for making assessments from inadequate data. Thus statistical hydrology and data simulation, sometimes called synthetic hydrology, came to the fore.

These modern methods dependent on the digital computer are also proving invaluable in water resource projects overseas in the developing countries, but the need for real measurements of rainfall and streamflow is still vital. The advent of the International Drinking Water Supply and Sanitation Decade, 1981–1990, is encouraging the international and national financing bodies to support the expansion of hydrometric schemes in the emerging nations to provide the necessary data bases for water resources accounting. The provision of potable water to all of the world's population is one of the main aims of the Decade.

19.1 Water Demand

The supply of wholesome water to a resident population is dependent on population size and rate of consumption. Additionally, there is the provision of water to industry according to its various needs, and for agricultural purposes of which irrigation is seasonal and weather dependent. Some of the largest users of water are electricity generating stations for cooling. Many industrial consumers organize their own resources and supply systems, but in the UK they must register their consumption with the Water Authorities. All those who require high-quality water generally obtain their supplies from public supply undertakings, water boards or water companies. The allocation of the total water resources of an area is the responsibility of the Water

Authority which must organize the licensing of all major abstractors, record their requirements and ensure that all demands can be met at all times as far as possible.

During this century, the general trend in demand for water has shown a continual increase due to the growth in populations and to higher standards of living. This applies in most countries. In England and Wales, the rate of population increase has now fallen. Fig. 19.1 shows the growth in the population of England and Wales since 1881. In 1965, assuming an increasing growth rate, the forecast population for 2001 was 66 millions. Subsequent forecasts take into account the reduction in the growth rate, e.g. in 1977, a forecast population for 2001 was 50 millions. A reduction of 16 millions in the possible population in 2001 requires a complete revision of projected water demands. However, there has also been an increase in per capita consumption of water resulting from the generally higher standard of living enjoyed by the population. In 1941, the per capita demand (including industrial use) on public water supplies was 150 l/head/day. This had risen to 250 l/head/day in 1965, but thereafter the growth rate has declined and since 1973, the per capita consumption has remained steady at 307 – 308 l/head/day. The present overall picture of water demands (15 400 Ml/day) in England and Wales is for a constant or only slightly increasing supply in the

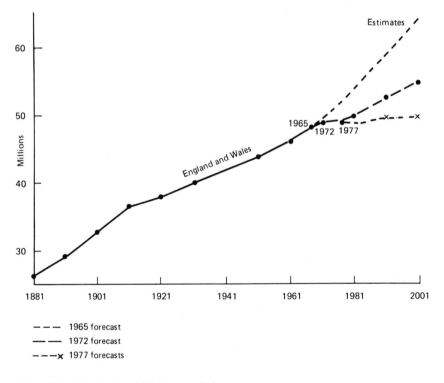

Fig. 19.1   England and Wales population.

near future. The major user of water, the Central Electricity Generating Board, has cut consumption drastically to match reduced power demands by rationalizing the number of power stations in operation, and the general effect of the present (1982) recession has been a decrease in water demand from other industries. Total abstractions registered by the Water Authorities fell from 15 583 Mm^3 in 1971 to 12 852 Mm^3 in 1978 (WDU, 1980).

Assessing the present pattern of water demand in England and Wales, it would appear that there is little need for the development of further water resources. However, if the quality of life were to improve and industrial demand increase to double the per capita consumption, then the water demand could rise to the present average consumption of 642 l/head/day in cities in the USA. In most developing countries, the water demand is still very much on the increase and development of resources is of prime importance.

19.2   Water Availability

Once the needs of an area have been established, be it for a single town's domestic supply, the irrigation water for a commercial plantation, or the total demands of a whole country, and some estimates of continuing requirements in the future made, then the engineer must investigate the availability of the resources. It is at this stage that the hydrological measurements described in the chapters of Part I come to be applied. Hydrological analyses such as those outlined in the chapters of Part II are usually needed in making assessments of water availability. When there are few measuring stations and data records are limited in length, then the analyses become more involved and statistical probabilities less assured.

From the general location and the physical features of the area to be developed, the sources of water are first identified. The average annual rainfall and its average seasonal distribution are first indicators of possible water availability. The existence of perennial streams and rivers is the next guide to available sources, but should these be lacking, then a search for suitable groundwater aquifers has to be made. Once a suitable source has been found, then in most climates of the world, in order to guarantee continuous supplies, some form of storage is essential. If the supplies are to come from surface waters, then failing an adequate natural lake, a storage reservoir must be constructed. Water from groundwater sources is usually already in suitable storage, but the aquifer capacity needs to be fully investigated before a steady supply can be relied on for the design life of a development scheme.

When countrywide water resources are being evaluated, it is necessary to note regional differences. For example, over England and Wales, from an average annual rainfall of around 900 mm, 440 mm are lost through evaporation leaving on average 460 mm of residual rainfall. Over the area of 151 100 km^2 this represents a volume of available water of about 69 500 Mm^3, which is some five times the annual abstractions. However, the depth of residual rainfall varies from over 2000 mm in the wet western mountains to under 100 mm in the driest parts of south-eastern England (WRB, 1973). Since the major centres of demand are in central and southern

England, the predominance of the water resources in the north and west poses problems of storage and distribution.

19.3   Selection of Storage Sites

When the most promising source region has been identified, suitable locations for the building of a storage reservoir must be investigated. A feasibility study for evaluating a possible reservoir site includes the following operations (Cole, 1975):

(a) topographical survey to determine proposed water area and water volume;
(b) geological survey of the dam site;
(c) survey of proposed aqueducts or pipelines to supply;
(d) hydrological assessments of the catchment river flows at the dam site for impounding reservoirs, or at the river site for pumping water to an off-channel storage, and of the design flood for the dam spillway;
(e) appraisal of the necessary land use changes and the future amenity value of the reservoir, and
(f) estimated cost of the scheme.

In carrying out feasibility studies of several possible catchments, it is necessary to make estimates of the *yields* in relationship to the amount of storage that would be available at each proposed reservoir site. The yield of a reservoir is the volume of water regularly available over a unit period of time. Approximate yield estimates for catchments in the UK can be obtained from the Lapworth Chart (Fig. 19.2). This was derived from the low-flow runoff records of 17 gauged catchments over a period which included the notable drought of 1933–34 (IWE, 1969). Corrections to these estimates are required for catchments with a large proportion of permeable rocks. To use the chart, a value of the average annual runoff (mm) is estimated for the catchment area to the proposed dam site for an impounding reservoir, and knowing the effective storage capacity from the topographical survey, the yield in mm per annum over the catchment can be read from the ordinate. This can then be converted into volumetric units of $m^3 \, day^{-1}$ or $M \, l \, d^{-1}$.

When all the governing factors have been considered in relationship to the feasibility studies and a final choice of site has been made, then the detailed planning and designing of the storage scheme can be advanced.

19.4   Reservoir Yield

At the detailed design stage of a water resource development involving an impounding reservoir, there are two important hydrological studies to be made: more precise evaluations of the yield of the catchment area, and the choice of a design flood for the dam spillway. The latter, which is assessed in close co-operation with the engineers designing the dam, has been considered in detail in Chapter 17.

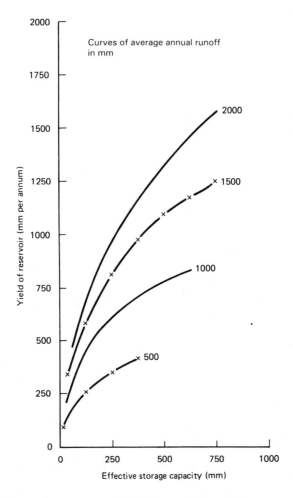

Fig. 19.2   Reservoir storage and yield (Lapworth chart). Storage (Mm^3) = ESC (mm) × Catchment area (km^2) × 10^3. Yield (m^3 day^{-1}) = Yield (mm/a) × Catchment area (km^2) × 2.74. (Reproduced, with modifications, from Institution of Water Engineers (1969) *Manual of British Water Engineering Practice, 4th edn., Vol. II,* by permission.)

In studying in greater detail the yield of a reservoired catchment area, it is important to appreciate more fully the implications of the term with respect to the variabilities in the catchment runoff. A definition of yield suggested by the Institution of Water Engineers and modified by Law (1955) is as follows:

'The uniform rate at which water can be drawn from the reservoir *throughout a dry period of specified severity* without depleting the contents to such an extent that withdrawal at that rate is no longer feasible'.

Thus given a fixed storage capacity in the reservoir, over a drought period of greater severity than that of the design, the regular amount of water available for supply, the yield, would have to be reduced. The occurrence of any such drought governs the yield for a given storage. Alternatively, if a constant yield is required beyond the capabilities of a reservoir designed to fulfil requirements over a drought of specified severity and duration, the capacity of the storage would need to be increased to meet the demand. To ensure adequate storage for a constant yield or to evaluate the yield of an existing reservoir, the study of low flows is fundamental.

Reservoir yield analysis has attracted considerable attention among researchers, particularly in the field of statistics, a subject essential for the assessment of drought probabilities (WRA, 1966). Important contributions to the subject have also been made in determining yields from inadequate discharge data using the techniques of synthetic hydrology. Only a selection from the numerous methods can be outlined here; for a comprehensive treatment of the subject the reader is referred to McMahon and Mein (1978).

19.4.1 Historic Droughts

When there is a sufficiently long representative series of discharge records for the stream to be impounded, the occurrence and frequencies of periods of critically low flows can be determined directly. Some regard must be paid both to the proposed life of the water resource scheme in accepting the representativeness of the data, and to the possibility of an increased yield being required in the future. The expected loss by evaporation from the reservoir must also be taken into account. Using the runoff time series, there are several straightforward methods of analysis.

Mass Curve (Rippl) Method.   From the historic record, the sequences of months having the lowest flows are abstracted and for each sequence the cumulative amounts plotted against time. The technique is demonstrated fully in Fig. 12.3, the reservoir being assumed full at the commencement of a drought period. The testing of all drought periods is needed before deciding on a design drought on which to assess the yield given a specified storage. A chronicle of some of the notable droughts in the UK is given in Chapter 10.

The Residual Mass Curve Method is an extension of the mass curve technique with the advantage of having smaller numbers to plot and hence increased accuracy on the ordinate scale. Each flow value in the record is reduced by the mean flow (mean monthly or mean annual according to the duration studied) and the accumulated residuals plotted against time (Fig. 19.3). A line such as AB drawn tangential to the peaks of the residual mass curve would represent a residual cumulative constant yield that would require a reservoir of capacity CD to fulfil that yield starting with the reservoir full at A and ending full at B. The largest deficit between this residual yield line and the residual mass curve gives the minimum storage required to maintain the yield. In Fig. 19.3, the residual mass curve has been plotted for the same data used in Fig. 12.3. The mean rate of flow over the 5 year period is 1855 Ml per month. The slope of the residual yield line (AB) is -240 Ml month^{-1}, which when adjusted by the mean rate gives $1855 - 240 = 1615$ Ml

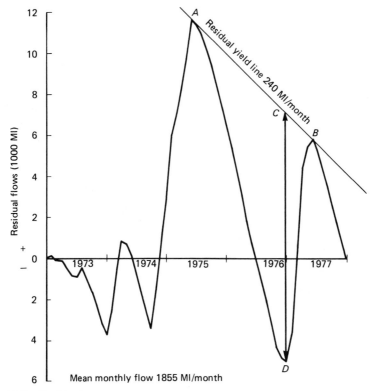

Fig. 19.3   Residual mass curve.

month^{-1} or 52 Ml day^{-1} for the actual yield rate for the period $A \rightarrow D \rightarrow B$. The storage required is then 12 000 Ml given by the line CD. This result compares favourably with the answer obtained in Fig. 12.3 for the necessary storage to sustain a 52 Ml day^{-1} yield over the particularly dry period represented by the period $A \rightarrow D \rightarrow B$ in Fig. 19.3.

The Minimum Flows over Various Durations e.g. 6, 12, 18, 24, 30, etc. consecutive months, can be abstracted from the runoff record and their flow volumes plotted against the corresponding durations (Fig. 19.4). A required yield line at the relevant slope is drawn from the origin of the plot, and a parallel line tangential to the plotted droughts curve identifies both the critical drought duration and the amount of storage, S, needed to provide the yield. In practice, it is recommended to add a year's required supply to the storage as a safety factor.

Considering also the probability of occurrence of a specified low flow, the design drought may be evaluated from the historic record by extreme value analysis of the low flows over a selected time period. Several durations, e.g. 6, 12, 18, 24, etc., months can be tested. The method of analysis by the

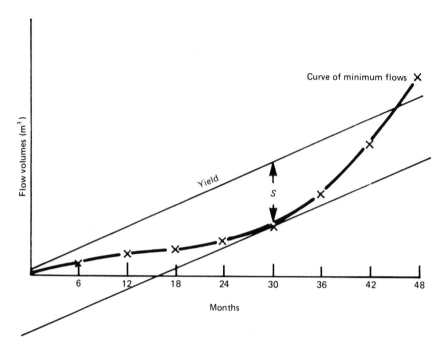

Fig. 19.4   Reservoir yield from minimum flows.

Gumbel EVIII extreme value distribution is described in Chapter 12. The lowest flow over a time period can be read from the curve for the expected return period (design period) of the scheme. The minimum value obtained gives an estimate of water availability from the catchment that can be compared with the reservoir storage capacity to give the reservoir yield.

19.4.2   Synthetic Droughts

Many decisions on reservoir yield and storage capacities necessarily have been made on very limited series of river flow records compared with the planned life of a water resources project. Generally, large safety factors for margins of error are allowed for, particularly when the return periods of the measured sequences are in doubt. During periods of expansion in water demand, over-designing of a storage reservoir is acceptable, but when increases in demand begin to fall off, water engineers must seek to refine their yield – storage assessments and pay due regard to the relative costs of different storage capacities.

In this respect, the developments in synthetic hydrology are making a valuable contribution to water resource planning. The techniques of stochastic hydrology (Chapter 15) allow the hydrologist to generate long series of runoff data matching the statistical properties of the short records available. With the random component incorporated into such generated series, the effects of occurrences of droughts of specified severity can be more precisely assessed. Using the synthetic data series, reservoir yield and related storages can be evaluated by applying the methods adopted for the historic records.

19.4.3   The Hurst Method

A distinctive property, the range, of the residual mass diagram was studied by the distinguished physicist, H.E. Hurst, who was concerned with reservoir storages on the River Nile. Definition of the range, R is shown in Fig. 19.5. Cumulative departures of sequential yearly discharges from the mean annual discharge are plotted against time. The range, R, is the difference between the maximum and minimum values of the residual curve. The range represents the amount of storage which would be required to sustain a discharge (yield) equivalent to the mean flow throughout the period of record. From several long annual series of river discharges and other natural phenomena such as rainfall and temperature, Hurst found the following relationship:

$$R/s = (N/2)^K \tag{19.1}$$

where R is the range previously defined, s is the standard deviation of the time series of data, N is the number of years of record and K is an exponent. If a data series is truly random, statistical analysis shows that $K \rightarrow 0.5$ for large N, but Hurst found a range of values for K from 0.46 to 0.96 with a mean of 0.72 (Hurst et al., 1965). This departure of K from 0.5 is known as the Hurst

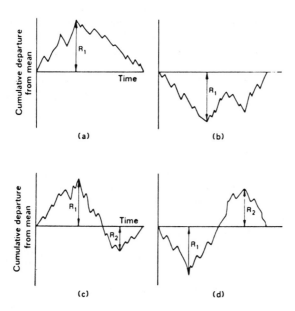

Fig. 19.5   Definition of range (R). For (a), $R = R_1$; (b) $R = R_1$; (c) $R = R_1 + R_2$; (d) $R = R_1 + R_2$. (Reproduced from T.A. McMahon & R.G. Mein (1978) *Reservoir Capacity and Yield*, by permission of Elsevier.)

phenomenon, and has stimulated many researchers in the representation of the phenomenon in modelling hydrologic time series. It is generally considered that a value of $K > 0.5$ represents a measure of persistence in natural time series, e.g. if a sequence of 2 or 3 wet years occurs, there is a tendency for the following year to be wet.

To evaluate required storages, S, for different yields, B, less than the mean discharge, M, Hurst calculated values of S/R for several data series with different values of $(M - B)$. He obtained the following empirical relationships:

$$\log_{10}(S/R) = -0.08 - 1.00\,(M - B)/s$$

or

$$S/R = 0.97 - 0.95\,[(M - B)/s]^{\frac{1}{2}}$$

for the data plots shown in Fig. 19.6. It is to be noted that there is a rapid reduction in the storage, S, required as the yield reduces from the mean, M. For application to reservoir yield, the hydrologist is recommended to derive comparable relationships from pertinent river flow records.

19.5   Reservoir Contents

Closely connected with reservoir yield analysis discussed in the previous section is the probabilistic treatment of reservoir contents, in particular the probabilities of being full or empty. In operating a reservoir for water supply, it is essential to be able to estimate, for a pattern of demand rates, the probability of the storage becoming empty. Thus the long term probability distribution of the reservoir contents under a known operating procedure is a main requisite.

Although the operation of a reservoir is a continuous process, it is convenient to consider it on a discrete time basis. Monthly, seasonal or annual time intervals may be used. The contents of the reservoir at any time $(t + 1)$, C_{t+1}, is given by:

$$C_{t+1} = C_t + X_t - M_t \tag{19.2}$$

where C_t is the reservoir contents at time t, X_t is the volume of inflow from t to $t + 1$ and M_t is the demand (which combines supply and compensation water), released during the same time interval. There are constraints on C_{t+1} when the reservoir is at or near its extremes of full or empty.

It is also convenient to denote all water quantities in terms of integer units of volume. Let the capacity of the reservoir when full be K such units of volume, and in the simplest case, let the demand be constant, i.e. $M_t = M$ units for all time intervals.

The assumed mode of operation of the reservoir governs the application of the discrete water balance equation, Equation 19.2. If the inflow, X_t, is sudden and precedes the demand, M, and the contents are considered *after* the demand, then the boundary constraints on the use of Equation 19.2 are given by:

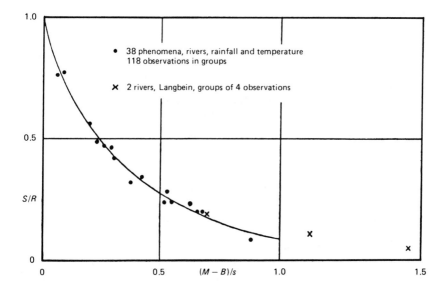

(b) Relation between draft (B) and maximum deficit (S)

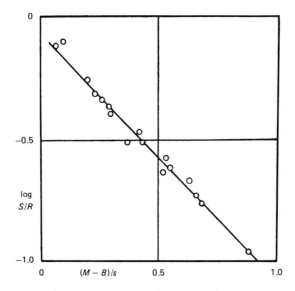

(a) Relation between maximum deficit (S) and draft (B):
     Rivers, rainfall and temperature

Fig. 19.6   (Reproduced from H.E. Hurst, R.P. Black & Y.M. Simaike (1965)
Long-Term Storage: An Experimental Study, by permission of Constable & Co.)

$$C_{t+1} = 0$$
if $(C_t + X_t)$

$$= M - \text{demand just satisfied}$$
$$< M - \text{demand not satisfied} \tag{19.3}$$

$$C_{t+1} = (K - M)$$
if $(C_t + X_t)$

$$> K - \text{reservoir spills excess before } M \text{ withdrawn}$$
$$= K - \text{reservoir just fills before } M \text{ withdrawn} \tag{19.4}$$

Thus, after the demand, the reservoir contents could be in any one of the states:

$$C_{t+1} = 0 \text{ (empty)}, 1, 2, 3, \ldots, (K - M)$$

A *transition probability matrix* may then be set up to give the probabilities of the storage finishing in a particular state after the demand, having started at any of the possible states after the previous demand. For this matrix, all the probabilities of transition from $C_t = 0, 1, 2, \ldots, (K - M)$ to $C_{t+1} = 0, 1, 2, \ldots, (K - M)$ need to be found.

Away from the extremes of storage (empty or full), the probabilities that $C_{t+1} = j$ given that $C_t = i$ is equivalent to the probability that $X_t = (j - i + M)$ (from Equation 19.2). The probability distribution of the X_t values may be found from the historic X_t record. An element $(t_{i,j})$ in the transition probability matrix is given by $\text{pr}\{C_{t+1} = j \mid C_t = i\}$ which is thus equal to $\text{pr}\{X_t = j - i + M\}$.

The state $C_{t+1} = j$ might be reached from any one of the n states $C_t = 0, 1, 2, \ldots i, \ldots (K - M)$ (where $n = K - M + 1$). Thus:

$$\text{pr}\{C_{t+1} = j\} = \text{pr}\{C_t = 0\}.t_{0,j} + \text{pr}\{C_t = 1\}.t_{1,j} + \ldots + \text{pr}\{C_t = i\}.t_{i,j} + \ldots + \text{pr}\{C_t = (K - M)\}.t_{n,j}$$

There are n such equations for $j = 0, 1, 2, \ldots, (K - M)$ and they can be written conveniently in the shorthand matrix form:

$$\underline{P}_{(t+1)} = \underline{T} \times \underline{P}_t \tag{19.5}$$

where $\underline{P}_{t+1}$ and $\underline{P}_t$ are column vectors, with the elements (p), giving the probability distributions of reservoir contents at times $(t + 1)$ and t respectively: and $\underline{T}$ is the $n \times n$ transition probability matrix with elements $t_{i,j} = \text{pr}\{X_t = j - i + M\}$ found from the X_t record and specified M value, care being needed when the contents are near empty or full.

As Equation 19.5 is applied recursively over successive intervals, it is found that after ten or so intervals, $\underline{P}_{t+1}$ and $\underline{P}_t$ both converge to an identical asymptotic distribution, $\underline{P}_s$ no matter what the initial distribution, $\underline{P}_0$ at time $t = 0$. Thus:

$$\underline{P}_s = \underline{T} \times \underline{P}_s \tag{19.6}$$

Equation 19.6 represents n simultaneous equations, which should be just sufficient to solve for the n unknown elements of $\underline{P}_s$. However, the n equations are not fully independent since all the columns of $\underline{T}$ (i.e. all the transition probabilities from any one initial state to all possible final states) must sum to unity and so one of the rows of $\underline{T}$ (i.e. one of the n equations) depends on the other rows (equations). However, any one of the rows (equations) can be replaced by the extra information that $\sum_{t=0}^{n-1} p_i = 1.0$ (i.e. all the probabilities of the asymptotic steady state vector should sum to unity) and so a solution is possible.

In any actual application to find the probability (p) of the reservoir contents being in any of the n states, K would be 10 or more, but a simple example with $K = 3$ will demonstrate the method. (For larger K, solutions are found using computer matrix routines.)

Consider a reservoir of capacity $K = 3$ units with constant demand $M = 1$ unit and the inflow, X_t, having the probability distribution:

X_t 0   1   2   3
Prob.   0.25   0.4   0.2   0.15   (Total = 1.0)

Then $n = K - M + 1 = 3$ so that $\underline{T}$ is a 3×3 matrix and $\underline{P}_s$ has 3 elements corresponding to the probabilities of the reservoir contents being 0, 1 or 2 units after deduction of the demand.

Applying Equation 19.2 to the different combinations of C_t, C_{t+1} and the four values of X_t with M always 1 unit when possible, the $t_{i,j}$ values are set down in the expanded form of Equation 19.6:

Final	p_0	$\dagger$ (0.25 + 0.4)	0.25	0		p_0 Initial
State	p_1 =	0.2	0.4	0.25	×	p_1 State
Probabilities	p_2	0.15	(0.2 + 0.15)	(0.4 + 0.2 + 0.15)		p_2 Probabilities

$\dagger$ Reservoir fails to meet demand
$*$ Reservoir spills after inflow and before demand

(19.7)

Note that the transition probability matrix then becomes:

$$\underline{T} = \begin{array}{ccc} 0.65 & 0.25 & 0 \\ 0.2 & 0.4 & 0.25 \\ 0.15 & 0.35 & 0.75 \end{array}$$

and the $n = 3$ simultaneous equations are:

$p_0 = 0.65\, p_0 + 0.25\, p_1$
$p_1 = 0.2\; p_0 + 0.40\, p_1 + 0.25\, p_2$
$p_2 = 0.15\, p_0 + 0.35\, p_1 + 0.75\, p_2$

To solve for the contents probabilities, p_i, the equations are rearranged and the last equation is replaced by $\Sigma p_i = 1$:

$0 = -0.35\, p_0 + 0.25\, p_1$
$0 = \;\;\;0.20\, p_0 - 0.60\, p_1 + 0.25\, p_2$
$1 = \qquad\quad p_0 \;\; + \;\; p_1 \;\; + \;\; p_2$

The solution is: $p_0 = 0.202$, $p_1 = 0.282$ and $p_2 = 0.516$. Thus the reservoir would be empty after the demand on virtually one in 5 occasions. From the values of $t_{i,j}$ in Equation 19.7 will be seen that the reservoir would fail to meet the demand with a probability of $0.25 \times 0.202 = 0.051$, roughly one in 20 occasions. On those occasions when it holds 2 units after the demand (just more than one in 2), it also spills with a probability of:

$$0.15 \times 0.282 + (0.2 + 0.15) \times 0.516 = 0.223$$

nearly one in 5 occasions.

The reason for this frequent spill is that the average inflow is:

$$0 \times 0.25 + 1 \times 0.4 + 2 \times 0.2 + 3 \times 0.15 = 1.25 \text{ units}$$

whereas the demand is only 1 unit. Nevertheless, despite this excess of average inflow over demand, the reservoir can be emptied and even fail to meet the demand.

Fuller details of this transition matrix technique will be found in McMahon and Mein (1978).

There have been many extensions to the probabilistic treatment of reservoir yield. Instead of assuming independent inflows, more realistic serially correlated inflows are used and allowances can also be made for seasonal differences in patterns of inflow. The probability of failure under given conditions has also received much attention. For further enlargement of the scope of the probabilistic methods, which stem from the work of Moran (1959), the contributions of Gould (1961) in Australia, and White (1966), Lloyd (1963, 1966) and Cole in the UK among others, should be consulted.

19.6   Alternative Surface Storage

The foregoing sections have been mainly concerned with the planning and operation of impounding reservoirs relying on the inflow from a surrounding catchment area. When the yield of a reservoir becomes insufficient to meet a rising demand, further resources may be available from neighbouring catchments. *Catchwaters* or *leats*, which are small canals, may be constructed with a gentle slope following the contours round the intervening hillside from intakes on the streams of adjoining catchments. Controlled flows from such catchwaters can result in a marked enhancement of available inflows into a reservoir. The design of catchwaters with regard to assessing their yield has been described by Mansell-Moulin (1966). There are many examples of such catchwaters supplementing the resources of impounding reservoirs throughout the UK.

When the siting of a reservoir means that it cannot obtain its supplies naturally by gravity flow, water has to be pumped up from a river to supply the reservoir. Such a reservoir is known as a *pumped storage reservoir* and it is most likely to be found in areas which do not have adequate residual rainfall. In the English lowlands, the new extensive Empingham Reservoir, covering a tenth of the old county of Rutland and with a gross capacity of 123 Mm^3, gains most of its water by pumping from the Rivers Welland and Nene. More obvious examples of pumped storage reservoirs are the numerous Thames Water Authority's storages for London's water supply, e.g. the reservoirs adjoining Staines and Datchet, into which the supply is pumped from the River Thames.

At the even lower end of a river system, the feasibility of operating *estuary storages* in tidal reaches has been studied in depth in England and Wales. Of the many schemes proposed, those of Morecambe Bay and the Wash

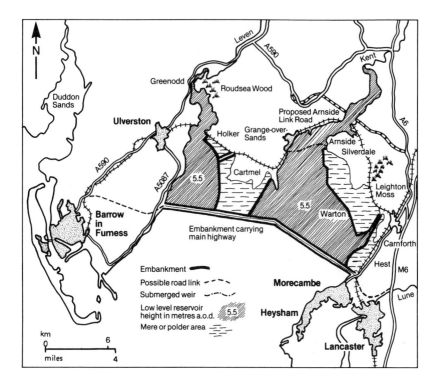

Fig. 19.7   The full barrage. (Reproduced from Water Resources Board (1972) *Morecambe Bay: Estuary Storage*, by permission of the Controller, Her Majesty's Stationery Office. © Crown copyright.)

received the most attention. The Morecambe Bay study suggested several alternative sites for a barrage across the bay to contain the freshwater flows from the Rivers Leven and Kent (Fig. 19.7). The environmental consequences of the proposed barrage ranged from the hydraulic modifications of the tidal flows affecting Heysham harbour to the destruction of the sea marshes, the natural home of many wild fowl. Apart from the storage of 328 Mm^3 of freshwater with a net yield of 2.2 Mm^3 day^{-1}, the principal advantage to be gained from a full barrage was the provision of a more direct motorway to Barrow-in-Furness from Lancaster and the South. The Morecambe Bay storage was to serve the growing needs of Manchester and other parts of Northern England (WRB, 1972).

Estuary storage in the Wash was proposed to serve the water demand of the increasing population in south-east England and the London area. The method of construction of the reservoir here was by building containing embankments to form 'bunded' reservoirs which would be filled by pumping from the rivers above the tidal limit. Thus there would be no interference with the facilities of the several small river ports and the enclosure of areas of the shallows of the Wash would avoid encroachment on valuable agricultural land for reservoir space. A small trial embankment was constructed in the Wash and civil engineers were able to learn a lot about the techniques of working in tidal waters on a scale which had not been attempted before in this country. The reports of these feasibility studies by consulting firms were made available by the clients, the Water Resources Board.

19.7   Groundwater Storage

Major aquifers, (the chalk, limestones and sandstones of the sedimentary rocks of Lowland Britain) have been valuable sources of water for many years. Such groundwater sources have provided high quality supplies for many large centres of population. London relied heavily on the artesian supplies from the underlying chalk. The lowering of regional water tables by the over-development of the resources has led to a new appreciation of aquifers as water storages.

An investigation into the feasibility of recharging the aquifers was proposed by the Water Resources Board for three areas in the London Basin, the Lee Valley, Leyton — Dagenham and Wandle - Ravensbourne. The designated trial areas are shown in Fig. 19.8. In the Leyton - Dagenham area it was also hoped that recharging of the chalk aquifer would stop all water intrusion caused by the dewatering of freshwater by overpumping for supplies. The Lee Valley scheme got under way in 1976 with the drilling of new boreholes and the first recharge tests were made in 1977. A typical dual-purpose borehole is shown in Fig. 19.9. Over a 5-month period, a recharge of 5800 Ml resulted in an area of about 80 km^2 showing a rise of over 1 m in the water table. The recharge water used is raw water from the River Thames treated to potable standards so that there is no danger of polluting the relatively good quality groundwater. On an operational basis, the recharging is done in the winter months when there is ample raw water available and there is surplus capacity at the treatment works. The replenishment of the groundwater storage makes it a reliable long term reservoir to serve extremely dry periods. An account of the Lee scheme with accompanying considerations of the economics of the groundwater recharge is given in Hawnt et al. (1981).

19.8   Conjunctive Use Schemes

In some regions with a concentrated high demand requiring major water resource development, the engineer may have a choice of sources between surface water from nearby amply watered mountainous catchments and groundwater from a thick water-bearing stratum of good transmissivity. It may be technically expedient, (with regard to the reliability of yields) and economically viable to develop both sources and use them jointly to best effect. This would be called a *conjunctive use scheme*. One of the advantages of having the two contrasting sources of supply is that the variations in the availability of surface water do not usually coincide with those of groundwater; thus an increased and steadier supply can be maintained by switching

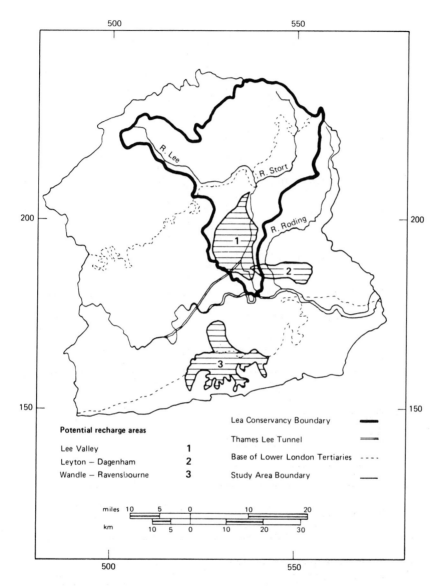

Fig. 19.8   Areas favourable for artificial recharge. (Reproduced by permission of the Thames Water Authority.)

from one source to the other. For example, in the UK, groundwater is a more reliable source in summer when available surface water tends to be at a minimum, yet the surface water can be drawn upon first in winter when residual rainfall is at a maximum, while allowing the groundwater storage to be replenished naturally by infiltration and percolation.

One of the first such conjunctive use schemes was that of the Fylde Water

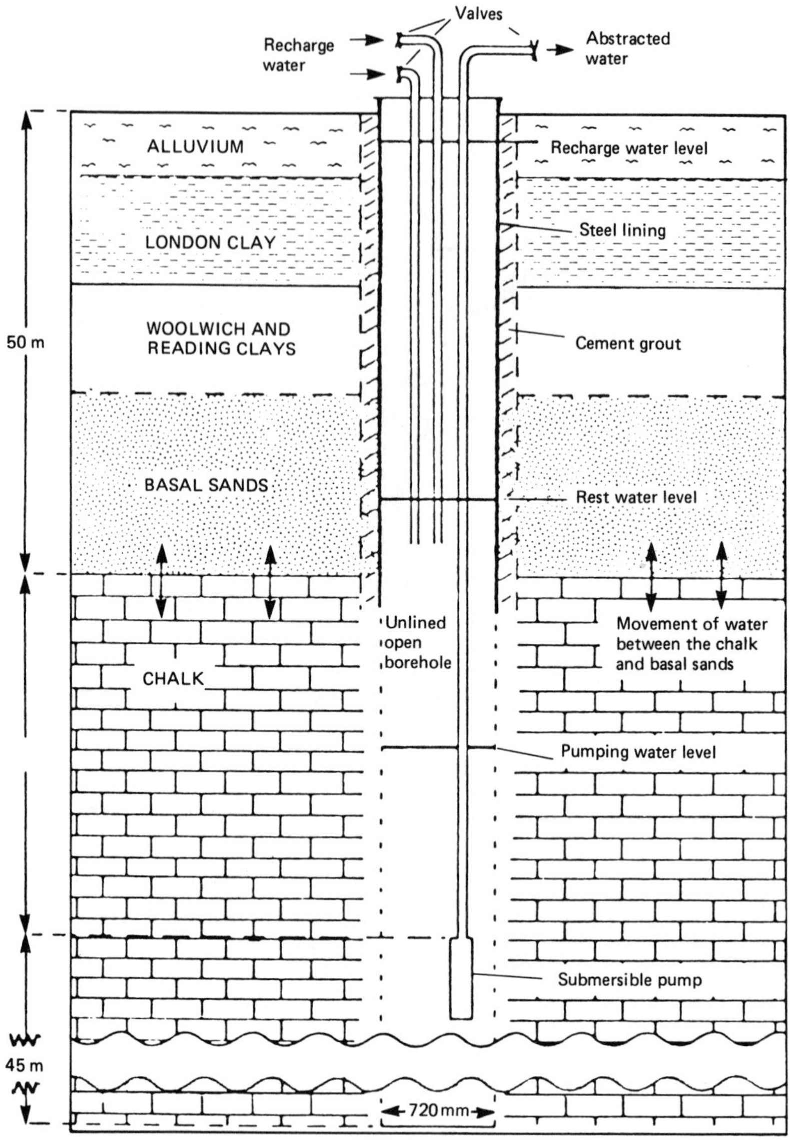

Fig. 19.9   Typical dual-purpose borehole (not to scale). (Reproduced from R.J.E. Hawnt, J.B. Joseph & R.J. Flavin (1981) *J. Inst. Water Eng. Sci.* **35**(5), 437–451, by permission.)

Board in north-west England (Law, 1965). The surface water resources are the Hodder, Dunsop and Barnacre catchments (Fig. 19.10) and the second source of supply is the groundwater of the Bunter Sandstone aquifer (Fig. 19.11). The main centres of demand are holiday resorts along the

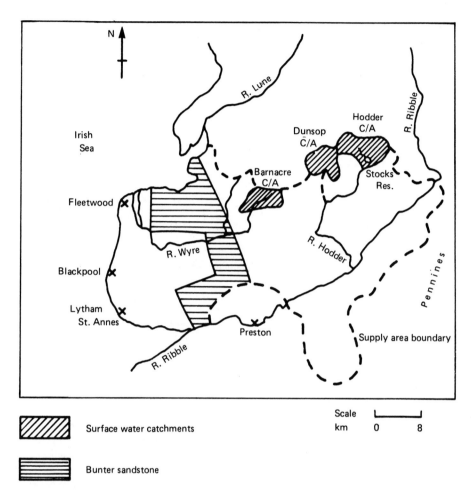

Fig. 19.10   Fylde Water Board. (Reproduced from F. Law (1965) *J. Inst. Water Eng.*, **19**, 413–436, by permission.)

coast; in the summer, the influx of visitors causes a large increase in water consumption. Although there was already provision for overdrawing of the upland resources, the belt of Bunter Sandstone in between the uplands and the coast was ideally placed for developing an alternative source to save the reduced surface waters in summer. This conjunctive use scheme has been the subject of considerable study in devising control rules for the management of the two sources (Walsh, 1971). Computer modelling of the supply and demand system to predict possible future water shortages and to assess economic aspects of the scheme (Burrow, 1971), and of the Bunter Sandstone aquifer behaviour to determine yields and natural recharges (Oakes & Skinner, 1975) has added greatly to prospective success in optimising management methods.

Conjunctive use schemes can also include the optimum operation of other

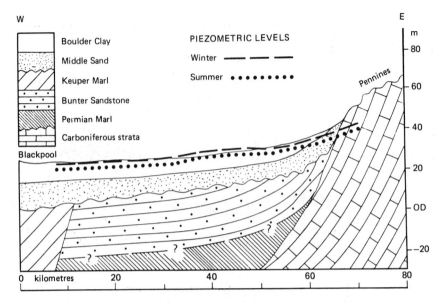

Fig. 19.11   Fylde Water Board generalized east – west cross-section showing range of natural water level fluctuation. (Reproduced from D.B. Oakes & A.C. Skinner (1975) *The Lancashire Conjunctive Use Scheme Groundwater Model*, TR 12, by permission of the Water Research Centre.)

combinations of water resources. For example, two surface reservoirs depending on inflows from different catchments with different yields may be operated conjointly to maximise the available water. A single reservoir may be used in connection with a reliable direct abstraction from another river, or a surface water source may be supplemented by the output from a desalination plant.

Applying the techniques of dynamic programming with the masses of data such as inflows, storage capacities, pumping rates, etc., handled by digital computer, the optimum yields can be evaluated for any chosen combination of sources. If operating costs are known, e.g. cost of pumping from groundwater or relative costs of necessary treatment of water from different quality sources, then these can be incorporated into the computer programs to provide the scheme which incurs the minimum costs for the yields needed to satisfy given demands. The cost of making good any water deficits during the operation of a resource system can also be evaluated.

19.9   Regional Planning

A recent tendency in the management of water resources has been for the growth of large executive authorities. Indeed this move can be seen in many aspects of modern life, ever larger companies and local authorities responsible for ever more and wider activities. One of the advantages of this outlook for water resources is that areas of plenty can be joined to areas of habitual shortages and the available water can be shared to everyone's advantage. With modern techniques of winning water, either by surface water control or by groundwater exploitation, joined to modern system management methods of linking sources to demand centres and including cost analysis, water resources should be considered on a regional scale.

The problems of irregular and scarce resources in south-east England were tackled by planning an overall regional solution. The development of the regional water resources envisaged the construction of major new source works and a network of pipelines and aqueducts to transmit the water to the demand centres. It has been estimated that it takes about 10 years to plan, promote, design and construct either a conventional surface storage scheme or a groundwater development scheme. Careful planning is therefore needed. It is particularly important to assess precisely the future demands of a region; it has already been seen that this is not easy now in the UK owing to the smaller increase in population and certain reductions in demand. However the regional plan for south-east England, taking into account existing sources of supply and service networks, was made by the engineers of the Water Resources Board, the former central authority set up under the Water Resources Act, 1963. This has been described by Armstrong and Clarke

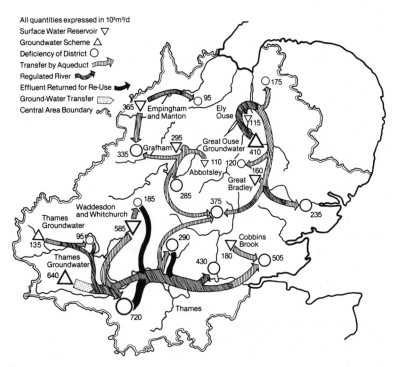

Fig. 19.12   Example programme — allocation of water in the year 2001. (Reproduced from R.B. Armstrong & K.E. Clarke (1972) *J. Inst. Water Eng.*, **26**(1) 11–47, by permission.)

(1972). A computer model was assembled, comprising several sub-programs for handling resources allocation, distribution and costing, to simulate the whole operation of assessing resources and distributing them to the main centres of demand. By varying the data inputs to the model, answers could be obtained to the questions of changing demands under a range of hydrological conditions, especially under extreme droughts. An example of an output from the computer model is the development plan for the allocation of water in parts of south-east England in the year 2001, shown in Fig. 19.12. It estimated that new resources totalling 3.845 Mm^3 day^{-1} would be required of which 0.85 Mm^3 day^{-1} could come from reuse of treated effluent. Of particular interest is the transfer of water from the Ely Ouse River via an aqueduct to the Essex rivers. The transfer of water from one catchment to another and the concept of river regulation will be considered further in the next chapter. Greater details of the optimum links between sources and demand centres for a regional planning model are to be found in a companion paper (O'Neill, 1972). The examples quoted here serve to demonstrate the methods of analysis and synthesis of water resource systems that can be usefully carried out to assist in regional planning. The results shown are based on data from the 1960s and are therefore now somewhat outdated.

References

Armstrong, R.B. and Clarke, K.E. (1972). 'Water resource planning in South-East England.' *J. Inst. Wat. Eng.*, **26**(1), 11–47.

Burrow, D.C. (1971). 'Conjunctive use of water resources (Computation)'. *J. Inst. Water Eng.*, **25**(7), 381–396.

Cole, J.A. (1975). 'Assessment of surface water sources.' *ICE Engineering Hydrology Today*, 73–85.

Gould, B.W. (1961). 'Statistical methods for estimating the design capacity of dams.' *J. Inst. Eng. Australia.*, **33**(12), 405–416.

Hawnt, R.J.E., Joseph, J.B. and Flavin, R.J. (1981). 'Experience with borehole recharge in the Lee Valley'. *J. Inst. Wat. Eng. Sci.*, **35**(5), 437–351.

Hurst, H.E., Black, R.P. and Simaike, Y.M. (1965). *Long-Term Storage: An Experimental Study*. Constable, 145 pp.

Institution of Water Engineers (IWE). (1969). *Manual of British Water Engineering Practice*. 4th ed. Vol II, 655 pp.

Law, F. (1955). 'Estimates of the yield of reservoired catchments'. *J. Inst. Wat. Eng.*, **9**(6), 467–487.

Law, F. (1965). 'Integrated use of diverse resources'. *J. Inst. Wat. Eng.*, 413–436.

Lloyd, E.H. (1963). 'A probability theory of reservoirs with serially correlated inputs.' *J. Hydrol.*, **1**(2), 99–128.

Lloyd, E.H. (1966). 'Probability of emptiness. I. Review of transition matrix techniques', in *Proceedings Reservoir Yield Symposium, September, 1965*. Water Research Association.

Mansell-Moulin, M. (1966). 'The application of flow-frequency curves to the design of catchwaters.' *J. Inst. Wat. Eng.*, **20**(6), 409–433.

McMahon, T.A. and Mein, R.G. (1978). *Reservoir Capacity and Yield.* Elsevier, 213 pp.

Moran, P.A.P. (1959). *The Theory of Storage.* Methuen.

Oakes, D.B. and Skinner, A.C. (1975). *The Lancashire Conjunctive Use Scheme Groundwater Model.* Water Research Centre, TR. 12, 36 pp.

O'Neill, P.G. (1972). 'A mathematical-programming model for planning a regional water resource system.' *J. Inst. Wat. Eng.*, **26**(1), 47–61.

Walsh, P.D. (1971). 'Designing control rules for the conjunctive use of impounding reservoirs.' *J. Inst. Wat. Eng.*, **25**(7), 371–380.

Water Data Unit (WDU). (1980). *Water Data 1978.* Dept. of the Environment, Water Data Unit, 61 pp.

Water Research Association (WRA). (1966). *Proceedings of the Reservoir Yield Symposium, September 1965*, 2 volumes.

Water Resources Board (WRB). (1972). *Morecambe Bay: Estuary Barrage.* HMSO.

Water Resources Board (WRB). (1973). *Water Resources in England and Wales.* HMSO. 2 volumes.

White, J.B. (1966). 'Probability of emptiness II. A variable season model', in *Proceedings Reservoir Yield Symposium, September, 1965*, Water Research Association.

Further Reading

Buras, N. (1972). *Scientific Allocation of Water Resources.* Elsevier.

Hall, W.A. and Dracup, J.A. (1970). *Water Resources Systems Engineering.* McGraw-Hill.

Kuiper, E. (1971). *Water Resources Project Economics.* Butterworths, 447 pp.

Maass, A. *et al.* (1962). *Design of Water Resource Systems.* Macmillan, 620 pp.

Okun, D.A. (1977). *Regionalisation of Water Management.* Applied Science, 377 pp.

Parker, D.J. and Penning-Rowsell, E.C. (1980). *Water Planning in Britain.* Allen & Unwin, 277 pp.

Wiener, A. (1972). *The Role of Water in Development.* McGraw-Hill, 483 pp.

20
River Basin Management

Rivers exercise contrasting influences on the activities of people in different parts of the world. To some extent, the influences exerted depend on the character of the river, but more so on the nature of the surrounding catchment area. Some large rivers have a unifying effect, attracting settlement along their banks; the Nile in Egypt and the great rivers of China are in this category. Others are barriers, forming frontiers between nations, although at the same time being used for communications; the Rhine and Danube are examples of this second group. In very few instances in the older civilizations does a single administrative authority have jurisdiction over a whole river basin. More generally, the benefits of rivers are shared among different communities, and some level of agreement between them has to be reached before major regulating works can be accomplished.

The advancement of civilizations and the spread of populations over the relatively newly settled continents such as the Americas and Australasia led to the establishment of large national territories that contain major river catchments within their boundaries. The inclusion of whole drainage basins within a single country meant that the development of a river could be under the control of one government. Of course this situation pertained in the vast regions of the USSR and the subcontinent of India, but the early status of these countries precluded considerations of comprehensive river basin management. However, it was not until the 20th century that technical knowledge and organizational capabilities could be combined to establish effective management of a single river basin, and the lead came in the New World.

20.1   The Tennessee Valley Authority

The classical example of the foundation of comprehensive river basin management was the creation of the Tennessee Valley Authority (TVA) in the Appalachian Mountains of the eastern USA. The Tennessee River is a tributary of the Ohio River, which in turn flows into the Mississippi River. It drains an area of 106 000 km^2, roughly the size of England (Fig. 20.1). Although the river was originally being developed for hydroelectric power, with the first dam on the main stream completed in 1913, it was not until 1933 that the establishment of the TVA expanded the development into a multipurpose plan that involved a great regional and social experiment. Over half of the catchment area had been cleared of natural forest for farming development, and the cultivated land had suffered intense erosion. The whole basin was ripe for studied development and progressive management.

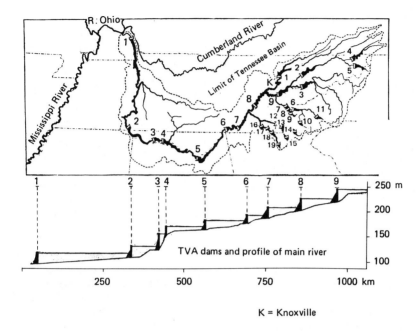

K = Knoxville

Fig. 20.1   Diagram of TVA river control. (Reproduced from L. Rodwell Jones &
P.W. Bryan (1946) *North America*, by permission of Associated Book Publishers Ltd.)

The main aims of the TVA can be summarized as follows:

(a) the provision of power and water for industry, agriculture and
   domestic use;
(b) the prevention of floods in the basin and the lessening of flood dangers
   in the Lower Mississippi;
(c) the provision of a navigational channel up to Knoxville;
(d) the reafforestation of denuded hillsides;
(e) the checking of erosion by contour strip farming and the practice of
   scientific crop rotations and modern methods of fertilizing; and
(f) the provision of recreational facilities on the reservoirs.

Some of these aims represented quite advanced thinking in conservation in
the 1930s.

In addition to the major dams on the main river (Table 20.1), the river
flow is controlled by numerous dams on the headwater tributaries. Of the
total drainage area down to Knoxville, 23 000 km², 89% is controlled, and
the storage capacity of the various reservoirs could sustain the long-term
average discharge of 362 m³s⁻¹ at Knoxville for 169 days if released at a
steady rate and assuming no other runoff (Hull, 1967). This high degree of
regulation of the river is a prime factor in achieving the aims of the
Authority. To manage the resources effectively, a network of rain gauge
stations gives prior warning of heavy falls, so that the reservoirs can be
operated to mitigate the effects of flood flows. Modern systems techniques

Table 20.1   Dams on the Tennessee River

	Height (m)	Location upstream of confluence (km)	HEP (kW)
1 Kentucky	48.8	36.0	160 000
2 Pickwick	34.4	332.6	216 000
3 Wilson	41.8	417.5	444 000
4 Wheeler	21.9	442.4	259 000
5 Guntersville	28.7	561.6	97 000
6 Hales Bar	25.3	693.8	50 483
7 Chickamauga	39.3	758.0	180 000
8 Watts Bar	29.6	852.8	150 000
9 Fort London	41.1	969.3	96 000

can be applied to the telemetered data for real-time control as well as the long-term planning of the optimum allocation of the water resources to meet the various demands (for further details, see Chow, 1964, Chapter 25).

River basin management in its widest sense, as exemplified by the TVA, has also been established in other parts of the USA; the Colorado River and the Sacramento River basins were initially developed on this basis, though their water resources gradually became incorporated into wider regional plans.

20.2   Southern Hemisphere Schemes

Other national schemes of renown are the Orange – Vaal in South Africa and the Murray – Darling in Australia (Fig. 20.2); both of these basin management schemes have been concerned with multipurpose management for flood mitigation, hydropower generation and water resources for irrigation, industry and domestic supply. The overall planning of the Orange – Vaal basin is in the hands of the Department of Water Affairs. The Australian scheme results from the planning of the Snowy Mountains Authority, named after one of the major source regions, and the River Murray Commission; the catchment extends over parts of four separate States. An interesting common feature of these two schemes is that they both seek to improve their resources by transference of water from the smaller well watered catchments on the eastern sides of the main watersheds, whose waters would run surplus to requirements into the Indian and Pacific Oceans, respectively. In both countries, major civil engineering works have been carried out and continue to be designed to effect the transference of supplies across the main divides and to improve the regulation of the main rivers. There are two current large developments, one in each of the two basins.

In South Africa, the most spectacular project is the Tugela – Vaal scheme. Water from the eastward flowing River Tugela, mean annual runoff 5073 Mm^3 from 29 137 km^2, is being pumped 500 m up the Drakensburg

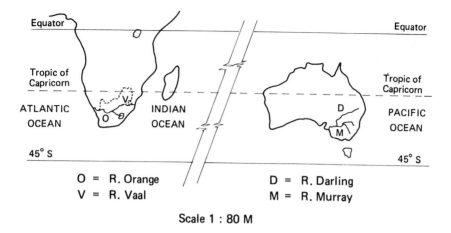

Fig. 20.2   Comparable river basin managements.

Mountains into the headwaters of the Vaal, mean annual runoff 3992 Mm³ from 136 234 km². One of the proposed strategies for the scheme envisages the transference of 350 Mm³ per annum. The water is stored in a special receiving reservoir, Sterkfontein, which is unique in having no flood spill-way. The surface area of the reservoir forms such a large proportion of its catchment area that it can easily absorb any storm, and the outlet works are large enough to discharge the severest floods. In such an event, the normal pumped supply of 5 m³s⁻¹ from the Tugela would be interrupted.

In Australia, the River Murray Commission, having reviewed increased demands for water from South Australia and promised additional supplies to New South Wales and Victoria, agreed to the construction of another large storage in the Murray Basin. The new Dartmouth Dam at 180 m on the Mitta Mitta tributary will be the highest rock-fill dam in Australia. A hydro-electric power station is incorporated into the scheme. The flood spillway is designed to pass 2755 m³s⁻¹, although the design inflow flood peak is assessed at 4246 m³s⁻¹, the large pondage behind the dam reducing the peak flow from the catchment.

20.3   Composite Management

The management of rivers whose catchment areas extend into several countries is a delicate complex matter requiring skilful diplomacy as much as technical expertise. The classic example of a river basin under composite management is the River Rhine, which rises in Switzerland, receives tributaries from France, Germany and Luxemburg, and flows through the Netherlands to the North Sea. It is a truly international river and an impor-tant highway for barge traffic inland from the North Sea ports, in the future to be extended via connecting canals to the River Danube and the Black Sea. The management body is the River Rhine Commission comprising representatives from all the riverine countries. The problems to be solved are mainly concerned with control of the quantity and quality of the flow. The Dutch authorities are particularly sensitive to the quality of the river water, one of their main sources of supply, since the Rhine has received effluent from several heavy industrial areas in Germany and France, and continuous pressure must be put on authorities to reduce the toxicity of such waste waters. The French and German authorities collaborate on flow control in the middle reaches. Where there is a sufficient head down a river reach, hydroelectric generating stations are paired with navigation locks across the river. At the head of the river in Switzerland is found the best quality water and the steepest gradients for power generation. The Swiss authorities bear responsibility for the main discharge measurements and the assessments of snow melt flood dangers that must be transmitted to the downstream authorities.

This gradual historical development of mutual cooperation in the management of a river basin contrasts markedly with the attempts to develop the potential resources of the River Mekong in south-east Asia. The Mekong rises in the Tibetan Plateau, flows through southern China before forming the eastern boundary of Thailand and flowing into the South China Sea near Saigon. As a scheme to aid developing countries, a United Nations Development Programme focused attention on the Lower Mekong with hydrological studies linked to engineering projects for the better use of the water resources. The political unrest in the area in the 1970s severely hampered progress in this ideal opportunity to apply beneficial techniques of river basin management.

20.4   River Management in England and Wales

Developments in the years up to the 1970s leading to the emergence of the present 10 Regional Water Authorities covering the whole of England and Wales were based on natural catchment boundaries. The River Boards which became River Authorities in 1963 were composed of single or groups of natural river basins. The Water Act, 1973, which joined adjacent River Authorities together to form the larger Regional Water Authorities, gave these Authorities responsibility for the whole of the land phase of the hydrological cycle within their areas. The evaluation of the water resources, their allocation, the disposal of waste water, and the problems of land drainage and flood protection all became the responsibility of the Water Authorities. The previous legislation, the Water Resources Act, 1963, had already stimulated research into management of quantities in river basins, but from 1973, there was added the requirement to include water quality into river basin management.

It was seen in Chapter 19 that regional planning of water resources is to be encouraged when sharing of resources is desirable. On a national scale, the transference of water from the wetter west to the demand centres in the east was proposed, but it appears now that most of the autonomous Water Authorities are large enough to be self-sufficient in resources. In consequence, the management of single river basins, or two or three joined together, is now receiving more attention. River basin management requires the regulation of the river flows to provide the maximum satisfaction of all demands on the river.

20.5   River Regulation

The regulation of the quantity and quality of water flowing in a river entails controlling the river discharge and effluent sources by appropriate river engineering and treatment works. The building of dams across the river, forming considerable storage capacities, enables the engineer to regulate the flow to satisfy design requirements. Thus regulating reservoirs are operated to ensure that there is adequate water in the river for all downstream abstractions. The river serves as the aqueduct and reservoir releases take account of expected tributary inflows downstream of the dam to provide some of the abstraction demands. In times of drought, larger quantities of water stored in the reservoir can be released to augment the low flows in the river. In this respect, there is often a minimum acceptable flow designated at an effluent outfall in order to ensure adequate dilution to maintain quality standards in the river.

20.5.1   The River Dee

One of the first designed management schemes for a river basin in England and Wales was for the Dee catchment in North Wales. Rising in the Cambrian Mountains, the River Dee flows eastwards before turning north to enter a wide estuary near Chester. The aims of the scheme were primarily to control high flows (up to 650 m^3s^{-1}) and to prevent widespread flooding along the middle course of the river (Collinge, 1967). The application of a catchment model to the Dee was described in Chapter 14. Radar measurements of storm rainfalls, calibrated by tipping bucket rain gauge measurements telemetered into a control station at Bala, were used in the model to forecast the flows at strategic points along the river. Although the natural Lake Bala acts as a reservoir storage, the linked regulating reservoir, Llyn Celyn, plays a part in the management of the system. In true multipurpose fashion, Llyn Celyn in addition to providing major supplies of water to Liverpool also generates hydropower for the national grid. In low-flow situations (down to 1.4 m^3s^{-1} at Erbistock in extreme droughts) the regulating reservoir can help to ensure that there is sufficient water in the river at downstream abstraction points at Chester for the Liverpool supplies and via the Llangollen canal for the Mid-Cheshire supplies. It will be seen from Fig. 14.11, that all the major tributaries (including the Alwen reservoir) play a part in the overall management of the River Dee (Jamieson & Wilkinson, 1972).

20.5.2   The River Severn

A second example of river regulation in the UK is given by the River Severn (Fig. 20.3). Of the two reservoirs shown on the catchment map, Lake

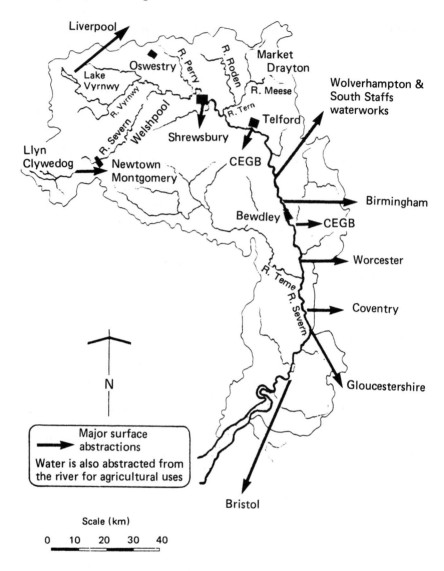

Fig. 20.3   Principal users of the River Severn. (Reproduced by permission of the Severn Trent Water Authority.)

Vyrnwy is an old single-purpose reservoir built in the last century for direct pipeline water supply to Liverpool, a function it still fulfils. The original aim in the building of the second reservoir, Llyn Clywedog, was to ensure further water supplies to the English Midlands. However, during the planning and design stages in the 1960s, the concept of river regulation was beginning to be accepted, and the proposed functions of the new dam at Clywedog were expanded (Fordham *et al.*, 1970). Flood mitigation for the vulnerable centres of Newtown and Welshpool became additional desirable aims. From the reservoir with a capacity of 49.94 Mm^3 (catchment area 50 km^2) the operation of the releases to serve the specified purposes must comply with the following provisions:

(a) minimum compensation water of 18 160 m^3 day^{-1} downstream of the dam;

(b) a minimum flow of 726 000 m^3 day^{-1} at Bewdley, the main river gauging station on the Severn;

(c) an adequate volume of water to satisfy the abstractors at the several points on the river downstream; and

(d) discharge releases between November and May to establish storage capacity in the reservoir to attenuate flood flows.

River regulation of the Severn began in 1968, and since then the operational procedures have continued to develop with the introduction of automated recording instruments and telemetering systems. The basin of the River Severn is the first to be served by radar-measured storm rainfall on a regular operational basis in a flood-warning procedure.

20.5.3   The River Thames

The River Thames may be considered a 'lowland' river. It does not rise in high mountains with plentiful annual rainfall; the sources of its headwaters in the Cotswold Hills and its tributaries from the Chilterns are fed mainly from groundwater aquifers. There are no suitable sites for impounding reservoirs to store water, but river regulation can be effected by pumping from groundwater. The Thames Groundwater Scheme is shown in Fig. 20.4a (Hardcastle, 1978). Water from boreholes in the Chalk and in the Jurassic Limestones (Oolites) (Fig. 20.4b) is pumped up and released into the nearest appropriate stream to augment the flow in the Thames in times of drought in order to maintain required rates of abstraction for the major urban centres. In the planning stage, it was estimated that 845 000 m^3 day^{-1} of groundwater output could increase the surface water resources by 455 000 m^3 day^{-1} on average for 6 to 8 months (Hardcastle, 1978). In operating such a scheme, due attention must always be paid to rates of natural recharge of the aquifers, and the pumping must not be allowed to interfere with private abstractions by unduly depleting the aquifer. Several years of careful study have been made in the valley of the River Lambourn (a tributary of the River Kennet), which has its sources in the unconfined Chalk strata forming the Berkshire Downs. Following successful results from a pilot scheme, the boreholes of Stage 1 are now operational and their effectiveness on the flows in the River Lambourn at the gauging station, Shaw, during the drought year 1975–76 is clearly seen in Fig. 20.4c. Extension of the scheme to development of the other areas, Stages 2 to 4, should provide further benefits in the ordered augmentation of the discharges of the River Thames during periods of drought.

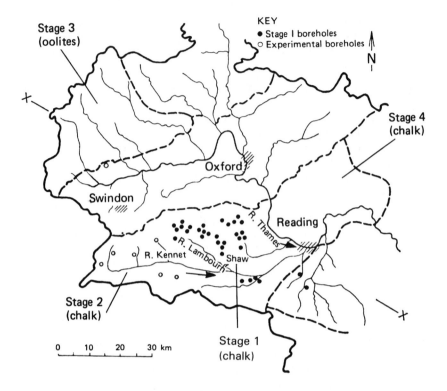

(a) Thames Ground Water Scheme: stages and aquifer units

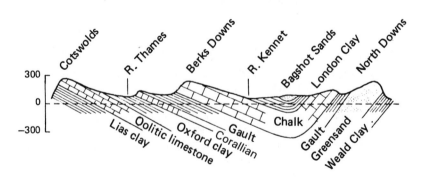

(b) Section X – X

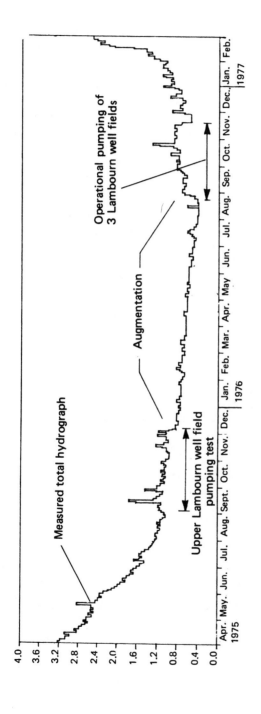

(c) River Lambourn at Shaw: flow augmentation in 1975 – 1976

Fig. 20.4 (Reproduced from B.J. Hardcastle (1978) *From Concept to Commissioning Thames Groundwater Scheme*, by permission of the Institution of Civil Engineers.)

20.5.4   The Kielder Water Scheme

In the limited confines of the UK it is only the largest river basins that can be considered as practical units for successful multipurpose water management schemes. Hence the Kielder Scheme in north-east England, designed to manage the water resources of the region, will be regulating the flow of three rivers, the Tyne, the Wear and the Tees (Fig. 20.5).

The need for the scheme arose from the evaluation of the regional water resources following the Water Resources Act, 1963, when it was estimated that the total water deficiencies would reach about 800 000 m³ day⁻¹ by the year 2001. Many solutions to the water supply problem were proposed, but it was finally agreed that the big, ambitious Kielder Scheme would be the most practical and most economical in the long run. The construction programme over 8 years costing approximately £163 million (1981 prices) (Coats & Ruffle, 1982) was completed in 1982 with the formal opening of Kielder Water.

The reservoir, Kielder Water, at the head of the River North Tyne with a storage of 188 Mm³ (one of the largest man-made lakes in Europe), serves initially to regulate the flow of the River Tyne. The 52-m high earth dam has a catchment area of 241 km², and the outflow spillway has been designed to

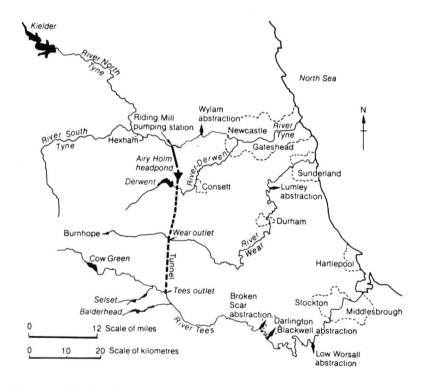

Fig. 20.5   Kielder Water scheme. (Reproduced from D.J. Coats & N.J. Ruffle (1982) *Proc. Inst. Civ. Eng. Part I*, **72**, 135–147, by permission.)

take 425 m^{3}s^{-1} outflow from a peak inflow of 570 m^{3}s^{-1}. However, sufficient dam freeboard allows for higher discharges without causing damage to the dam structure and the overflow works will pass even the Probable Maximum Flood attenuated outflow of 1010 m^{3}s^{-1} designated by the Institution of Civil Engineers following the *Flood Studies Report* (Coats & Roche, 1982). Concern for the aesthetic appearance of the area at the head of the reservoir during periods of severe drawdown led to the building of a smaller subsidiary dam 12.5 m in height (Bakethin Dam) to maintain water levels.

The regulation of the Wear and Tees is made by the transference of water taken from the Tyne at Riding Mill (Fig. 20.5). The transfer works comprise a weir across the river and a pumping station with a final capacity of 1.092 Mm^3 day^{-1}, which takes the water up to a headpond (storage 200 000 m^3), from which there is a tunnelled gravity feed to the Wear and Tees. The 40 km of tunnels, mostly 2.9 m in diameter, allow for a transference of up to 1.18 Mm^3 day^{-1}.

In operating the combined system for the three rivers to serve the domestic and industrial needs of the populated areas, the yield of Kielder Water together with existing storages would be about 1.3 Mm^3 day^{-1}. With 418 000 m^3 day^{-1} coming from the combined smaller sources, the net yield from Kielder approaches 900 000 m^3 day^{-1}, above the estimated deficiencies for 2001. Revised estimates in 1982, at a time of industrial recession, showed that the 1963 estimates were too high. Thus the implementation of the Kielder Water Scheme means that north-east England, a development area, will have amply adequate water supplies into the next century, and is in a position to be able to attract water-consuming industries to the region.

In the design and execution of the scheme, considerable attention has been paid to its multipurpose functions and to aspects of environmental conservation. Provision is made for hydroelectric generation, but river regulation will be given priority. Water quality has been studied, and steps taken to mitigate the adverse effects of thermal stratification in the reservoir. Tests have shown that the quality of the waters in the three rivers is similar, and there should be no adverse effects in their mixing. Loss of dissolved oxygen in the water in temporary storage in the tunnels between periods of transfer is remedied by aerating the water on outlet to the augmented rivers. During the construction period, it was found expedient to provide facilities for the general public to view the operations and these have been expanded on completion to include a wide range of amenities and provision for water sports. Kielder Water will become a tourist attraction and a recreation area for the neighbouring urban populations.

20.6   Real-Time Forecasting

The management of a river basin implies that a continuous watch must be kept of the natural events occurring in the catchment and on man's activities affecting water movements. The elements of the hydrological cycle and the state of the water storages (natural and artificial) must be monitored so that any changes can be assessed. Consequent remedial actions can then be taken to preserve acceptable conditions in the catchment or ensure constant supplies. For a beneficial management scheme, it is helpful to be able to fore-cast changes in the river flows or in the state of the catchment. This applies more particularly to day-to-day management rather than to the monthly or annual organization of the water resources. The need for prior knowledge of rainfall or runoff an hour or a day ahead is fulfilled by short-term or real-time forecasting. The forecasting of rainfall is mainly the concern of the meteoro-logists. When the rainfall forecast is received, the hydrologist can apply the expected input to a model of the river basin system to provide a real-time forecast of the subsequent behaviour of the system. With this facility, the water resources engineer can initiate control actions and subsequently allow for the effects of the actual rainfall by appropriate initial operations and subsequent adjustments of the control structures, e.g. dam releases or sluice gate closures. In the event of a toxic spillage into a stream, the real-time fore-casting of the arrival of the pollution at a water supply intake can ensure that abstraction is halted before the arrival time.

The application of microcomputers to the rapid processing of telemetered measurements enables real-time forecasting and subsequent control of a river system to form the foundation of river basin management. The operations may be carried out by the rapidly developing techniques of systems engineering. In conjunction with these methods can be applied the principles and mathematical tools of control engineering, which has its applications in many areas (IWES, 1981).

20.6.1   Example – A Water Resource System

The practical application of such methods can be demonstrated by consider-ing the simple system portrayed in Fig. 20.6 (O'Connell, 1980). The flow in a main river is regulated by a multipurpose reservoir. The water resources of the area are developed to serve the needs of two urban demand centres. The first town draws supplies from the river at A_1 and from the boreholes in an unconfined aquifer. Waste water from the town passes through treatment works before being returned to the river downstream of A_1. The second demand centre lower down the catchment receives its water supplies entirely from a pumped storage reservoir mainly fed from the river by pumping at the point A_2, but also having some runoff contributed by its own small tributary catchment. Treated effluent from the second town is disposed of into a neigh-bouring river system.

At the points marked L on the various streams, the discharges have been assigned a statutory lower limit below which the river flow must not be allowed to fall for the following reasons:

L_1: an adequate volume of water is required at all times to receive the intake of effluent from the first town and still maintain an acceptable water quality standard in the river; the flow at L_1 must also be sufficient to serve the abstraction needs at A_2 and maintain L_2;

L_2 and L_3: the flow remaining at these points must be able to maintain water quality standards in the river and the tributary and be adequate to sustain the amenities of the river downstream;

L_4: a gauging station at this point with a prescribed lower flow limit is needed to ensure that pumping of the aquifer is not harmful to the stream fed from the aquifer.

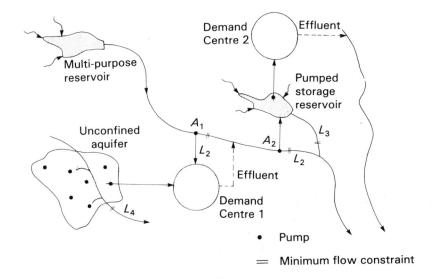

Fig. 20.6   Water resource system in a river basin. (Reproduced from D.G. Jamieson (1976) *Formulating Operational Policy for a River Basin Management Authority*, by permission of Martinus Nijhoff Publishers B.V.)

The problems for the water resources engineer operating this system would stem primarily from irregularities in hydrological inputs. It is to be expected that the resources of the system would have been assessed correctly at the planning stage so that there is adequate storage to meet demands over a severe period of drought. In really extreme drought events, normally unacceptable water conservation measures may have to be introduced, e.g. curtailment of piped supplies. However, heavy rainfalls may cause trouble in any part of the catchment. Local storms over the towns may upset the smooth running of a waste water treatment works if there is a combined storm water/sewerage system, and a widespread storm over the whole area might severely tax the capacity of the system to contain the volume of water without flooding riverine areas.

The operating of the system when there are no extreme events occurring then turns on the economics of the day-to-day meeting of water demands and the efficient running of the water treatment works. If there is enough surface water available, saving the cost of pumping groundwater may be considered; during a wet period the pumped storage reservoir may be receiving enough runoff from its small catchment to meet the needs of demand centre 2 without pumping from the main river at A_2. More timely savings could be made in electricity charges by pumping during off-peak hours and shutting down during the highest cost periods of peak electricity demand.

The operational balancing of the supplies and demand of water would be carried out from a system control centre as depicted in Fig. 20.7. The diagram relates directly to the supposed requirements of the simple system in Fig. 20.6. In addition to the rainfall information coming in from the

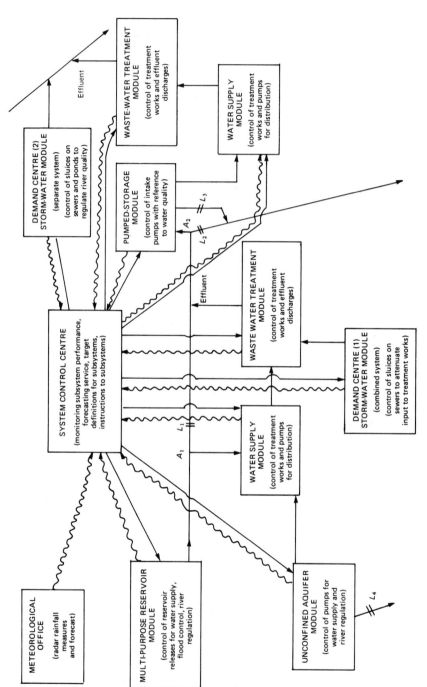

Fig. 20.7 River basin system control scheme. (Reproduced, with modifications, from D.G. Jamieson (1976) *Formulating Operational Policy for a River Basin Management Authority*, by permission of Martinus Nijhoff Publishers B.V.)

Meteorological Office, telemetered rainfall measurements would also be received at the control centre from the Authority's own hydrometric network.

The various modules represent the operational functions of the subsystems; the multipurpose reservoir, a water treatment works or a waste water treatment works are subsystems that fit into the overall scheme. At each subsystem element, a microcomputer would process the data pertaining to that module, e.g. at the impounding reservoir, rainfall, evaporation and river flow data would be processed and reservoir levels, compensation water, released water, etc. would be monitored continuously and relevant records compiled. Determination of runoff by rainfall – runoff modelling would be carried out at this location. The data would be transmitted to the system control centre served by a large computer. Here, the state and current performance of all the subsystems can be appraised and modifying operational instructions can be given to any subsystem according to real-time forecasting of any changes in circumstances. Such changes are naturally thought of as being most likely due to a storm warning, but they may result from an irregular increased demand for water in one of the towns or a breakdown of a waste water treatment plant which could endanger river water quality. At the control centre, the resulting effects of such aberrations in the smooth running of the system can be forecast in real-time by systems models and direct mitigating controls of the affected subsystems can be transmitted. On a short time basis, feedback information on the control change effects can be fed into the system model and any control adjustments made. The length of the time steps in the control – feedback interchanges depends on the dynamic response time of the system (or the pertiment subsystem if only one module is involved).

20.6.2   Example – Flood Forecasting

The introduction of areal rainfall measurements by radar into real-time flood forecasting has been a major development in the past decade. While the initial applications of radar rainfall in the UK were in flow forecasting in single river basins (Dee and Severn), this example demonstrates how radar rainfall has been incorporated into the flood forecasting for a whole region, that of the North West Water area (Noonan, 1985).

A Regional Communications Scheme (RCS) was established to assemble rain gauge and river level measurements made every 15 minutes and transmitted by a telemetry system. The radar rainfall data automatically obtained every 15 minutes at the Hameldon Hill weather radar and converted into catchment rainfalls are also relayed to RCS. The pattern of the data collection is shown in Fig. 20.8. From the Meteorological Office, forecast rainfalls are provided by the FRONTIERS system which uses radar and satellite imagery for very short-range precipitation forecasting. The main data receiving centre is at FRANK LAW where two mainframe computers store the data, one acting as stand-by for the other in case of a failure. Then river level and flow forecasts are produced automatically every 15 minutes for the significant flood risk areas.

The river flows are determined from the input data by rainfall-runoff

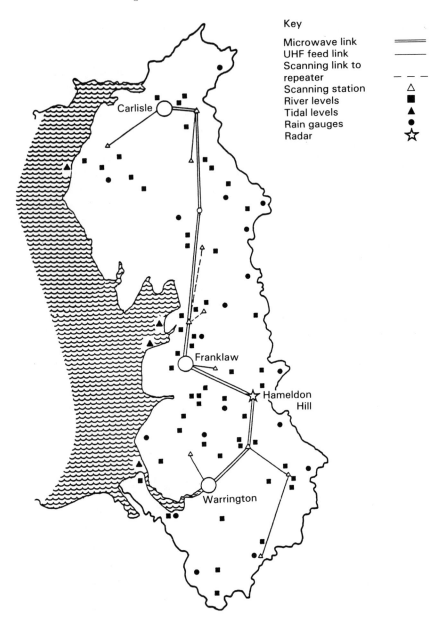

Fig. 20.8   North West Water regional communications scheme. (Reproduced from Collinge, V.K. and Kirby, C., eds, *Weather Radar and Flood Forecasting*. John Wiley & Sons, 1987.)

models programmed in the computer. For operational purposes the important model features are forecast accuracy, a suitable lead time (the time between the forecast and the flooding) and the resources necessary for its operation. Several suitable models have been investigated and two groups have been identified.

(1) Models that can operate continuously and automatically in the RCS system and produce forecast levels for a short time ahead. The system at present operates on the Inflow Storage Outflow model developed by North West Water from the earlier Lambert model. It is represented by:

$$Q_{T+1} = aQ_T + bR_T + Q_B$$

where Q_B is the baseflow at the start of an event and Q_{T+1}, Q_T and R_T are river flows and rainfall at appropriate times T. The parameters a and b and suitable time increment are derived from several events to give the best fit for flood flows. The simple Peak Level Correlation method is used to route flood levels downstream from the river gauging station to danger areas. Times of travel and correlation equations are determined from previous events. These parameters and simple models are easily handled automatically in the computer and they can be updated with added experience.

(2) Many sophisticated models requiring larger computers could, with the necessary rainfall forecasts, give flow forecasts with longer time leads, which would be advantageous for the large catchments. Such models are developments of ARMA, Muskingum-Cunge and FLOUT but without automatic real-time correction and needing skilled operators, they are not at present so attractive for operational purposes. Future research and development will no doubt make the advanced models more amenable to real-time usage. The 15-minute forecasts are transmitted from Franklaw to the operations headquarters at Warrington where the information is displayed in graphical and tabular forms (Fig. 20.9). There is always an officer on duty, and on receipt of an alert either a heavy rainfall forecast from the Meteorological Office at Manchester Airport or an alarm triggered off at Franklaw by an abnormal reading above preset levels either from a rain gauge or river level monitor, the Flood Forecasting system becomes operational. The duty officer may monitor the situation from his home on a portable terminal linked to Franklaw and Warrington. For each gauged catchment, two standby water levels have been fixed the first for the Water Authority, the second a little higher for the Police who are responsible for making public flood warnings and then a series of flood levels according to severity. Public warnings are only made when a flood level has been forecast.

The forecast river levels for up to 4 hours ahead are printed out and plotted on graphs and when the appropriate levels are forecast the information is sent out and if necessary warnings are given.

Although much of the procedure is automated, a team of trained duty officers is needed to be able to assess the flood risk zones from the radar screen showing the rainfall intensities and to concentrate forecast print-outs on the rivers in danger of flooding. In monitoring the forecasts, adjustments to the rainfall-runoff models may be necessary. Training sessions on the flood forecasting system are held regularly and when new developments or modifications have been made all duty officers must be informed.

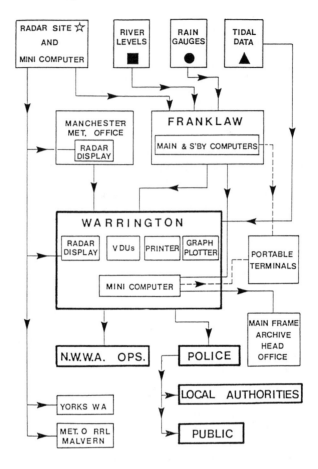

Fig. 20.9   Flow diagram for data and forecast information. (Reproduced from Collinge, V.K. and Kirby, C., eds, *Weather Radar and Flood Forecasting*. John Wiley & Sons, 1987.)

Many of the hydrological modelling techniques described in this text are applicable in the developments now taking place in real-time forecasting and river basin management. Technological advances soon make any data processing or operational system out-of-date; constant renewal of procedures (hardware and software) is inevitable in operational hydrology.

References

Chow, V.T. (1964). *Handbook of Applied Hydrology*. McGraw-Hill.
Coats, D.J. and Roche, G. (1982). 'The Kielder Headworks'. *Proc. Inst. Civ. Eng. Part I*, **72**, 149–176.

Coats, D.J. and Ruffle, N.J. (1982). 'The Kielder Water scheme.' *Proc. Inst. Civ. Eng. Part I*, **72**, 135–147.

Collinge, V.K. (1967). 'Research on river management', in *River Management*, ed. Isaac, P.C.G., Maclaren, 258 pp.

Fordham, A.E., Cochrane, N.J., Kretschmer, J.M. and Baxter, R.S. (1970). 'The Clywedog Reservoir project.' *J. Inst. Water Eng.*, **24**(1), 17–76.

Hardcastle, B.J. (1978). 'From concept to commissioning,' in *Thames Groundwater Scheme*. ICE, London, 241 pp.

Hull, C.H.J. (1967). 'River regulation', in *River Management*, ed. Isaac, P.C.G., Maclaren, 258 pp.

Institution of Water Engineers and Scientists (IWES). (1981) *Micro-electronics in the Water Industry*. 118 pp.

Jamieson, D.G. (1976). *Formulating Operational Policy for a River Basin Management Authority*. NATO Advanced Study Institute on Systems Analysis and Reservoir Management, 12 pp.

Jamieson, D.G. and Wilkinson, J.C. (1972). 'River Dee research program.' *Water Res. Res.*, **8**(4), 899–920.

Noonan, G.A. (1987). 'An operational flood warning system' in Collinge, V.K. and Kirby, C. (eds.), *Weather Radar and Flood Forecasting*, John Wiley.

O'Connell, P.E. (ed). (1980). *Real-Time Hydrological Forecasting and Control*. Institute of Hydrology, 264 pp.

Rodwell Jones, Ll. and Bryan, P.W. (1946). *North America*, 7th ed. Methuen, 580 pp.

Bibliography

General

Chow, Ven Te (1964). *Handbook of Applied Hydrology*. McGraw-Hill, New York.
Fleming, G. (1975). *Computer Simulation Techniques in Hydrology*. Elsevier, New York.
Gray, D.M. (ed.) (1970). *Handbook on the Principles of Hydrology*. I.H.D. Canadian National Committee.
Linsley, R.K., Kohler, M.A. and Paulhus, J.L.H. (1982). *Hydrology for Engineers*, 3rd ed. McGraw-Hill, New York.
Nemec, J. (1972). *Engineering Hydrology*. McGraw-Hill, New York.
Raudkivi, A.J. (1979). *Hydrology: an Advanced Introduction to Hydrological Processes and Modelling*. Pergamon, Oxford.
Rodda, J.C. (ed.) (1976). *Facets of Hydrology*. Wiley, London.
Rodda, J.C. (ed.) (1985). *Facets of Hydrology*, Volume 2. Wiley, Chichester.
Rodda, J.C., Downing, R.A. and Law, F.M. (1976). *Systematic Hydrology*. Newnes-Butterworths, London.
Viessman, W., Harbaugh, T.E. and Knapp, J.W. (1977). *Introduction to Hydrology*. 2nd ed. Harper & Row, New York.
Ward, R.C. (1975). *Principles of Hydrology*. 2nd ed. McGraw-Hill, London.
Wilson, E.M. (1983). *Engineering Hydrology*. 3rd ed. Macmillan, London.

Specialist

Hydrometeorology

Donn, W.L. (1975). *Meteorology*, 4th ed. McGraw-Hill, New York.
Mason, B.J. (1975). *Clouds, Rain and Rainmaking*, 2nd ed. Cambridge U.P.
McIlveen, J.F.R. (1986). *Basic Meteorology*. Van Nostrand Reinhold (UK), Wokingham, England.
Riehl, H. (1979). *Climate and Weather in the Tropics*. Academic Press.
Wiesner, C.J. (1970). *Hydrometeorology*. Chapman & Hall, London.

Evaporation and Soil Moisture

Hillel, D. (1980). *Fundamentals of Soil Physics*. Academic Press.
Marshall, T.J. and Holmes, J.W. (1979). *Soil Physics*. Cambridge U.P.
Monteith, J.L. (1973). *Principles of Environmental Physics*. Arnold, London.

Groundwater

Freeze, R.A. and Cherry, J.A. (1979). *Groundwater.* Prentice Hall.
Heath, R.C. and Trainer, F.W. (1981). *Introduction to Groundwater Hydrology.* Water Well Journal Pub. Co., Worthington, Ohio.
Raudkivi, A.J. and Callander, R.A. (1976). *Analysis of Groundwater Flow.* Arnold, London.
Todd, D.K. (1980). *Groundwater Hydrology*, 2nd ed. Wiley.

River Flow

Ackers, P., White, W.R., Perkins, J.A. and Harrison, A.J.M. (1978). *Weirs and Flumes for Flow Measurement.* Wiley.
Henderson, F.M. (1966). *Open Channel Flow.* Macmillan, New York.
Herschy, R.W. (ed.) (1978). *Hydrometry, Principles and Practices.* Wiley.
Brandon, T.W. (ed.) (1987). *River Engineering – Part 1, Design Principles.* IWES for the Institution of Water and Environmental Management.

Hydrological Modelling

Clarke, R.T. (1973). *Mathematical Models in Hydrology.* FAO Irrigation and Drainage Paper 19, Rome.
Fleming, G. (1979). *Deterministic Models in Hydrology.* FAO Irrigation and Drainage Paper 32, Rome.
Haan, C.T. (1977). *Statistical Methods in Hydrology.* Iowa State U.P.
Kottegoda, N.T. (1980). *Stochastic Water Resources Technology.* Macmillan, London.
McCuen, R.H. and Snyder, W.M. (1986). *Hydrologic Modelling: Statistical Methods and Applications.* Prentice-Hall.
Yevjevich, V. (1972). *Probability and Statistics in Hydrology.* Water Resources Publications, Fort Collins, Colorado.

Urban Hydrology

Hall, M.J. (1984). *Urban Hydrology.* Elsevier Applied Science, London.
Helliwell, P.R. (ed.) (1978). *Urban Storm Drainage.* Proc. Intern. Conf. Southampton, Pentech Press, London.
Watkins, L.H. and Fiddes, D. (1984). *Highways and Urban Hydrology in the Tropics.* Pentech Press, London.

Problems

Although allocated to specific chapters, the problems may require material from more than one chapter. Numerical answers are given where appropriate. For questions requiring subjective judgement or a trial and error method of solution, i.e. where there is no single answer, one numerical answer is provided as a guide to the result.

Chapter 1   The Hydrological Cycle

1.1 97% of the Earth's water is stored in the oceans. Describe the natural physical processes involved in the translation of oceanic water to an upland stream flowing into an impounding reservoir.

1.2 A water engineer is required to assess the quantity of water available in a particular catchment in the Scottish mountains. What phases of the hydrological cycle would he have to consider, and what measurements would you recommend him to make?

1.3 Compare and contrast the synoptic situations resulting from waves in the easterlies and waves in the westerlies. Indicate how they affect precipitation.

1.4 The following aerological observations were taken from a radiosonde ascent:

Pressure (mb)   1006 920 800   740 700 660 600 500 400
Specific humidity (g kg^{-1})   14.0 13.4 10.2 9.4 7.2 6.6 5.6 4.0 1.8

If 60% of the precipitable water is produced from the air up to the 600 mb level to form precipitation at ground level, what would be the depth of the rainfall? [*Ans.*   25.4 mm]

1.5 Define and explain, with diagrams where appropriate, three of the following: anticyclone, occlusion, albedo, aerosols, monsoon.

1.6 Hurricanes and typhoons are extreme meteorological phenomena. Describe the initial conditions and stages in their development. Give briefly some of the serious hydrological consequences of a hurricane/typhoon affecting a land area.

1.7 The rainfall events portrayed on the given hyetograms (Fig. P.1) from a rainfall recorder were caused by two distinctive meteorological situations. In *A*, the rainfall was from a cold front and in *B* from an occlusion. How do these two weather patterns help to account for the differing rainfall traces?

  From the charts, obtain the average rainfall intensity for each storm. Identify the period of maximum intensity in each event and evaluate these maximum rates of rainfall. [*Ans.*   Av. int.: CF 3.2 mm h^{-1}; Occ. 1.8 mm h^{-1}. Max. int.: CF 15 mm h^{-1}; Occ. 5 mm h^{-1}]

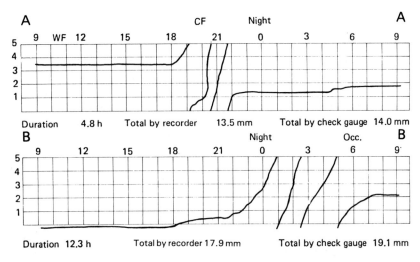

Fig. P1

Chapter 2   Hydrometric networks

2.1 Select a catchment of about 1000 km² with the upper reaches of the river draining a mountainous area. Define the catchment boundary and design a multipurpose precipitation network for the catchment.

A river-regulating reservoir is to be constructed in one of the tributary valleys. Select a suitable valley; plan and specify the instrumentation required to measure the precipitation before reservoir construction and indicate the network of gauges necessary for the efficient operation of the reservoir after completion.

2.2 You are to be sent as an engineering hydrologist to a developing country to assist in the establishment of a Hydrometric Division of the Ministry of Public Works. List those aspects of background research and planning to which you would pay particular attention in designing a multipurpose network of precipitation stations for the country.

2.3 Compare and contrast the factors to be considered in designing a precipitation network for catchments in the following pairs of hydrological situations:

(a) An equatorial region and a north-west European type of climate.
(b) An urban catchment and a rural catchment.

Chapter 3   Precipitation

3.1 A standard instrument for the daily measurement of precipitation in the UK is the Meteorological Office Mark II rain gauge. Draw an annotated sketch diagram of this instrument installed in the ground. Explain the importance of its special features. In a separate diagram, show a recommended installation for an exposed moorland site.

3.2 With the aid of a diagram, explain the principle of the mechanism of the tipping-bucket rain gauge. What are the advantages of this type of instrument to the engineering hydrologist?

3.3 The sequential data (mm at 30-min intervals) shown in Table P.1 have been compiled from a transmission via the satellite METEOSAT from a counter attached to a recording rain gauge on 28 December 1979 at 0512 h. The first record is for 0530 h on 27 December 1979 and the counter is returned to zero at midnight. Describe, with sketch diagrams where appropriate, the mechanism of the type of rain gauge used for these measurements and a recommended installation for the instrument.

Table P.1

0006	0006	0006	0007	0008	0009	0011	0011
0012	0013	0015	0016	0018	0020	0022	0025
0027	0029	0031	0034	0038	0041	0043	0046
0047	0048	0055	0068	0072	0076	0079	0081
0085	0088	0090	0091	0092	0092	0000	0000
0000	0001	0001	0001	0001	0001	0001	0001

From the data, give the total fall for the rain day of 27 December, assuming no further rain fell on the 28th after transmission. Plot the data, identify the period of heaviest rainfall and give the maximum intensity value. [*Ans.*   82 mm, 26 mm h^{-1}]

3.4 The hyetograms shown in Fig. P.1. come from a Dines Tilting Syphon Rain Recorder. Draw a diagram to show the mechanism of the instrument and, making reference to the charts, describe how the recorder works. Outline the difficulties encountered in its use and some of the modifications that may be made to overcome them.

Chapter 4   Evaporation

4.1 Describe the standard US evaporation pan, and indicate by sketch diagrams how it is installed at a climatological station. In using a measurement of evaporation from this pan, explain the need for a pan coefficient and give an example of such a coefficient.

4.2 Many types of tanks or pans have been used to obtain a measure of evaporation from an open water surface. Describe the UK, US and USSR standard instruments, and indicate with annotated sketch diagrams how they are installed at a climatological station. Comment on their ability to give a representative measure of evaporation loss from a reservoir.

4.3 In a catchment area in northern England, the measurements of evaporation (mm) shown in Table P.2 were made for 1958–62 and 1968–72. For the years 1968–72, measurements from the US Class A pan only were available. Estimate the volume of water lost each year in the later period from a reservoir of surface area of 1.4 km^2. [*Ans.*   1968, 652.4; 1969, 610.4; 1970, 721.0; 1971, 655.2; 1972, 596.4 10^{3}m^3]

Table P.2

Year	UK tank	US Class A pan	Year	US Class A pan
1958	351	491	1968	621
1959	536	713	1969	581
1960	502	653	1970	687
1961	437	586	1971	624
1962	486	612	1972	568

4.4 Distinguish between the two forms of evaporation, E_o and E_t, and outline the factors which affect evaporation. What is meant by potential evaporation?

4.5 Evaporation loss from a reservoir can be estimated by the water budget method. Identify the necessary items in a water budget and describe how these might be measured.

4.6 The calculation of evaporation from meteorological variables is often recommended. What are the measurements required for evaluating evaporation? Describe the instrumentation necessary for making the observations.

4.7 Describe the principal features of an automatic weather station. How are the measurements used for calculating evaporation?

Chapter 5   Soil Moisture

5.1 How is soil formed? Identify the major soil types according to their composition.

5.2 Define porosity and describe how water is contained within a soil.

5.3 Explain what is meant by 'field capacity' and 'permanent wilting point' by reference to sandy soils and clay soils.

5.4 Distinguish between sorption and desorption in the relationship of soil water content to the pore water pressure of a soil.

5.5 Describe the neutron scattering method for determining soil water content.

Chapter 6   River Flow

6.1 Describe the method of discharge measurement by the velocity – area method. Itemize the instruments required and any necessary structural installations.

6.2 State what is understood by the term 'control' in open-channel flow. Define channel controls and multiple controls, giving an example of each type.

6.3 Outline the characteristics required in choosing a current meter. How do the cup-type meters and propeller-type meters compare with the outlined specification.

6.4 The river gauging measurements in Table P.3 were carried out by wading on the Zhob River in Baluchistan. Calculate the flow by the mid-section method assuming the length units are metres. Comment on this assumption. [*Ans.*   47.79 m³s⁻¹]

Table P.3

Distance	50	60	70	80	90	100	110	120	130	140	150
Depth	0	0.65	0.71	1.42	1.18	1.33	1.48	1.75	1.62	1.31	0
Mean velocity	0	0.270	0.327	0.421	0.459	0.459	0.478	0.459	0.421	0.327	0

6.5 The details of a gauging carried out by the velocity – area method at Namasagali on the River Nile are as shown in Table P.4. Estimate the discharge using the mid-section procedure. [*Ans.*   548 m³s⁻¹]

Table P.4

Chainage (m)	0	50	100	150	200	250	300	350	400	450	470
Depth (m)	0	1.7	2.5	3.3	4.1	4.4	3.6	2.4	1.6	1.6	0
Mean velocity (m s⁻¹)	0	0.27	0.40	0.38	0.51	0.54	0.52	0.42	0.38	0.26	0

6.6 Current-meter measurements were taken at a section on a river as shown in Table P.5. Calculate the discharge by the mean-section method. [*Ans.*   155.9 m³s⁻¹]

Table P.5

Distance (m)	5	10	20	30	35	40	45	47
Depth (m)	0	1.8	3.7	9.0	12.6	10.1	5.3	0
Mean velocity (m s⁻¹)	0	0.1	0.2	0.6	1.1	0.8	0.5	0

6.7 State the advantages to be gained by plotting stage – discharge relationships on logarithmic scales, particularly with regard to gauging stations with multiple controls.

6.8 The data in Table P.6 relate to a chemical gauging using the constant-rate injection method. Estimate the rate of flow in m³s⁻¹. [*Ans.*   12.5 m³s⁻¹]

Table P.6

Injection solution flow rate	5.0 mℓ s⁻¹
Concentration of injection solution (when diluted by a factor of 2×10^6)	100 μg ℓ⁻¹
Background concentration of chemical in river	0
Concentration of chemical at sampling site	80 μg ℓ⁻¹

Chapter 7   Groundwater

7.1 Draw diagrams to explain a confined aquifer and a perched water table. What kind of rocks form aquicludes?

7.2 Explain Darcy's Law of groundwater flow, and define the term 'hydraulic conductivity'.

7.3 Define three of the following terms: anisotropic, storage coefficient, transmissivity, unsteady (transient) flow, flow net, piezometric surface.

7.4 Outline the principles involved in the application of electrical analogues in the solution of groundwater flow problems.

7.5 In investigating sources of groundwater in an undeveloped region, compare the application of the different survey methods.

7.6 The variation of drawdown with time at an observation well 46 m from a well that fully penetrates a confined aquifer being pumped at 26.5 ℓ s^{-1} is given in Table P.7. Using the Jacob approximation to the Theis equation, determine values of S and T for the aquifer. [*Ans.* $T = 289$ m^{2}day^{-1}, $S = 0.014$]

Table P.7

Time (h)	1.8	2.7	5.4	9.0	18.0	54.0
Drawdown (m)	0.55	0.73	1.10	1.31	1.77	2.47

7.7 A well tapping an unconfined aquifer having an initial saturated thickness of 8.2 m is pumped at a rate of 65 ℓ s^{-1} until a steady-state cone of depression is established. The drawdowns measured at two observation wells situated 15 m and 31 m from the pumped well are then found to be 1.80 m and 1.40 m, respectively. Determine the transmissivity of the aquifer. [*Ans.*   2016 m^2 day^{-1}]

7.8 An extensive horizontal confined aquifer having a constant thickness of 40 m is pumped steadily by a series of identical wells located at 300 m intervals in a N – S line. Each well completely penetrates the aquifer and delivers 20 ℓ s^{-1}. Field measurements a long way to the east of the well line show a hydraulic gradient of 1 in 175, and similarly to the west of the wells a hydraulic gradient of 1 in 500. Both piezometric surfaces slope in a westerly direction. Estimate the Darcy hydraulic conductivity of the aquifer. [*Ans.*   4.487 × 10^{-4} m s^{-1}]

7.9 A well penetrates 33.5 m below the static water table. After pumping at 26.5 ℓ s^{-1}, the drawdown in observation wells 17 m and 45 m from the pumped well is found to be 3.65 m and 2.25 m, respectively. What is the Darcy permeability (K) of the aquifer? [*Ans.*   9.94 × 10^{-4} m s^{-1}]

Chapter 8   Water Quality

8.1 It is required to assess the water quality of a small stream. List the characteristics of water that need to be investigated and outline a suitable sampling procedure for these investigations.

8.2 Water quality standards for a particular industry should be related to river flow characteristics, existing water use and environmental considerations rather than being fixed at uniform national or international levels. Discuss.

8.3 Explain why river pollution is difficult to define, and categorize the main sources of such pollution.

8.4 Explain what is meant by the term 'water hardness', and indicate its relevance to water usage.

8.5 What are the advantages and disadvantages of using a single biological index as an indicator of overall water quality for river classification?

8.6 Explain the significance of four of the following as indicators of water quality: pH, turbidity, biochemical oxygen demand, colour, dissolved oxygen, temperature, chloride content.

Chapter 9   Data Processing

9.1 Show by means of a flow chart the basic steps in calculating the daily mean discharge of a river. Discuss how computer processing might be used to improve the procedure. What modifications might need to be made to the field instrumentation?

9.2 Construct a flow chart to show the stages required in a computer program to effect quality control of a series of daily rain gauge measurements. State the form of input to be used and indicate how the various outputs produced might be used by a practising hydrologist.

9.3 The average annual rainfall is an important rainfall statistic. Explain how it is derived and outline some of its uses in hydrological analysis.

9.4 Consecutive annual rainfall totals and their percentages of normal are provided for two rainfall stations in Table P.8. Derive a satisfactory average annual rainfall value for each station giving reasons for the judgement used during the procedure. [*Ans.*   Station A, 790.2 mm; Station B, 683.3 mm]

Table P.8

Station A										
Years	1	2	3	4	5	6	7	8	9	10
Rain (mm)	858.0	560.1	812.3	683.3	693.7	605.8	896.9	1032.5	759.0	642.0
%	107.5	91.8	105.5	90.5	94.4	69.5	114.0	126.5	100.0	78.3

Station B										
Years	1	2	3	4	5	6	7	8	9	10
Rain (mm)	722.9	731.5	652.0	665.0	787.4	744.7	629.1	642.2	622.9	558.2
%	105.0	106.5	96.5	100.0	112.0	106.5	94.0	111.0	112.5	103.3

9.5 Define the daily mean stage and show how it is obtained for flashy streams. Given a stage – discharge relationship for such a stream, indicate how the daily mean discharge should be calculated.

9.6 Identify sources of errors that may exist in a digitally recorded series of stage measurements.

Chapter 10   Precipitation Analysis

10.1 The precipitation falling on a catchment area of 50 km² is sampled by six rain gauges. From the measurements in Table P.9 deduce the areal rainfall for 1978. If a dam is built at the catchment outfall and a statutory minimum discharge of 0.1 m³s⁻¹ is maintained throughout the year in the river downstream, assess the volume of water available for supply. (Assume the drainage basin is water-tight and there is a total evaporation loss of 400 mm). [*Ans.*   66.75 Mm³ in the year]

Table P.9

Rain gauge	1	2	3	4	5	6
1978 rainfall (mm)	2052	1915	1868	1723	1640	1510
Thiessen polygon area (km²)	7.8	8.3	10.2	11.5	5.4	6.8

10.2 The frequencies of occurrence of monthly rainfall (mm) for Manchester from 1786–1971 are grouped as shown in Table P.10. What kind of frequency distribution do these data represent, and how does it differ from the frequency distributions of annual rainfall totals? Construct a cumulative percentage frequency diagram for the data. From a smooth cumulative frequency curve drawn through the group mid-points on the diagram, find the median value of monthly rainfall and the interquartile range about the median. [*Ans.*   62 mm, 54 mm]

Table P.10

0.0 – 20.0	121	140.1 – 160.0	90
20.1 – 40.0	304	160.1 – 180.0	33
40.1 – 60.0	448	180.1 – 200.0	18
60.1 – 80.0	447	200.1 – 220.0	7
80.1 – 100.0	358	220.1 – 240.0	5
100.1 – 120.0	243	240.1 – 260.0	1
120.1 – 140.0	156		

10.3 From a topographical map of Northeast England (scale 1:250,000) define the catchment boundary of the River Wear. The total area is of the order of 1180 km². Locate the rain gauge stations given in Table P.11 and plot their average annual rainfalls. Determine the average annual rainfall over the total catchment using (a) the arithmetic mean (b) the Thiessen polygon methods. Compare and comment on the results obtained. [*Ans.*   871 mm, 843 mm]

10.4 The areas (Table P.12) contained within given isohyets were obtained for a 2-day storm over Norfolk in August 1912 when the maximum point total measured was 210 mm. Calculate the areal rainfall within each isohyet and plot the depth – area values on semi-logarithmic paper. Draw a smooth curve through the plotted points. On the same graph, draw a curve for the 24 h depth – area relationship when the

Table P.11

Rain gauge	N.G. Ref.	Average annual rainfall (mm)
Grassmeres	NY(35)825372	1517
Rookhope	NY(35)938425	1096
Waskerley	NZ(45)022444	903
The Grove	NZ(45)066299	883
Waterhouses	NZ(45)189413	766
Washington Glebe	NZ(45)313560	663
Ryhope P. Stn.	NZ(45)404524	659
Mill Hill Res.	NZ(45)413425	704
Shildon S. Wks.	NZ(45)243253	644

maximum rainfall total was 185 mm. What assumptions have been made in drawing this second curve?

Table P.12

Isohyet (mm)	200	175	150	125	100
Area (km²)	50	630	1800	2600	5200

10.5 The 30-year averages shown in Table P.13 are provided for a climato-logical station in the Northern Hemisphere. Describe the precipitation regime for which the data are representative and itemize some of the hydrological problems of such an area.

Table P.13

	J	F	M	A	M	J	J	A	S	O	N	D
Temperature (°C)	– 19	– 15	– 9	3	10	15	18	17	11	5	– 6	– 14
Rainfall (mm)	18	18	17	21	34	57	53	45	33	19	19	18

Chapter 11   Evaporation Calculations

11.1 The following two general formulae are used to evaluate evaporation from open water:

(i) $E_o = f(u)\,(e_s - e_d)$

(ii) $Q_{E_o} = Q_s - Q_{rs} - Q_L - Q_c \pm Q_G \pm Q_v$

Explain the principles of the two different methods. Identify and name the terms in each equation.

11.2 A practical realization of the vapour flow equation for open water evaporation is given by:

$$E_o = 0.291\, A^{-0.05}\, u(e_s - e_d)$$

where E_o is in mm day^{-1}, A in m^2, u in m s^{-1}, e_s and e_d in mb. Calculate the loss of water over the summer months (April – September) from Kielder Water, surface water area 10.86 km^2, given the mean wind speed is 9 knots and the values of e_s and e_d are 15.3 and 11.2 mb, respectively. [*Ans.*   4.869 Mm^3]

11.3 The Penman equation for the evaluation of potential evaporation can be expressed as follows:

$$PE = \frac{\Delta/\gamma H + E_{aT}}{\Delta/\gamma + 1}$$

where:

$$H = 0.75\, R_I - \sigma T_a^4 (0.47 - 0.075\sqrt{e_d})\,(0.17 + 0.83\, n/N)$$

and

$$E_{aT} = 0.35\,(1 + u_2/100)\,(e_a - e_d)$$

Define the individual terms, and explain the physical basis behind the derivation of the formulae.

11.4 Calculate the potential evaporation for a rural station at latitude 53° N for the two months March and June from the data in Table P.14, using the Penman formula. Comment on the variation in the two results. [*Ans.*   30.7 mm, 75.0 mm]

Table P.14

	March	June
Mean temperature (°C)	5.3	14.7
Mean relative humidity (%)	79	77
Sunshine (h)	119.5	97.4
Mean wind speed (knots)	7.1	8.1

11.5 The data in Table P.15 for two climatological stations in the London area, Kew and Kingsway (both latitude $51\frac{1}{2}°$ N) are given for the month of April. Kew may be considered a rural location and Kingsway a smoky area. Explain the difference between the two values of potential evaporation calculated by the Penman formula. [*Ans.* Kew, 56.7 mm; Kingsway, 51.6 mm]

Table P.15

	Kew	Kingsway
Mean Temperature (°C)	8.1	8.6
Mean vapour pressure (mm)	6.0	6.5
Sunshine (h day^{-1})	4.82	4.67
Wind (knots)	7.4	6.2

11.6 Records of monthly rainfall and potential evaporation are available for a climatological station in the northern hemisphere as shown in Table P.16. At the beginning of the year the soil is quite dry. Evaluate the actual evaporation (E_a) for each month assuming the soil moisture can reach a field capacity of 100 mm. Plot the data and values of E_a on a water balance diagram and indicate the periods of water surplus, soil moisture use and soil moisture recharge. In which climatic region is the station situated? Calculate the water available for the year from a catchment of 150 km². [*Ans.* $49.65 \times 10^6 \text{m}^3$]

Table P.16

	J	F	M	A	M	J	J	A	S	O	N	D
Rainfall (mm)	5	19	47	91	140	158	186	299	251	112	32	10
PE (mm)	154	173	203	200	154	146	130	97	90	119	120	130

Chapter 12   River Flow Analysis

12.1 The monthly average discharges are given in Table P.17 for two rivers, the Yenisei (USSR) and the Severn (UK) whose mean annual discharges are 17 800 and 61.5 m³s⁻¹, respectively. For each river, plot bar graphs of the monthly values as ratios of the mean annual discharges. Identify the two river regimes and explain the main features of the patterns obtained with special reference to the climatic conditions of their drainage basins.

Table P.17

	J	F	M	A	M	J	J	A	S	O	N	D
Yenisei (10^3 m³s⁻¹)	4.8	4.5	4.2	4.0	30.2	76.0	27.9	18.9	17.9	14.7	6.1	4.9
Severn (m³s⁻¹)	110	107	69.9	51.5	34.7	28.2	22.4	29.1	42.0	52.4	92.6	99.0

12.2 Distinguish between gauged river flows and naturalized river flows, and explain some of the causes of differing values for the discharge at a river gauging station.

12.3 Account for the difference in the enveloping curves for peak discharges in the UK and the USA.

12.4 The series of annual maxima for a 24-year record of instantaneous peak discharge (m^3s^{-1}) for a river is given in Table P.18. From a suitable histogram representing these data, estimate the return period of an annual maximum discharge exceeding 12.5 m^3s^{-1}. [*Ans.*   2.66 years]

Table P.18

15.35	14.20	12.25	10.24	12.11	11.23
10.05	6.55	13.05	12.10	21.00	10.00
12.60	4.90	13.80	11.35	6.72	9.65
15.20	6.95	13.98	10.60	11.20	9.05

12.5 Define the flow duration curve, and describe how it is constructed from the records at a river gauging station. What information does the flow duration curve give on the nature of the catchment?

12.6 Define the following terms, illustrating the definition where appropriate: annual maximum series, partial duration series, return period, EVI probability distribution.

12.7 Floods on a particular river conform to the Extreme Value Type I distribution given as:

$$1 - P_m(y) = \exp\left[-e^{-\gamma(y-a)}\right]$$

where y is the flood magnitude, $P_m(y)$ is the probability of y being exceeded in the annual maximum series and γ and a are parameters. If the 100-year flood has a magnitude of 260 m^3s^{-1} and the 10-year flood has a magnitude of 170 m^3s^{-1}, what is the probability of having a flood as great or greater than 240 m^3s^{-1}. If the 240 m^3s^{-1} flood is to be adopted as the design flood for a spillway, what is the probability that the design flood will be exceeded at least once in a 5-year period of construction? [*Ans.*   0.017; 0.082]

Chapter 13   Rainfall – Runoff Relationships

13.1 Describe the 'Rational' method of designing surface water drainage systems. Define fully any terms used in the description. Discuss the assumptions made in applying the rational method, and comment upon the procedure used to predict the peak rate of runoff for a given return period.

13.2 The principal dimensions of a catchment are shown in Fig. P.2. If all isochrones at intervals of 1 min are arcs of circles centred at the outfall and the velocity of flow is assumed constant at 1 m s^{-1}, draw the time area diagram for the catchment. Ignore times of entry. A design storm has the following sequence of 1-min intensities:

5.1, 6.4, 8.1, 10.7, 14.1, 19.6, 30.3, 43.8, 28.0, 20.8, 12.3 and 4.5 mm h^{-1}

Assuming a runoff coefficient of 0.25, estimate the peak rate of flow at the outfall. [*Ans.*   117.5 ℓ s^{-1}]

13.3 The 1-h 1 mm unit hydrograph for a small catchment is given in Table P.19. Determine the peak of the hydrograph that should result from the following storm: 5 mm in the first hour, no rain in the next 30 min, and 8 mm of rain in the next final hour. Assume a loss rate of 3 mm h^{-1} in the first hour and 2 mm h^{-1} for the remainder of the storm. [*Ans.*   36.8 m^{3}s^{-1}]

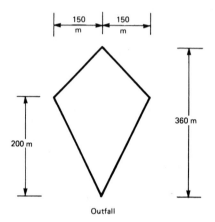

Outfall

Fig. P2

Table P.19

Time (h)	0	0.5	1	1.5	2	2.5	3	3.5	4	4.5
Flow (m^{3}s^{-1})	0	0.9	3.0	5.5	4.3	3.0	1.9	1.0	0.4	0

13.4 During a notable storm in July 1968, rainfall measurements were made at five stations in the River Exe catchment area as shown in Fig. P.3. Given the ordinates of the 5-h (1 mm) unit hydrograph shown in Table P.20, derive the outflow hydrograph of the storm at Thorverton gauging station assuming 80% of the total precipitation is lost at a constant rate. [*Guideline Ans.*   Peak flow 80.5 m^{3}s^{-1} after 15 h]

Table P.20

Time (h)	0	5	10	15	20	25	30	35	40
u(m^{3}s^{-1})	0	0.8	5.0	7.5	5.0	2.7	2.0	1.6	1.3

Time (h)	45	50	55	60	65	70	75	80
u(m^{3}s^{-1})	1.1	0.9	0.8	0.6	0.4	0.3	0.1	0

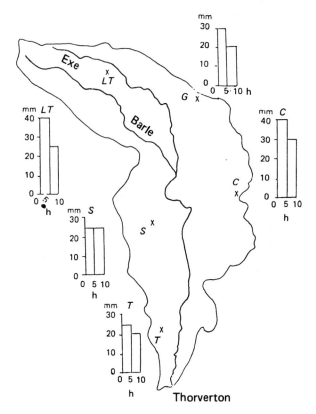

Fig. P3   River Exe catchment area (600 km²).

13.5 An acceptable 1-h unit hydrograph (10 mm) has been derived for a catchment. Its ordinates are shown in Table P.21. What is the approximate area of the catchment? Determine the peak flow that would result from a storm whose effective rainfall, assumed over the whole catchment, is given in Table P.22. [*Ans.*   34 km²; 95 m³s⁻¹]

Table P.21

Time (h)	0	1	2	3	4	5	6	7
$u(t)$ (m³s⁻¹)	0	12	35	24	16	8	3	0

Table P.22

Time (h)				0	1	2	3
Total accumulated effective rainfall (mm)				0	6	26	35

13.6 The ordinate of the 1-h unit hydrograph of a catchment area are summarized in Table P.23.

Table P.23

Time (h)	0	1	2	3	4	5	6	7
TUH ordinate $(m^3s^{-1}\ mm^{-1})$	0	25	125	358	465	405	305	220
Time (h)	8	9	10	11	12	13	14	15
TUH ordinate $(m^3s^{-1}\ mm^{-1})$	170	130	90	60	35	20	8	0

(a) Derive the S-curve for the catchment area.
(b) Use the S-curve to obtain the 2-h unit hydrograph; and
(c) Forecast the peak runoff that would result from a storm in which
the effective rainfall totals in two consecutive 2-h periods were
20 mm and 5 mm. [*Ans.*   Peak 9910 m^3s^{-1}]

Chapter 14   Catchment Modelling

14.1 Outline the principle of the Nash model, and describe how it can be
used to obtain the instantaneous unit hydrograph for an ungauged
catchment.

14.2 Distinguish between lumped and distributed catchment models, and
give examples of both.

14.3 Explain why the runoff-routing model is non-linear. Draw a diagram
to show how the runoff-routing model is applied to a catchment.

14.4 Define a conceptual model, illustrating the answer with a simple
example, and explain the function of parameters.

14.5 Describe the fitting of a simple conceptual catchment model. Itemize
the data required for the model, and indicate the functions of the parameters in the fitting process.

14.6 What is a component model? Discuss the usefulness of component
models in solving hydrological problems.

Chapter 15   Stochastic Hydrology

15.1 Identify the several components of a hydrological time series. Suggest
causes for the appearance of a catastrophic looking peak in a recorded
series of river flow data.

15.2 Compile sample time series for rainfall and runoff data of annual,
monthly and daily durations for measuring stations in your home area.

15.3 Define periodicity. Explain how it can be identified in a time series and
modelled.

15.4 Define the lag k autocorrelation coefficient, explaining clearly what it
measures.' Show how the information contained in the lag k autocorrelation coefficient can best be presented.

15.5 For a sample time series of length n, explain how you would compute the range of cumulative departures from the sample mean, thence define Hurst's Law and explain briefly why it is significant in stochastic hydrology.

15.6 What is autocorrelation analysis? What salient features of a time series would such an analysis reveal? Sketch the correlograms for the following:
   (i)   a simple sine – cosine wave;
   (ii) a first-order Markov process given by $Z_t = \rho Z_{t-1} + \epsilon_t$, $0 \leqslant \rho \leqslant 1$.
   (iii) a population of uncorrelated random variables.

15.7 Discuss the application of stochastic modelling of hydrological data in solving engineering problems. When are stochastic techniques most useful, and what safeguards must be applied before results are adopted for design purposes?

15.8 Compare and contrast the modelling of short-duration rainfall and river flow data.

Chapter 16   Flood Routing

16.1 Using a finite difference formulation of the Muskingum equation, derive a recurrence relation for outflow from a river reach in terms of inflow to the reach and outflow at the preceding time interval. The Muskingum constants, K and x, are estimated for a given river reach to be 12 h and 0.2. Assuming an initial steady flow, determine the peak discharge at the downstream end of the reach for the inflow hydrograph shown in Table P.24. [*Ans.*   55.9 m³s⁻¹]

Table P.24

Time from start of rise (h)	0	6	12	18	24	30	36	42
Inflow (m³s⁻¹)	25.0	32.5	58.0	65.0	59.4	49.5	42.5	35.0

16.2 The crest of a 20-m wide reservoir spillway consists of two 10-m wide gates, one of which is kept 0.5 m lower than the other. The flow, Q (m³s⁻¹), over each gate is given by $Q = C_d\, b\, h^{\frac{3}{2}}$ where b (m) is the gate width and h (m) is the head on the gate. The coefficient of discharge, C_d, which does not vary with gate position, can be taken as 2.0 (m$^{\frac{1}{2}}$s⁻¹) for each gate. Initial flows into and out of the 1 ha reservoir are equal at 10 m³s⁻¹. The inflow into the reservoir is to be increased steadily to 20 m³s⁻¹ over 1 h. Find the outflow at the end of that time, assuming the gate positions remain unchanged (1 ha = 10⁴m²). [*Ans.*   12.9 m³s⁻¹]

16.3 A lake, having steep banks and a surface area of 6 km², discharges into a steep channel which is approximately rectangular in section, with a width of 50 m. Initially conditions are steady with a flow of 170 m³s⁻¹ passing through the lake; then a flood comes down the river feeding the lake, giving rise to the inflow hydrograph shown in Table P.25. Compute the outflow hydrograph and plot it on the same graph with the inflow hydrograph. Note the difference in magnitude and time between the two peaks. (Critical flow exists at the lake outlet, $g = 9.81$ ms^{-1}.) [*Ans.*   20 m^{3}s^{-1}; 1 h]

Table P.25

Time from start (days)	0	1	2	3	4	5	6
Inflow (m^{3}s^{-1})	170	230	340	290	220	180	170

16.4 Discharge measurements at two gauging stations for a flood flow on the Macquarie River in Australia are given in Table P.26. It is assumed that there are no tributaries to the river between the upstream station B and the downstream station W. Apply the Muskingum method of flood routing to derive a model for calculating sequential outflows at the downstream end of the reach from measured flows at B. From the given inflows, derive the computed peak outflow discharge and its time of arrival at W. [*Ans.*   4493 m^{3}s^{-1} at 20 h]

Table P.26

Time (h)	0	4	8	12	16	20	24
B (m^{3}s^{-1})	595	1699	3837	5636	4305	3059	2271
W (m^{3}s^{-1})	130	496	1189	2209	3087	3823	3781
Time (h)	28	32	36	40	44	48	
B (m^{3}s^{-1})	1756	1359	1062	830	637	504	
W (m^{3}s^{-1})	3285	2393	1841	1416	1147	850	

16.5 Analysis of past records for a certain river reach yielded the values of the Muskingum coefficients $x = 0.22$ and $K = 1.5$ days. Route the flood shown in Table P.27 through the reach. The inflow continues at 14.2 m^{3}s^{-1} after day 12. (*Guideline Ans.*   Peak 217 m^{3}s^{-1} on day 9.)

Table P.27

Day	1	2	3	4	5	6
Inflow (m^{3}s^{-1})	14.2	76.7	129.4	166.8	171.9	113.8
Day	7	8	9	10	11	12
Inflow (m^{3}s^{-1})	187.8	264.2	152.1	75.0	45.0	14.2

Chapter 17   Design Floods

17.1 Define the term 'design flood', and examine what criteria must be considered in its determination.

17.2 Outline some of the various flood protection works that could be designed to mitigate flooding in a town at the confluence of a small tributary with a main river. Indicate the possible effects of the protection works on downstream flooding.

17.3 What are the advantages and disadvantages of using the unit hydrograph method to determine the design flood for a small catchment?

17.4 Describe the difficulties in quantifying the benefits of a flood protection scheme.

17.5 With reference to a particular example, explain how a water supply reservoir can be used for flood control.

Chapter 18   Urban Hydrology

18.1 Table P.28 shows the averages of bright sunshine and the mean annual temperatures for three climatological stations. Suggest which of these stations is representative of the centre of a large city, which is the suburban station and which is located in a rural area. Discuss the reasons governing the identification of the stations.

Table P.28

	A	B	C
Average of bright sunshine (h day^{-1})	3.95	3.60	4.33
Mean annual temperatures (°C)	10.3	11.0	9.6

18.2 The urbanization of a catchment area has a marked effect on its drainage characteristics. List the changes to be expected in regard to the quantity and quality of water in the stream.

18.3 The 1-mm, 0.5-h unit hydrograph of a small rural catchment is approximated by a triangle of height 0.66 m^{3}s^{-1} at time 2 h and of base length 8 h. Following construction of a major housing development the time to peak of the unit hydrograph is reduced by 25% and the base length by 3 h.

A design storm has mean intensities of 8.0, 22.0, and 12.0 mm h^{-1} for three consecutive 0.5-h intervals. If the ratio of storm flow volume to total rainfall volume for this storm is 0.2 before development and 0.5 after development, and a proportional rainfall loss is assumed, by what factor is the peak storm runoff increased? [*Ans.* 2.8]

18.4 Owing to urban development subsequent to the installation of a storm water drainage system, a length of road on the outskirts of a new town has become subject to frequent flooding. Within this length of road, there are three pipes AB, BC, CD with the outfall at D. The details of the pipes and the areas they serve are shown in Table P.29. Assuming a time of entry of 2 min, use the Rational Method to determine which pipe length has insufficient capacity to carry away the 2-year storm for which the intensity, i (mm h^{-1}) is related to the duration, t(min) by the expression $i = 825/(t + 8)$.

Table P.29

Pipe	Diameter (mm)	Length (m)	Gradient (1 in)	Area (ha)	Runoff coeff.
AB	533	450	225	2.9	0.8
BC	610	390	238	2.9	0.7
CD	686	225	250	2.4	0.75

Chapter 19   Water Resources

19.1 Obtain several years population data for your own country, local authority or town. Evaluate the per capita consumption of water for those years and assessing the possible changes in demand over the next few years, estimate the likely water requirements in the year 1990/2000.

19.2 Identify the water resources for your home region. Describe the location of the water source, and explain how its availability is assessed and how it has been developed for supply.

19.3 Choose a catchment in the UK suitable for a storage reservoir. Use the Lapworth chart to determine the effective storage capacity required to yield 200 000 $m^3 \, day^{-1}$.

19.4 An upland catchment of 80 km^2 produced the minimum flows shown in Table P.30 over selected durations during a very dry year. Assess the storage required in a reservoir to yield 110 000 $m^3 d^{-1}$. [*Ans.* 2.45 Mm^3]

Table P.30

Days	1	10	30	60	90	180
Q_{min} ($m^3 s^{-1}$)	0.065	0.101	0.378	0.857	1.013	1.464

19.5 A reservoir of storage capacity $3 \times 10^9 m^3$ is to be used jointly for power generation and flood control. The monthly inflows may be assumed to be independent and can be described by the following discrete probability distribution:

p_0	p_1	p_2	p_3
$\frac{1}{6}$	$\frac{1}{3}$	$\frac{1}{3}$	$\frac{1}{6}$

where p_i is the probability that the monthly inflow is $(i \times 10^9 m^3)$. If the monthly inflow exceeds the available storage capacity, the excess is released through spillways. Power generation requires the release of $2 \times 10^9 m^3$ at the end of each month; if less than $2 \times 10^9 m^3$ is available, the water remaining in storage is released. Determine the distribution of reservoir storage after release. [*Ans.*   Empty, $\frac{3}{4}$; $1 \times 10^9 m^3$, $\frac{1}{4}$]

19.6 A small irrigation reservoir of storage capacity 3 units is required to supply 2 units of water yearly, if this is feasible, to neighbouring agricultural lands. This demand occurs during the summer months with the annual inflows occurring in the previous winter. The annual inflows, which may be considered independent, are characterized by the following discrete probability distribution:

p_0	p_1	p_2	p_3	p_4
0.1	0.3	0.3	0.2	0.1

where p_i is the probability that the annual inflow is i units. If water in excess of the 'target release' of 2 units plus the storage capacity of the reservoir is available, the excess will also be released. By observing the reservoir state at the end of each year, construct the transition probability matrix and indicate how this would be used to evaluate the probability of emptiness. [*Ans.* $p_0 = 0.57$]

Chapter 20   River Basin Management

20.1 Derive a flow chart to indicate the necessary considerations in the formulation of a river basin management policy.

20.2 Examine the differences between river basin management policies for flood-prone areas and for drought-prone areas.

20.3 To what extent does river basin management involve a compromise between conflicting demands for water in a developing country.

20.4 Discuss how a river management scheme is dependent upon the gauging network. Illustrate with examples.

Appendix: Statistical Formulae

A series of measurements x_1, x_2, x_3, ... x_n form a 'sample' data set of n values taken from a 'population' of *all* such possible values.

A.1 Measures of Location

The *arithmetic mean* or *average*:

$$\bar{x} = \frac{x_1 + x_2 + x_3 + \ldots + x_n}{n} = \sum_{i=1}^{n} \frac{x_i}{n}$$

$\bar{x}$ is the sample mean which provides an estimate, $\hat{\mu}$, of the population mean, μ.

The *median* is the middle value of a sample data set arranged in order of magnitude.

The *mode* is the value of a sample data set that occurs with the greatest frequency.

A.2 Measures of Spread

Mean absolute deviation (MAD) is the average absolute deviation from the sample mean:

$$\text{MAD} = \frac{|x_1 - \bar{x}| + |x_2 - \bar{x}| + \ldots + |x_n - \bar{x}|}{n}$$

$$= \sum_{i=1}^{n} \frac{|x_i - \bar{x}|}{n}$$

The *variance* of the sample:

$$s^2 = \frac{(x_2 - \bar{x})^2 + (x_2 + \bar{x})^2 + \ldots + (x_n - \bar{x})^2}{(n-1)}$$

$$= \sum_{i=1}^{n} \frac{(x_i - \bar{x})^2}{(n-1)}$$

s^2 is the sample variance, which provides an estimate, $\hat{\sigma}^2$, of the population variance, σ^2.

The *standard deviation* of the sample:

$$s = \sqrt{\left[\sum_{i=1}^{n} \frac{(x_i - \bar{x})^2}{(n-1)} \right]}$$

A more convenient form for calculating s is:

$$s = \sqrt{\left[\frac{\{(\Sigma x_i^2) - n\bar{x}^2\}}{(n-1)} \right]}$$

The *coefficient of variation* is a measure of the spread of the sample in relative terms:

$$C_v = \frac{s}{\bar{x}}$$

Quartiles and *percentiles* may be considered as measures of spread about the median (the 50 percentile value). With the sample values arranged in ascending order of magnitude, the *lower quartile* is the value at the first quarter of the data series (the 25 percentile). The *upper quartile* marks the beginning of the top quarter of the data (the 75 percentile). Thus the interquartile range would contain the middle 50% of the data.

A.3   Discrete Probability Distributions

If a sample of discrete random variables x_1, x_2, x_3, ... x_n has the individual probabilities of occurrence p_1, p_2, p_3, ... p_n, then $p_1 + p_2 + p_3 + ... + p_n = 1$ and $p_i \geq 0$, for all i, defines a discrete *probability distribution* for values of x, and is denoted by $p(x)$. The average or *expected value* of the population of x, $E(x)$ is given by:

$$E(x) = \sum_{i=1}^{n} x_i p_i \quad \left(\sum_{i=1}^{n} p_i = 1 \right)$$

The variance:

$$\sigma^2 = \sum_{i=1}^{n} (x_i - \mu)^2 p_i$$
$$= E(x_i - \mu)^2$$

Binomial Distribution.   In a series of n independent trials, a success can occur with probability p, and hence a failure occurs with probability $(1 - p)$. Then:

$$[p + (1 - p)]^n = 1$$

The population mean μ or $E(x) = np$
The population variance $\sigma^2 = np(1 - p)$

Values of the binomial distribution are available in tables.

Poisson Distribution

$$p(x) = \frac{e^{-\mu}\mu^x}{x!} \qquad (x = 0, 1, 2, \ldots \mu > 0)$$

$$\sum_{x=0}^{\infty} p(x) = 1$$

Poisson probabilities are also tabulated.

$$\text{Mean} = \text{Variance} = \mu = \sum_{x=0}^{\infty} xp(x)$$

Truncated Negative Binomial Distribution (TNBD)

$$p(x = k) = \binom{k + r - 1}{k} \frac{p^r q^k}{1 - p^r} \qquad k = 1, 2, \ldots$$

with $0 < p \leqslant 1$, $r \geqslant -1$ and $q = 1 - p$

Geometric Distribution (GD).   This is given by the TNBD with $r = 1$. Thus:

$$p(x = k) = pq^{k-1} \qquad k = 1, 2, \ldots$$

A.4   Continuous Probability Distributions

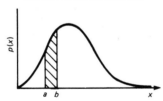

Probability density function (PDF)

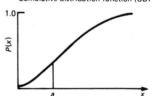

Cumulative distribution function (CDF)

$$\text{Prob.} \ (a \leqslant x \leqslant b) = \int_a^b p(x)\,\mathrm{d}x$$

$$\int_{-\infty}^{\infty} p(x)\,\mathrm{d}x = 1$$

$$\text{Prob.} \ (x \leqslant a) = P(x) = \int_{-\infty}^{a} p(x)\,\mathrm{d}x$$

$$\text{Mean} \ E(x) = \mu_x = \int_{-\infty}^{\infty} x\,p(x)\,\mathrm{d}x$$

$$\text{Variance} \ \sigma_x^2 = E(x - \mu_x)^2$$

$$= \int_{-\infty}^{\infty} (x - \mu_x)^2\,p(x)\,\mathrm{d}x$$

Normal or Gaussian Distribution

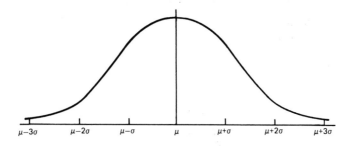

$$p(x) = \frac{1}{\sigma\sqrt{2\pi}} \exp\left[-(x-\mu)^2/2\sigma^2\right] \qquad -\infty < x < +\infty$$

The curve of the PDF is symmetrical about $x = \mu$, and the shape of the normal distribution depends on the standard deviation, σ.

$$\int_{\mu-\sigma}^{\mu+\sigma} p(x)\, dx = 0.68$$

$$\int_{\mu-2\sigma}^{\mu+2\sigma} p(x)\, dx = 0.95$$

$$\int_{\mu-3\sigma}^{\mu+3\sigma} p(x)\, dx = 0.997$$

Thus the PDF is very small for values of x more than 3σ away from the mean, μ.

The Standard Normal Distribution.   Values of the variable x are standardized to give a series of values, z:

$$z = \frac{(x-\mu)}{\sigma}$$

The standardized variable has a mean, $\mu_z = 0$ and a variance and standard deviation, σ_z^2 and σ_z, $= 1$.

A standardized variable, z, normally distributed, is denoted by $N(0, 1)$, and the normally distributed original variable by $N(\mu, \sigma^2)$.

Tables of the standardized cumulative distribution function of the normal distribution are given in most statistical texts.

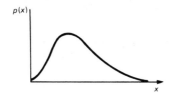

Log – Normal Distribution.   If $y = \ln(x)$, then the new values, y, are normally distributed. The variable is denoted by $N(\mu_y, \sigma_y^2)$, where:

$$\mu_x = \exp\left(\mu_y + \tfrac{1}{2}\sigma_y^2\right)$$
$$\sigma_x = \mu_x \sqrt{[\exp(\sigma_y^2) - 1]}$$

Gamma Distribution

$$p(x) \;=\; \frac{1}{a\Gamma k}\left(\frac{x}{a}\right)^{k-1} \exp\left(-x/a\right)$$

where $a = \sigma^2/\mu$ and $k = \mu^2/\sigma^2$.

Weibull Distribution

$$p(x) \;=\; \begin{cases} \dfrac{\gamma}{\theta}\; x^{\gamma-1} \exp\left(-\dfrac{x}{\theta}\right)^{\gamma} & x \geqslant 0 \\[2mm] 0 & x < 0 \end{cases}$$

where γ and θ are shape and scale parameters, respectively. When the shape parameter $\gamma < 1$, the curve of the PDF is the reversed J-shape. When $\gamma = 1$, the Weibull distribution reduces to the exponential distribution.

Exponential Distribution

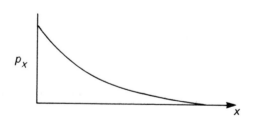

$$p(x) \;=\; \begin{cases} \lambda \exp\left(-\lambda x\right) & x \geqslant 0 \; (\lambda > 0) \\ 0 & x < 0 \end{cases}$$

$$\text{The CDF } P(x) \;=\; \begin{cases} 1 - \exp\left(-\lambda x\right) & x \geqslant 0 \\ 0 & x < 0 \end{cases}$$

J-Shaped Distribution.   This is also a bounded distribution with no values for x < 0, and it resembles the exponential distribution. However at $x = 0$, the probability $p(x)$, greatly exceeds the probability of occurrence of the positive x values. If the total number of occurrences is N, the cumulative distribution of $N - F$ is considered, where F is the value of $P(x)$ for a given x. Then

$$N - F = \alpha \exp(- \beta x)$$

and

$$\ln(N - F) = k - \beta x.$$

in using logarithms to base 10:

$$\log(N - F) = a - bx$$

and if the differential $- \dfrac{d\{P(x)\}}{dx} = y$, the form of the equation of the curve

becomes $y = c\, d^x$

A.5   Probability Distributions for Extreme Values

The data analysed are usually annual maximum/minimum values. Thus a series of annual maxima (or minima) may be considered as $X_1, X_2, X_3, \ldots, X_i,$ $\ldots, X_N$ for a sample of N years. From the cumulative distribution function of a series, $F(X)$ represents the probability of an annual maximum $X_i \leqslant X$.

The General Extreme Value distribution (GEV) has the form:

$$F(X) = \exp\left(-\left(1 - \frac{k(X - u)}{\alpha}\right)^{1/k}\right)$$

According to the value of k, three types of distribution are defined:

when $k = 0$, the GEV becomes the EV1 with no limit,
when k is negative, the GEV becomes the EV11 with a lower limit,
and when k is positive, the GEV becomes the EV111 with an upper limit.

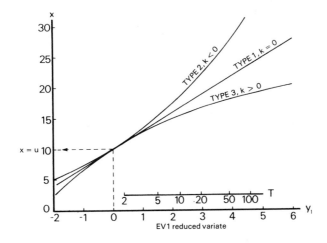

If the three types are plotted as functions of the EV1 reduced variate $y_1 = (x_1 - u)/\alpha$, the differences in shape are shown by the relation

$$x = u + \alpha(1 - e^{-ky_1})/k$$

EV11 is also known as the Fréchet distribution, and EV111 with x negative is a Weibull distribution.

Gumbel Extreme Value Type I distribution (EVI) with no defined upper limit:

$$F(X) = \exp\left[- e^{-b(X-a)}\right] = \exp\left[- e^{-y}\right]$$

where y is called the reduced variate and is equal to $b(X - a)$. a and b are two parameters related to μ_X and σ_X^2:

$$a = \mu_X - \frac{\gamma}{b} \qquad (\gamma = 0.5772)$$
$$b = \pi/\sigma_X\sqrt{6}$$

Probability of exceedence of X, $P(X)$ is given by $1 - F(X)$.

Gumbel Extreme Value Type III Distribution (EVIII) with the variate redefined has a lower limit and is used for extreme minimum values.
    If $P(X)$ = Prob. (annual minimum $\leqslant X$):

$$P(X) = \exp\left[- \{(X - \epsilon)/(\theta - \epsilon)\}^k\right]$$

where ϵ, θ and k are three parameters. ϵ is the lower limit $\geqslant 0$, θ is defined by $P(\theta) = \exp(-1) = 0.368$
    If the lower limit is made 0, the formula reduces to a two-parameter distribution:

$$P(X) = \exp\left[- (X/\theta)^k\right]$$

The following PDFs are given for information only. Before they are applied, reference to Matalas (1963) and advanced statistical texts should be made.

3-Parameter Log – Normal Distribution

$$p(X) = \frac{1}{\sqrt{(2\pi)}\sigma(X - \epsilon)}\exp\left[- \frac{1}{2\sigma^2}\left\{\ln(X - \epsilon) - m\right\}^2\right]$$

m and σ are mean and standard deviation of $\ln(X - \epsilon)$.

Pearson Type III Distribution

$$p(X) = \frac{1}{a\Gamma(b + 1)}\left(\frac{X - m}{a}\right)^b \exp\left[-\left(\frac{X - m}{a}\right)\right]$$

with parameters a, b and m.

Pearson Type V Distribution

$$p(X) = \frac{1}{a\Gamma(b + 1)}\left(\frac{X - m}{a}\right)^{-b} \exp\left[-\left(\frac{a}{X - m}\right)\right]$$

with parameters a, b and m.

Bibliography: Statistics

Brooks, C.E.P. and Carruthers, N. (1953). *Handbook of Statistical Methods in Meterology*. HMSO, M.O. 538, 412 pp.

Chatfield, C. (1970). *Statistics for Technology*. Penguin Books, 359 pp.

Chatfield, C. (1980). *The Analysis of Time Series: An Introduction*. Chapman & Hall, 2nd ed, 268 pp.

Elderton, W.P. and Johnson, N.L. (1969). *Systems of Frequency Curves*. Cambridge University Press, 216 pp.

Ezekiel, M. and Fox, K.A. (1967). *Methods of Correlation and Regression Analysis*. John Wiley, 3rd ed, 548 pp.

Gumbel, E.J. (1958). *Statistics of Extremes*. Columbia University Press, 375 pp.

Loveday, R. (1969). *Statistics*. Cambridge University Press, 2nd ed, 206 pp.

Matalas, N.C. (1963). *Probability Distribution of Low Flows* US Geol. Surv. Prof. Paper 434-A, 27 pp.

Natural Environment Research Council (NERC) (1975). *Flood Studies Report* Vol. 1.

Author Index

Subject Index